ECOLOGICAL ENGINEERING:
AN INTRODUCTION TO ECOTECHNOLOGY

ECOLOGICAL ENGINEERING

An Introduction to Ecotechnology

Edited by

WILLIAM J. MITSCH

School of Natural Resources
The Ohio State University
Columbus, Ohio

and

SVEN ERIK JØRGENSEN

Department of Environmental Chemistry
Institute of Chemistry AD.
The Royal Danish School of Pharmacy
Copenhagen, Denmark

WILEY

A WILEY-INTERSCIENCE PUBLICATION

JOHN WILEY & SONS

New York · Chichester · Brisbane · Toronto · Singapore

To Howard T. Odum,
our teacher and colleague
and a pioneer in ecological engineering

Library of Congress Cataloging in Publication Data:

Ecological engineering : an introduction to ecotechnology / edited by
William J. Mitsch and Sven Erik Jørgensen
       p.   cm.
  Includes index.
  ISBN 0-471-62559-0
  1. Environmental engineering. 2. Environmental engineering—Case
studies. 3. Human ecology. 4. Human ecology—Case studies.
I. Mitsch, William J. 1947–   . II. Jørgensen, Sven Erik, 1934–   .
TD153.E28 1989
628—dc19   86-23576
                                                   CIP

CONTRIBUTORS

JOHN CAIRNS, JR., Department of Biology and University Center for Environmental and Hazardous Materials Studies, Virginia Polytechnic Institute and State University, Blacksburg, Virginia

MILADY A. CARDAMONE, Wetlands Research, Inc., Chicago, Illinois

CHUNG-HSIN CHUNG, Institute of Spartina and Tidal Land Development, Biology Department, Nanjing University, Nanjing, China

ROBERT COSTANZA,* Center for Wetland Resources, Louisiana State University, Baton Rouge, Louisiana

M. SIOBHAN FENNESSY, School of Natural Resources, The Ohio State University, Columbus, Ohio

DONALD L. HEY, Wetlands Research, Inc., Chicago, Illinois

LIEF ALBERT JØRGENSEN, The Engineering College of Copenhagen, Copenhagen, Denmark

SVEN ERIK JØRGENSEN, Department of Environmental Chemistry, Institute of Chemistry AD, The Royal Danish School of Pharmacy, Copenhagen, Denmark

MARGARETE KALIN, Boojum Research Limited, Toronto, Ontario, Canada

DAVID M. KLARER, Old Woman Creek State Nature Preserve and National Estuarine Research Reserve, Huron, Ohio

AKIRA KURATA, Lake Biwa Research Institute, Otsu, Shiga, Japan

MA SHIJUN, Center of Eco-environmental Studies, Academia Sinica, Beijing, China

JUDITH MAXWELL, School of Natural Resources, The Ohio State University, Columbus, Ohio

WILLIAM J. MITSCH, School of Natural Resources, The Ohio State University, Columbus, Ohio

HOWARD T. ODUM, Department of Environmental Engineering Sciences and Center for Wetlands, University of Florida, Gainesville, Florida

DAVID ORVOS, Department of Biology and University Center for Environmental and Hazardous Materials Studies, Virginia Polytechnic Institute and State University, Blacksburg, Virginia

* Current affiliation, Chesapeake Biological Laboratory, University of Maryland, Solomons, Maryland.

DAVID PIMENTEL, Department of Entomology, Cornell University, Ithaca, New York

BRIAN C. REEDER, School of Natural Resources, The Ohio State University, Columbus, Ohio

J. HENRY SATHER, Institute for Environmental Management, Western Illinois University, Macomb, Illinois

MASARU SATOUCHI, Shiga Prefectural Junior College, Otsu, Shiga, Japan

YAN JINGSONG, Nanjing Institute of Geography and Limnology, Academia Sinica, Nanjing, China

YAO HONGLU, Jiangsu Provincial Freshwater Fisheries Research Institute, Nanjing, China

SERIES PREFACE
Environmental Science and Technology

The Environmental Science and Technology Series of Monographs, Text-books, and Advances is devoted to the study of the quality of the environment and to the technology of its conservation. Environmental science therefore relates to the chemical, physical, and biological changes in the environment through contamination or modification, to the physical nature and biological behavior of air, water, soil, food, and waste as they are affected by man's agricultural, industrial, and social activities, and to the application of science and technology to the control and improvement of environmental quality.

The deterioration of environmental quality, which began when man first collected into villages and utilized fire, has existed as a serious problem under the ever-increasing impacts of exponentially increasing population and of industrializing society. Environmental contamination of air, water, soil, and food has become a threat to the continued existence of many plant and animal communities of the ecosystem and may ultimately threaten the very survival of the human race.

It seems clear that if we are to preserve for future generations some semblance of the biological order of the world of the past and hope to improve on the deteriorating standards of urban public health, environmental science and technology must quickly come to play a dominant role in designing our social and industrial structure for tomorrow. Scientifically rigorous criteria of environmental quality must be developed. Based in part on these criteria, realistic standards must be established and our technological progess must be tailored to meet them. It is obvious that civilization will continue to require increasing amounts of fuel, transportation, industrial chemicals, fertilizers, pesticides, and countless other products, and that it will continue to produce waste products of all descriptions. What is urgently needed is a total systems approach to modern civilization through which the pooled talents of scientists and engineers, in cooperation with social scientists and the medical profession, can be focused on the development of order and equilibrium in the presently disparate segments of the human environment. Most of the skills and tools that are needed are already in existence. We surely have a right to hope a technology that has created such manifold environmental problems is also capable of solving them. It is our hope not only that this Series in Environmental Sciences and Technology will serve to make this challenge more explicit to the established professionals, but also that it will help to stimulate the student toward the career opportunities in this vital area.

ROBERT L. METCALF
WERNER STUMM

PREFACE

Human beings have had and will always have an impact on nature. Our book proposes a new approach in which we can deal with these impacts, while at the same time benefiting from those natural systems. We propose the terms *ecological engineering* and *ecotechnology* to mean the design of human society with its natural environment for the benefit of both. As with other technologies, ecological engineering and ecotechnology have their basis in scientific theory. The principles of ecology and environmental science have been well developed over the past 80 years, particularly over the last 20 years; the time is ripe for application of these principles. Furthermore, tools such as ecological modeling have been developed and refined with the advent of high-speed computers to offer the possibilities of quantification of ecosystems and an understanding of how they operate as a whole.

As we approach an age of diminishing resources, we need to conserve the remaining resources and use them wisely. Ecological engineering and ecotechnology offer a strategy to accomplish these goals. Our belief is that it is equally as unwise to allow the environment to deteriorate by pollution to the point where it is not functioning property as it is to isolate the environment from humankind or even bring it back to the stage where it was before human settlement. The tradeoff between these two extremes is ecological engineering, using the tools of ecotechnology, where humans are considered *a part* of nature rather than *apart* from nature. Our approach is a cooperation between humans and nature.

The book and its outline are the results of many hours of discussion between the editors. Originally one of us (Jørgensen) proposed the idea about a book on this new and emerging technology, but only through our many discussions and considerations did the idea of the book become streamlined. We have contibuted equally to these discussions and considerations to attempt to make a coherent book out of the many authors' fine but diverse contributions. The other one of us (Mitsch) carried the load of refining the language and editing the chapters as the contributions became reality.

The book is divided into two parts. Part One, written partly by the editors and partly by invited authors, introduces the basic definitions, concepts, and principles. Chapter 1, the introduction, defines ecological engineering and ecotechnology and points out the perspectives of this new discipline. The second chapter presents a few examples of ecological engineering. Chapter 3 identifies a series of ecological engineering principles, i.e. the rules that we must follow in managing natural ecosystems. Because ecological modeling is one of the primary quantitative tools in ecological engineering, Chapter 4 describes the general methodology for this approach. Chapters 5 and 6 describe economic, ecological, and educational aspects of ecological engineering, and Chapter 7 describes the general application of ecotechnology in an agricultural setting.

Part Two includes twelve chapters of case studies of ecological engineering from around the world. These chapters were written by invited scientists and engineers who either have considerable experience in what we consider ecological engineering and ecotechnology or have been, in fact, using these terms in their work. We purposely sought geographical diversity and have chapters from Denmark, China, Japan, the United States, and Canada. Most of the examples are applications in aquatic ecosystems, including hydrological modification; pollution control; wetland management; and lake, reservoir, and stream restoration. The absence of a significant number of terrestrial ecosystem examples was not intentional; we hope that future papers and books on this subject will correct that deficiency.

Chapters in Part Two have similar outlines, which include a survey of the problem or existing methodologies, a discussion of where and when these methods are ecologically sound (i.e., where they use proper ecological engineering techniques), and a case study to illustrate the proper use of ecological engineering in detail. If a simulation or conceptual model is used in the case study, it is included with the case study. Many of the case study chapters identify the state of the art of ecological engineering and further research needs for the selected topic.

This book will be appreciated by a wide variety of environmental and natural resource managers, with particular appeal to ecologists, engineers, and planners. It could also serve as a textbook for developing courses in applied ecology, environmental planning, and ecological engineering. These courses differ from traditional courses in ecology (where the book would not be appropriate) by emphasizing application of ecological principles rather than the development of theory, and from traditional courses in natural resource management by presenting an approach based on ecological science rather than empirical evidence. The book should be of considerable interest to practicing environmental managers because of its multidisciplinary approach to practical problems and opportunities. Although we stop short of describing the book as a major breakthrough in environmental management, we believe that it will spawn much discussion and improvement of present-day applied ecology and will encourage a symbiotic relationship between humans and their natural environment.

We could not have produced this volume without the assistance and support of a number of individuals. Mary Conway of John Wiley & Sons showed faith in our ideas by encouraging us in this project. Many of the chapters were retyped by the staff at the School of Natural Resources, The Ohio State University, under the direction of Jan Gorsuch. Ruthmarie H. Mitsch contributed her editorial talents to several chapters in the book. Judy Kauffeld and Julie Cronk assisted with graphics in many of the chapters. And finally, we appreciate the timely responses and good writing from our distinguished group of authors. It is they who will be defining ecological engineering.

WILLIAM J. MITSCH
SVEN ERIK JØRGENSEN

Columbus, Ohio
Copenhagen, Denmark
November 1988

CONTENTS

ECOLOGICAL ENGINEERING:
AN INTRODUCTION TO ECOTECHNOLOGY

PART ONE

BASIC PRINCIPLES

1

INTRODUCTION TO ECOLOGICAL ENGINEERING

William J. Mitsch

School of Natural Resources, The Ohio State University, Columbus, Ohio

and

Sven Erik Jørgensen

Department of Environmental Chemistry, Institute of Chemistry AD, The Royal Danish School of Pharmacy, Copenhagen, Denmark

1.1 WHAT ARE ECOLOGICAL ENGINEERING AND ECOTECHNOLOGY?

We are approaching an age of diminishing resources, the growth of the human population is continuing, and we have not yet found means to solve local, regional, and global pollution problems properly. About two decades ago, ambitious goals such as "zero discharge," defined as the complete elimination of pollution from entering the environment, were seriously discussed. There was a strong belief in the possibilities that environmental technology offered. Today we have finally recognized that we cannot achieve the complete elimination of pollutants owing to a number of factors. First, we have a finite quantity of resources to address to the problems of pollution control. This is particularly true for developing countries. Second, when we offer a technological alternative we are usually transferring the material from one medium (e.g., air) to another (e.g., water). We must seek additional means to reduce the adverse effects of pollution, while at the same time preserving our natural ecosystems and conserving our nonrenewable energy resources. Ecotechnology and ecological engineering offer such additional means to cope with pollution problems, by recognition of the self-designing properties of natural ecosystems.

We define *ecological engineering* and *ecotechnology* as the design of human society with its natural environment for the benefit of both (Mitsch, 1988). It is engineering in the sense that it involves the design of this natural environment using quantitative approaches and basing our approaches on basic science. It is a technology with the primary tool being self-designing ecosystems. The components are all of the biological species of the world.

H. T. Odum (Odum 1962; Odum et al., 1963; see also Chapter 6) was among the first to define ecological engineering as "environmental manipulation by man using small amounts of supplementary energy to control systems in which the main energy drives are still coming from natural sources. Formulae for ecological engineering may begin with natural ecosystems as a point of departure, but the new ecosystems which develop may differ somewhat. . . ." Odum (1971) further developed the concept of ecological engineering in his book *Environment, Power and Society* as follows: "The management of nature is ecological engineering, an endeavor with singular aspects supplementary to those of traditional engineering. A partnership with nature is a better term." He later states in his comprehensive work *Systems Ecology* (Odum, 1983) that "the engineering of new ecosystem designs is a field that uses systems that are mainly self-organizing." Ma (1985) also has contributed to the application of ecological principles in the concept of ecological engineering in China (see also Chapter 10).

Uhlmann (1983), Straškraba (1984, 1985), and Straškraba and Gnauck (1985) have defined ecotechnology as the use of technological means for ecosystem management, based on deep ecological understanding, to minimize the costs of measures and their harm to the environment. In this book we consider ecological engineering and ecotechnology synonymous.

1.2 CONTRASTS WITH OTHER TECHNOLOGIES

It may also be useful in discussing what ecological engineering and eco-technology are first to define what they are not. Ecological engineering is not the same as *environmental engineering*, a field that has been well established in universities and the workplace since the 1960s and that was around for decades before that under the name of sanitary engineering. The environmental engineer is certainly involved in the application of scientific principles (sometimes called *unit processes*) to clean up or prevent altogether pollution problems, and the field is a well-honored one. Environmental engineers are taught and use valuable environmental technologies (called *unit operations*) such as setting tanks, scrubbers, sand filters, and flocculation tanks.

Ecological engineering, by contrast, is involved in identifying those eco-systems that are most adaptable to human needs and in recognizing the multiple values of these systems. As do other forms of engineering and technology, ecological engineering and ecotechnology use the basic principles of science (in this case it is mainly the multifaceted science of ecology) to design a better living for human society. However, unlike other forms of engineering and technology, ecological engineering has as its raison d'être the design of human society with its natural environment, instead of trying to conquer it. And unlike conventional engineering, ecological engineering has in its toolbox all of the ecosystems, communities, organisms, that the world has to offer.

Ecological engineering and ecotechnology should also not be confused with *bioengineering* and *biotechnology*.* Some of the contrasts between the two approaches are shown in Table 1.1. Biotechnology, by its very design, involves the manipulation of the genetic structure of the cell to produce new strains and organisms capable of carrying out certain functions. Ecotechnology does not manipulate at the genetic level, but considers an assemblage of species and their abiotic environment as a self-designing system that can adapt to changes brought about by outside forces—by the changes introduced by humans or the natural forcing functions. Biotechnology has begun with much fanfare but now is beginning to realize the enormous costs and concerns arising from manipulation at this micro level. In contrast, eco-

---

* *Biotechnology* in this chapter is defined in the narrow sense as the development of new species and varieties through alteration of genetic structure. We recognize that the term *biotechnology* is used in a broad sense also to include manipulation of biological systems, without genetic changes. We would include these latter changes at the species level and higher as ecological engineering or ecotechnology.

Table 1.1 Comparison of Ecotechnology and Biotechnology

Characteristic	Ecotechnology	Biotechnology
Basic unit	Ecosystem	Cell
Basic principles	Ecology	Genetics; cell biology
Control	Forcing functions, organisms	Genetic structure
Design	Self-design with some human help	Human design
Biotic diversity	Protected	Changed
Maintenance and development costs	Reasonable	Enormous
Energy basis	Solar based	Fossil fuel based

technology introduces no new species that nature has not dealt with before, nor does it involve major laboratory development costs except for micro-cosm studies of new combinations of organisms.

1.3   THE VIEW OF ECOTECHNOLOGY

1.3.1   Self-design

Ecotechnology involves an acceptance of the concept of the self-designing capability of ecosystems and nature. We may even go so far as to say that the concept of a polluted ecosystem is an anthropogenic view, one that may not recognize the beauty of natural systems shifting, substituting species, reorganizing food chains, adapting as individual species, and ultimately de-signing a system that is ideally suited to the environment that is superim-posed on it. The ecosystem also does another wonderful thing—it begins to manipulate its physical and chemical environment to make it a little more palatable! It is this self-designing capability of ecosystems that ecological engineering recognizes as a significant feature, because it allows nature to do some of the "engineering." We participate as the choice generator and as a facilitator of matching environments with ecosystems, but nature does the rest.

So ecological engineering is not a license to pollute, but human society *and* ecosystems are viewed as an entity. Ecological and holistic viewpoints are emphasized.

1.3.2   Ecosystem Conservation

Just as an engineer depends on an abundance of tools and raw materials to design and build products and processes, so too does the ecological engineer depend on an abundance of species and ecosystems. Therefore it would be counterproductive to eliminate, drain, or even disturb natural ecosystems unless absolutely necessary. These are the environments that protect the biological diversity that the ecological engineer may call on one day. This means that the ecotechnology approach will lead to a greater environmental conservation ethic than has up to now been realized. It has been noted, for example, that when wetlands were recognized for their abiotic values of flood control and water quality enhancement in addition to the long understood values as habitat for fish and wildlife, then wetland protection efforts gained a much wider degree of acceptance and enthusiasm. Recognition of ecosystem values leads to conservation of ecosystems.

1.3.3   Solar Basis

Because ecosystems are solar-based systems, either directly or indirectly (with the possible exception of biological communities that have developed along vents in the abyss of the oceans), the concept of self-sustaining systems is basic to ecotechnology. Once a system is designed and put in place, it should be able to sustain itself indefinitely with only a modest amount of intervention. This means that an ecosystem, running on solar energy or the products of solar energy, does not have to depend on technological energies as much as would high tech solutions. If the system does not sustain itself, it does not mean that the ecosystem has failed us (its behavior is ultimately predictable). It means that we have not designed the proper interface between nature and the environment.

1.3.4   A Part of, Not Apart from Nature

Figure 1.1 demonstrates the difference between a society that treats natural ecosystems as something separate from its daily existence (Figure 1.1a) (except for recreational and aesthetic values, shown in the faint line in Figure 1.1a) and a society that derives significant value from the surrounding ecosystems by working in a symbiotic pattern with them (Figure 1.1b). The recreational and aesthetic values are still there, but so also are the tangible benefits of cleaner air and water and more nonrenewable energy to accomplish other tasks or to save for posterity.

1.4   ECOLOGY—THE BASIC SCIENCE OF ECOLOGICAL ENGINEERING

Management of ecosystems is not an easy task, but complex problems of complex systems should not be expected to have easy solutions. Most environmental problems require the application of environmental technology as well as ecotechnology to find an optimal solution. This again will require a deep ecological knowledge to understand the processes and reactions of

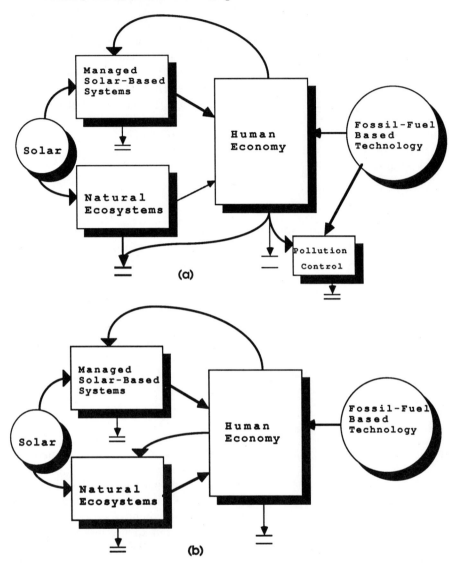

Figure 1.1 Comparison for interactions of economy and ecosystems (*a*) without and (*b*) with ecotechnology.

ecosystems to possible management strategies. Recognition of nature's ability to self-design to its forcing function is also necessary to find the right ecotechnological methods. Thus ecology becomes the basis of ecotechnology just as genetics is for biotechnology, as chemistry is for chemical engineering, and so on.

All new technologies have grown from their basic science. When the basis in science was sufficiently strong, the technology emerged. During the past two decades, the fields of ecology and environmental science have grown very rapidly. So the time is mature for ecotechnology to appear on the stage after waiting in the wings for several years. Ecological engineering must be based on sound ecological methods and a thorough understanding of eco-systems, their reactions, and their information subsystems (species), just as biotechnology requires understanding the cell and its information (genetic) subsystems.

1.5   THE NEED AND APPROPRIATE TIMING FOR ECOLOGICAL ENGINEERING AND ECOTECHNOLOGY

Pollution problems became widely recognized during the 1960s. Although many billions of dollars have been invested all over the world on solutions of these problems we are still far from an acceptable solution to many serious problems, which threaten the survival of *Homo sapiens*. Some even claim that we are further from a total solution of our environmental problems today than we were 10 to 20 years ago, owing to the continuous growth of pop-ulation and continuous disappearance of natural resources. We must ac-knowledge that there are 2 billion more people on earth and that the non-renewable resources are more limited today than 20 years ago.

We therefore need to find new ways. We have attempted to solve the problem by use of available technology. It has partially failed. Therefore we must think more ecologically and consider additional means. Ecotechnology is based on ecological considerations and attempts to optimize ecosystems (including limited resources) and man-made systems for the benefit of both. It should therefore afford additional opportunities to solve the crisis.

We have had two energy crises during the past 15 years and we know that new crises will appear in the future. Therefore we have to rely more on solar-based ecosystems, which are the basis for ecotechnology.

The limited resources and the high and increasing human population force us to find a trade-off between the two extremes of pollution and totally unaffected ecosystems. We cannot and we must not accept a situation of no environmental control, but we cannot afford zero-discharge policies either, knowing that we do not provide one-third of the world's population with sufficient food and housing.

Some of the ecotechnological methods presented in this book are now new and, in fact, some have been practiced for centuries (see, e.g., integrated fish ponds, Chapter 17). In earlier times these methods were considered as good, empirical approaches. Today, ecology has developed sufficiently to understand the scientific background of ecological engineering, to formalize usage of these approaches, and to develop new ones. We must understand not only how we can influence the processes in the ecosystem and how the ecosystem components are linked together, but also how changes in one ecosystem can produce changes in neighboring ecosystems.

1.6   ECOTECHNOLOGY AND MODELING

All applications of technologies, whether of biotechnology, chemical technology, or ecotechnology, require quantification. Because ecosystems are complex systems, the quantification of their reactions becomes complex. However, ecological modeling represents a well developed tool to survey ecosystems, their reactions, and the linkage of their components. Ecological modeling is able to synthesize the pieces of ecological knowledge, which must be put together to solve a certain environmental problem.

The application of ecological modeling emphasizes the need for a holistic view of environmental systems. Optimization of subsystems does not necessarily lead to an optimal solution of the entire system. There are many examples in environmental management where optimal management of several aspects of a resource separately does not optimize management of the resource; for example, in some governments water quality is controlled by one agency and water quantity by another, often leading to conflicting policies and regulations.

1.7   ECOTECHNOLOGY AND OUR EDUCATIONAL SYSTEM

Environmental problems are, by definition, multidisciplinary, but our educational system often is not. There is therefore a need in environmental management and ecological engineering for integration of disciplines, particularly between ecologists and engineers.

Ecological training in schools and universities is often out of date, owing to a rigid departmentalization that has centuries of inertia behind it. A far better integration of disciplines is urgently needed in the very near future for environmental issues. Our higher educational system has, for many years, encouraged specialization. Merits have been given to scientists or experts who fully mastered a very narrow problem or topic. Analyses are more appreciated than syntheses. But this system has created an isolation from other problems and topics and has paralyzed the interaction with other disciplines. The result has been that ecologists have often not understood the need for quantifying ecosystems and dealing with applications of ecology in a technological framework. On the other hand, engineers have not taken the demands and value of ecosystems into consideration in their development of technologies and planning of production. Therefore we have been slowed by our educational system in solving the environmental problems properly. A much more integrated education is urgently needed in the future, if we shall be able to find the right environmental solutions that consider human society *and* ecosystems as an entity.

This does not imply that we should develop educational systems and curricula that give all students a little knowledge about everything. That type of "cafeteria" educational approach has been discredited in many circles and we concur. Rather, we need to educate fundamentally sound generalists as well as specialists. Specialists must be taught to work together on multidisciplinary projects. Such collaboration will force specialists to understand other languages, it will prevent isolation, and it will provoke cooperation and coordination. Future students must devote some time to learning what other disciplines can offer.

Experiments with such multidisciplinary problem-oriented and project-oriented education have given very positive results (see, for example, suggested curricula on ecological engineering in Chapter 6). Courses of this type are offered at several universities all over the world, but more are needed.

1.8   FUTURE OF ECOTECHNOLOGY

Our experience with ecotechnology and its possibilities is limited today. Although the results we do have look very promising, we need to integrate the application of ecotechnology much more in our pollution control and environmental planning in the future. This will require a continuous development of ecology, applied ecology, ecological modeling, and ecological engineering. Ecotechnology offers us a very useful tool for better planning in the future. It is a real challenge to humankind to use this tool properly.

The science of ecology is expected to continue its progress and, owing to a rapid growth of computer technology, we shall be able to overview and solve more and more complex ecological problems. This is absolutely necessary if we are to apply ecological engineering approaches to the growing problems of humans. However, a better general ecological understanding is needed by our politicians and the entire population if we are to use ecological engineering successfully in the future. This understanding will not come with a more complex technology if our educational system lags behind. Therefore a better ecological education with multidisciplinary aspects is urgently needed. Ecology should be introduced as a basic and compulsory subject at all school levels including the elementary.

1.9   ROLE OF THIS BOOK

We expect this book to be a catalyst, not the major reactant, in the development of ecological engineering. Many fine scientists have preceded us with the development of the idea itself. Our collection of principles and case studies here is meant to provide a demonstration of the breadth of application of ecotechnology and to suggest the significance of these applications. It is up to environmental scientists in academia, government, and industry to continue to find innovative solutions to environmental problems through this "partnership with nature."

REFERENCES

Ma Shijun. 1985. Ecological engineering: application of ecosystem principles. *Environmental Conservation 12*(4):331–335.

Mitsch, W. J. 1988. Ecological engineering and ecotechnology with wetlands: Applications of systems approaches. In *Advances in Environmental Modelling: Proceedings of Conference on State of the Art of Ecological Modelling, June, 1987, Venice, Italy.* Elsevier, Amsterdam.

Odum, H. T. 1962. Man in the ecosystem. In *Proceedings Lockwood Conference on the Suburban Forest and Ecology.* Bull. Conn. Agr. Station 652. Storrs, CT, pp. 57–75.

Odum, H. T., W. L. Siler, R. J. Beyers, and N. Armstrong. 1963. Experiments with engineering of marine ecosystems. *Publ. Inst. Marine Sci. Uni. Texas 9:*374–403.

Odum, H. T. 1971. *Environment, Power and Society.* Wiley, New York, 331 pp.

Odum, H. T. 1983. *Systems Ecology: An Introduction.* Wiley, New York, 644 pp.

Straškraba, M. 1984. New ways of eutrophication abatement. In M. Straškraba, Z. Brandl, and P. Procalova, eds. *Hydrobiology and Water Quality of Reservoirs.* Acad. Sci., Cěské Budějovice, Czechoslovakia, pp. 37–45.

Straškraba, M. 1985. *Simulation Models as Tools in Ecotechnology Systems. Analysis and Simulation,* Vol. II. Academic Verlag, Berlin.

Straškraba, M. and A. H. Gnauck. 1985. *Freshwater Ecosystems: Modelling and Simulation.* Elsevier, Amsterdam, 305 pp.

Uhlmann, D. 1983. Entwicklungstendenzen der Ökotechnologie. *Wiss. Z. Tech. Univ. Dresden 32:*109–116.

2

CLASSIFICATION AND EXAMPLES OF ECOLOGICAL ENGINEERING

Sven Erik Jørgensen

Department of Environmental Chemistry, Institute of Chemistry AD, The Royal Danish School of Pharmacy, Copenhagen, Denmark

and

William J. Mitsch

School of Natural Resources, The Ohio State University, Columbus, Ohio

2.1 ECOTECHNOLOGICAL METHODS

Part Two of this book is devoted to examples and case studies of ecological engineering applications. The list of potential ecotechnological methods is

13

Table 2.1 Examples of Ecological Engineering Approaches for Terrestrial and Aquatic Systems

Ecological Engineering Approaches	Terrestrial Examples	Aquatic Examples
Nutrient recycling	Sludge disposal on agricultural land	Recycling in wetlands
Hydrologic modification	Artificial ponds and wetlands	Control of retention time of reservoirs
Ecosystem recovery	Coal mine reclamation	Restoration of lakes
Enhancement of ecological diversity	Tropical forest management	Biomanipulation in lakes
Ecological sound biotic harvest	Good forestry practices	Multispecies fisheries

long, however, and it is therefore not possible to present them all in one book. We have attempted to treat in detail the most important methods either from an environmental management or ecological point of view. Other methods are touched on more superficially in the text to give alternative illustrations of ecological engineering.

Table 2.1 gives an overview of the ecotechnological methods mentioned in this volume plus a few additional ones. The list is comprehensive without being complete, because it is not our goal to present all methods. Rather we propose to give the readers a good ecological understanding of how, where, and when ecotechnology can be applied to solve environmental problems, and to illustrate the approaches and their ecological soundness.

2.2 A CLASSIFICATION OF ECOTECHNOLOGY

The examples in Table 2.1 are classified according to the engineering approach used, such as nutrient recycling and hydrologic modification. The various approaches are based upon one or more of the following principles of application:

1. Ecosystems are used to reduce or solve a pollution problem that otherwise would be very harmful to other, or other types of ecosystems. Examples are sludge disposal and wastewater recycling in terrestrial ecosystems or wetlands.
2. Ecosystems are imitated or "copied" to reduce or solve a pollution problem, leading to artificial ecosystems that can cope "on behalf of" natural ecosystems. Examples are integrated fishponds and created wetlands.
3. The recovery of ecosystems is supported after significant disturbances. Examples are coal mine reclamation, methods for restoration of lakes

and rivers, and biomanipulation (such as the use of grass carp to reduce eutrophication).

4. Ecosystems are used for the benefit of humankind *without* destroying the ecological balance, that is, utilization of the ecosystem on an eco- logically sound basis. Typical examples are the use of agroecosystems and a sound ecological basis for the harvest of renewable resources (fish and timber, etc.).

2.3 EXAMPLES OF ECOLOGICAL ENGINEERING—A SYSTEMS APPROACH

Examples of the general types of ecological engineering projects that one might get involved in are presented in Table 2.1. The diversity of approaches, for both aquatic and terrestrial ecosystems, shows the breadth of ecological engineering. It can also be noted that ecological engineering encompasses parts of some applied fields such as forestry and fisheries when a systems approach is taken toward these activities.

Hypothetical details of some of these examples are presented below, al- though more detailed approaches are presented in Part Two of this book. By giving specific examples of eutrophication control, sludge disposal, drink- ing water treatment, and aquaculture, we hope to illustrate the wide range of possibilities for ecological engineering in contrast with more technological approaches.

2.3.1 Eutrophication Control

Figure 2.1 illustrates the difference between traditional environmental tech- nology and ecotechnology for the control of eutrophication. Eutrophication of a lake is considered to be a pollution problem, and environmental tech- nology can be used to reduce the input of nutrients (phosphorus, for instance, by chemical precipitation or nitrogen by ion exchange or denitrification). However, the reduction in nutrient input from wastewater is not sufficient to control eutrophication owing to the contributions from non-point sources. Furthermore, the retention time of the lake is long, so more rapid improve- ments would be advantageous. Here ecotechnology comes into the picture. A wetland is developed to trap the nutrients in the inflow, and nutrient-rich hypolimnetic water is siphoned off downstream of the lake. This example shows that environmental technology is not sufficient and ecotechnological methods can be used to supplement them.

2.3.2 Sludge Disposal

Figure 2.2 illustrates even more clearly the difference between the environ- mental technological and ecotechnological approaches. In an environmental

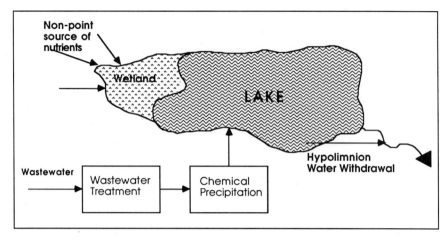

Figure 2.1 Control of lake eutrophication with a combination of chemical precipitation for phosphorus removal from wastewater (environmental technology), a wetland to remove nutrients from the inflow (ecotechnology) and siphoning off of hypolimnetic water, rich in nutrients, downstream (ecotechnology).

technology approach sludge from a wastewater treatment plant is incinerated, causing air pollution problems, and the slags and ash still have to be deposited. The ecotechnological solution, on the other hand, recognizes the sludge as a resource of nutrients and organic matter, and deposits it on agricultural land, where the resources can be utilized.

2.3.3 Drinking Water Treatment

Figure 2.3 gives a third example, concerned with nitrate removal from drinking water. Environmental technology would use either ion exchange or denitrification. The first method has high operating costs and produces a regeneration solution, which must be deposited. The second method is expensive in installation (4–8 hours retention time is required), and the water requires disinfection. The ecotechnological method uses an artificial wetland. To optimize the denitrification potential of the organic matter, cellulose (or bark-based ion exchange) is placed in a layer of about 1 m under the wetland. This causes a high rate of denitrification within the ion exchange, owing to the high accumulation of organic matter. Finally, a deeper layer of sandy soil of about 1 m further purifies the water by removal of microorganisms, organics, and so on. The wetland is designed to be a self-sustaining system with the production of surface vegetation supplying organic matter necessary for denitrification to take place.

2.3.4 Aquaculture

The examples given so far illustrate how pollution problems can be solved elegantly and at relatively moderate costs by use of ecotechnological meth-

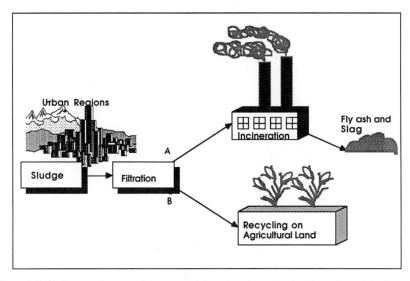

Figure 2.2 Sludge may be treated by use of (*a*) environmental technology through incineration or (*b*) ecotechnology through application on farmland.

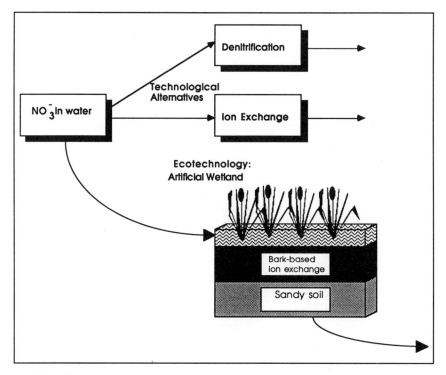

Figure 2.3 Nitrates can be removed from water by methods of environmental technology, such as denitrification and ion exchange, or ecotechnology, such as wetland treatment.

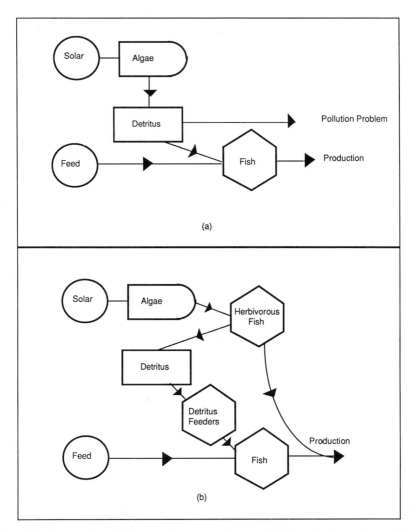

Figure 2.4 A comparison of (*a*) conventional aquaculture and (*b*) integrated aquaculture utilizing an ecotechnological approach.

ods. However, it is also considered to be ecotechnology to *design* our production systems in harmony (we may even use the words *symbiosis* or *partnership*) with nature. This is illustrated by integrated aquaculture (for more detail, see Chapter 17). Aquaculture is practiced in many developed countries by feeding fish and solving the problem of pollution from the production of detritus from dead algae and fish feces (Figure 2.4a). In integrated fish production, however, the cycling of energy and mass is imitated. Herbivorous fish are used to take care of algal production and detritus feeders utilize the detritus and serve as feed for other fish (Figure 2.4*b*). Thereby pollution is eliminated, or at least almost so.

2.4   APPLICATION OF ECOTECHNOLOGY

If a proper use of ecotechnology is achieved during the next decade or two, it will most probably imply that all planning of projects with related pollution control problems will have been subjected to the following examinations and considerations before they are launched.

1. The parts of nature that are directly or indirectly touched by the project with various pollution control alternatives must be determined.
2. Quantitative assessment (by use of models) of the environmental impact for all alternatives must be carried out.
3. The project is optimized taking into consideration the *entire* system, that is, human society *and* the affected ecosystem. Ecotechnical solutions to existing problems should be included in this step.
4. The optimization must include short- as well as long-term effects, and ecology as well as economy. In this context, the application of ecological economic models becomes a very strong tool.
5. The renewable and nonrenewable resources involved in the various alternatives should be quantified.
6. All such examinations have an uncertainty, which should be stated, and the uncertainty should be at least equally accounted for in ecological and economic considerations.

We have introduced the application of points 1 and 2 during the last two decades, but points 3–6 are open for introduction by ecological engineering. However, before we really can apply any ecological engineering approach to project planning, we need more experience in the application of ecotechnology, and this again requires that we reinforce ecological research in its broadest sense, including systems ecology, applied ecology, and ecological modeling. More resources must be allocated to this research in the coming years to assure a proper application of all the considerations mentioned above. Ecological engineering will not flourish without a strong base in ecology.

3

ECOLOGICAL ENGINEERING PRINCIPLES

Sven Erik Jørgensen

Department of Environmental Chemistry, The Royal Danish School of Pharmacy, Institute of Chemistry AD, Copenhagen, Denmark

and

William J. Mitsch

School of Natural Resources, The Ohio State University, Columbus, Ohio

A list of ecological engineering principles is given below. They are all based on a combination of our ecological knowledge and our experience from practical applications of ecotechnology. This should be considered a first attempt to list and discuss such principles. It is therefore not complete and more principles will be added to the list as we gain experience. The principles are given as prognoses; general (site) specific examples and illustrations are given for each principle.

1. Ecosystem structure and function are determined by the forcing functions of the system. Alterations of the forcing functions cause the most drastic changes in ecosystems.

The abiotic and biotic structure of the ecosystem is ultimately determined by the system's forcing functions (temperature, inflows of chemical compounds, including water, etc.). This is also reflected in ecological modeling, where relationships between forcing functions and state variables form the core of the model. The relationships between state variables can to a certain extent be considered pathways for the influence of the forcing functions.

Eutrophication is a typical example of how the forcing functions determine the state of the system including species composition. The input of nutrients is very closely related to the concentration of algae in a reservoir or a lake. Figure 3.1 illustrates this relationship between a forcing function (nutrient inflow) and the concentration of phytoplankton (the most central

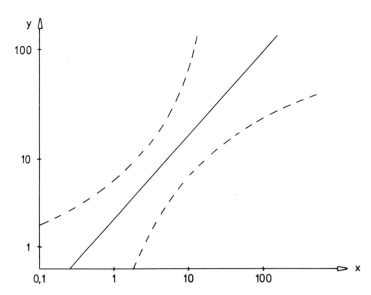

Figure 3.1 The chlorophyll *a* concentration (*x*) in μg/L versus orthophosphate concentration (*y*) in μg/L. This relationship is obtained from more than 100 different lakes. The dashed lines indicate the standard deviation.

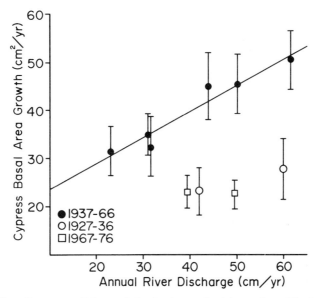

Figure 3.2 *Taxodium* tree radial growth in riparian wetland in southern Illinois versus river flooding (From "Ecosystem dynamics and a phosphorus budget of an alluvial cypress swamp in southern Illinois," by Mitsch et al., 1979; *Ecology 60*, 1116–1124. Copyright © 1979 by The Ecological Society of America. Reprinted by permission.).

state variable in eutrophication). Observations in Glumsø Lake in Denmark (see Chapter 16), where the phosphorus input has been reduced from 1.6 g/m²-yr to about 0.4 g/m²-yr, show a significant decrease in the algae concentration and corresponding increase in transparency. Simultaneously, a shift of the dominant algae species took place from *Scenedesmus* sp. to diatoms.

There are many examples of the role of forcing functions on ecosystem structure in wetlands (Mitsch and Ewel, 1979; Mitsch et al., 1979; Mitsch, 1988). Figure 3.2 illustrates the effect that flooding has an ecosystem function of a forested wetland. When wetlands are "subsidized" by flooding, for example, productivity is often higher owing to the imported nutrient-laden sediments. When wetlands are isolated from their surroundings (i.e., isolated from their forcing function), as is often done when dikes are built around wetlands for wildlife enhancement along the North American Great Lakes (see Chapter 8), the ecological manipulation does not optimize system function but maximizes for only one benefit. We would not necessarily characterize this as ecological engineering in the best sense of the term.

The role of forcing functions is even more pronounced when we consider discharge of toxic substances into ecosystems. This is an obvious case of a forcing function affecting an ecosystem, although in a negative way. If the lethal concentration of certain species is reached, it is obvious that the structure and function of the system will change radically.

2. Ecosystems are self-designing systems. The more one works with the self-designing ability of nature, the lower the costs of energy to maintain that system.

An ecosystem's regulation and feedback mechanisms give it the ability to adapt and self-design to the environment and minimize changes in the function of the ecosystem. In Chapter 6, Odum describes this phenomenon as *self-organization*, whereby ecosystems and ecological processes are used, not replaced, in ecological engineering. In Chapter 10, Ma and Yan describe much the same concept for the adaptability of aquatic systems as *self-purification*. Often, this self-designing capability may yield an important solution to a pollution problem by playing on these abilities of an ecosystem. For example, toxic wastes are causing great problems in many industrialized countries. Although it is forbidden to deposit such wastes today, 10 to 20 years ago it was not regulated. Consequently, in industrial countries many examples of soil contamination are found that are due to deposition some years ago of toxic substances. All the contamination soil could be removed to another, less harmful place but that is an expensive and unacceptable solution from an ecological point of view. From an ecotechnological view, however, nature can help us. If the toxic substance is organic, it will in many cases be possible to find microorganisms adapted to this toxic substance. This methodology has been applied on an individual plot in a suburb of Copenhagen. The soil was contaminated with xylol and related chemical compounds. According to the Danish Environmental Protection Agency, the soil is being purified by the use of adapted organisms.

Heavy metals can be removed from soil by the use of plants. The plants will adapt to the heavy metal concentration of the soil and take it up; they could after harvest be burned, the energy used, and the metal even recovered. Calvin (1983) described an ecological engineering application of certain species of the family Euphorbiaceae (*Euphorbia lathyris* and *Euphorbia esula*). It was found to be possible to obtain the equivalent of 5000 L of hydrocarbon fuel per hectare per year. The entire plant, not just the seeds, which contain approximately 37% hydrocarbons, was briquetted. Experiments have shown that the plants were able to remove several grams per square meter of such toxic heavy metals as lead and cadmium without any observable effect on the growth. Chapter 12 describes a *Typha* wetland that is used to control acid mine drainage, particularly iron, from a coalfield in eastern Ohio. Although the plant uptake of iron on an annual basis is low, the wetland environment is ideal for iron retention owing to enhancement of iron precipitation and other physiochemical processes.

3. Elements are recycled in ecosystems. Matching humanity and ecosystems in recycling pathways will ultimately reduce the effects of pollution.

Elements cycle in all ecosystems. An example of a nutrient cycle is shown in Figure 3.3. It is of great importance in the application of ecological en-

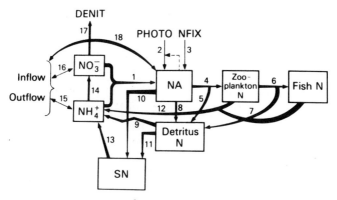

Figure 3.3 Model of the nitrogen cycle of a lake. The processes are: (1) uptake of NO_3^- and NH_4^+ by algae, (2) photosynthesis, (3) nitrogen fixation, (4) grazing with loss of undigested matter, (5–7) predation and loss of undigested matter by predation, (8) mortality, (9) mineralization, (10) settling, (11) settling of detritus, (12) settling, (13) release from sediment, (14) nitrification, (15,16,18) input/output, (17) denitrification.

gineering to understand the individual cycles in ecosystems and their rates. More importantly, ecological engineering approaches that utilize recycling can reduce pollution effects and save resources. For example, the use of terrestrial ecosystems for nutrient recycling can be a very effective means of protecting fragile ecosystems at a minimum cost and leads to recovery of a resource that is otherwise discharged to aquatic ecosystems (Figure 3.4). Chapters 8 and 9 document studies that have utilized wetlands to protect downstream ecosystems. If the wetland is then harvested for biotic resources, the cycle is complete. Chapter 10 describes how nutrients are recovered from a river by water hyacinths in China and then the plants are used as feed in a nearby fish farm, completing the recycling loop.

Models can be useful for optimizing this recycling. For example, manure is used as a natural fertilizer in agriculture in Denmark. It can, however, be a source of nutrient input to lakes and streams or cause nitrate contamination of the groundwater. If, however, you consider (1) the influence of the temperature on the nitrification and denitrification (storage capacity of manure must be available to assure right application time), (2) the adsorption capacity of the soil, (3) the hydraulic conductivity of the soil, (4) the slope of the field, and (5) the growth of the roots in your planning, then the nutrients will be used for the benefit of the plant growth and not lost to the environment, where they will cause pollution problems. This requires that a model be used to set up fertilization plans. Preliminary results in Denmark have shown that a 10–20% reduction in fertilizer can be achieved by careful use of a fertilizer model, without any effect on the yield but with a 20–35% reduction of the nitrogen discharges to the environment (internal research paper at the Danish Agriculture University, 1987).

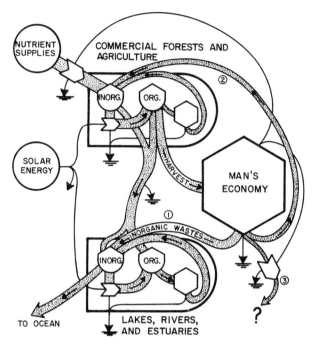

Figure 3.4 Schematic of chemical cycling in ecosystems and human activities, showing loss of nutrients to aquatic systems versus recycling to terrestrial systems. Pathway 1 is pollution of aquatic ecosystems; pathway 2 is proper recycling of chemicals (ecological engineering); pathway 3 is energy-costly environmental technology for pollution control.

4. Homeostasis of ecosystems requires accordance between biological function and chemical composition.

Table 3.1 gives the typical compositions of various living organisms. Their biochemical functions define their compositions, although these are not to be considered as fixed concentrations, but rather as ranges. Figure 3.5, which is an illustration of Shelford's law, shows that organisms are able to regulate only within these given ranges and this law should be considered in the use of ecotechnology.

The success of ecotechnological methods depends on the understanding of this principle. For example, some of the consequences of lake acidification caused by emission of sulfur dioxide from the use of fossil fuel can be explained by a violation of this principle. The low pH value, observed for lakes in Sweden, Norway, southern Canada, and the northeastern United States, causes a lack of inorganic carbon for photosynthesis. The lakes then turn superoligotrophic, the entire food chain is affected, and the lakes become biota poor.

Table 3.1 Elements in Plant and Animal Tissues (mg/kg Dry Matter)

Element	Plankton[a]	Brown Algae	Ferns	Bacteria	Fungi
Ag	0.25	0.28	0.23		0.15
Al	1000	62		210	29
As		30			
Au		0.012			
B		120	77	5.5	5
Ba	15	31	8		
Be					<0.1
Br		740			20
C	225,000	345,000	450,000	538,000	494,000
Ca	8000	11,500	3700	5100	1700
Cd	0.4	0.4	0.5		4
Cl		4700	6000	2300	10,000
Co	5	0.7	0.8		0.5
Cr	3.5	1.3	0.8		1.5
Cs		0.067			
Cu	200	11	15	42	15
F		4.5			
Fe	3500	690	300	250	130
Ga	1.5	0.5	0.23		1.5
H	46,000	41,000	55,000	74,000	55,000
Hg		0.03			
I	300	1500			
K		52,000	18,000	115,000	22,300
La		10			
Li		5.4			
Mg	3200	5200	1800	7000	1500
Mn	75	53	250	30	25
Mo	1	0.45	0.8		1.5
N	38,000	15,000	20,500	96,000	51,000
Na	6000	33,000	1400	4600	1500
Ni	36	3	1.5		1.5
O	440,000	470,000	430,000	230,000	340,000
P	4250	2800	2000	30,000	14,000
Pb	5	8.4	2.3		50
Ra	4×10^{-7}	9×10^{-8}			
Rb		7.4			
Re		0.014			
S	6000	12,000	1000	5300	4000
Se		0.84			2
Si	200,000	1500	5500	180	
Sn	35	1.1	2.3		5
Sr	260	1400	13		320
Ti	80	12	5.3		
U					0.25
V	5	2	0.13		0.67
W		0.035			
Y			0.77		0.5
Zn	2600	150	77		150
Zr	20		2.3		5

[a] Mainly diatoms.

Table 3.1 (continued)

Element	Coelenterata	Annelida	Mollusca	Crustacea	Insecta	Pisces	Mammalia
Ag	5?				≤0.07	11?	0.006
Al		340	50	15	100	10	<3
As	30	6	0.005	0.08		0.3	0.2
Au	0.007		0.008	0.0005		0.0003	<0.009
B		2.1?	20	15	20		<2
Ba			3	0.2			2.3
Bi	0.3?					0.04?	
Br	1000	100?	1000	400		400	4
C	436,000	402,000	399,000	401,000	446,000	475,000	484,000
Ca	1300	11,000	1500	10,000	500	20,000	85,000
Cd	1		3	0.15		3	
Ce							0.47
Cl	90,000		5000	6000	1200	6000	3200
Co	4?	5?	2	0.8	<0.7	0.5	0.3
Cr	1.3?					0.2	<0.3
Cs							0.06
Cu	50	4?	20	50	50	8	2.4
F			2	2		1400	500
Fe	400	630	200	20	200	30	160
Ga	0.5?					0.15?	
Ge	1.5?					0.3?	
H	45,000	59,000	60,000	60,000	73,000	68,000	66,000
Hg			1?			0.3?	0.05
I	15	160	4	1	0.9	1	0.43
K	3000	16,000	19,000	13,000	11,000	12,000	7500
La							0.09
Li			1?		≤7		<0.02
Mg	5500	6000	5000	2000	750	1200	1000
Mn	30?	0.06?	10	2?	10	0.8	0.2
Mo	0.7		2	0.6	0.6	1	<1
N	63,000	99,000	85,000	84,000	123,000	114,000	87,000
Na	48,000		16,000	4000	3000	8000	7300
Ni	26?	11?	4	0.4	9	1	<1
O	271,000	340,000	390,000	400,000	323,000	290,000	186,000
P	14,000	8100	6000	9000	17,000	18,000	43,000
Pb	35?		0.7	0.3	≤7	0.5	4
Ra			1.5×10^{-7}	7×10^{-9}		1.5×10^{-8}	7×10^{-9}
Rb			20				18
Re			0.006	0.0005		0.0008	
S	19,000	14,000	16,000	7500	4400	7000	5400
Sb	0.2					0.2	0.14
Sc							0.006
Se							1.7
Si		150	1000	300	6000	70	120
Sn	23?		15?	0.2		3?	<0.16
Sr		20	60	500			21
Th	0.03						
Ti	7		20	17	160	0.2	<0.7
U						≤0.06	0.023
V	2.3	1.2	0.7	0.4	0.15	0.14	<0.4
W			0.05	0.0005		0.0014	
Zn	1500?	6?	200	200	400	80	160

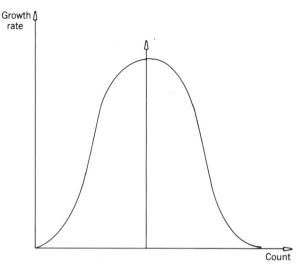

Figure 3.5 Illustration of Shelford's law of limitation. The growth rate is related to a concentration as shown, a normal distribution around an optimal concentration.

5. Processes in ecosystems have characteristic time scales that may vary over several orders of magnitude. Manipulation of ecosystems must be adapted to the ecosystem dynamics.

Examples of wrong environmental management due to lack of time scale considerations can be found in the lack of recognition of prey–predator relationships. Humans have often considered predators as harmful animals and therefore have killed carnivorous predators to protect the herbivorous prey. But this strategy has often failed because a short-time benefit was often realized at the expense of a long-term pattern; that is, the different time scales were not considered. When the number of predators was reduced, the number of prey increased very rapidly. But herbivorous animals have a time scale different from that of their food source, and they have increased above their carrying capacity, with the result of overgrazing and starvation in great numbers (Figure 3.6). The observations at the Kaibab Plateau of Arizona during the years 1918–1940 exemplify these statements (Knight, 1965). In 1918 the plateau supported about 4000 deer, an aggregation below the carrying capacity of the area (about 30,000 deer). However, many of the deer's natural enemies, including wolves, coyotes, and pumas, were killed in the early 1920s. This reduction in predation pressure resulted in a maximum of 100,000 deer (more than three times the carrying capacity) during the winter of 1924–1925. The food supply was consequently depleted, causing the starvation of 60,000 deer during the next two years. The depletion of the initial food supply resulted in a continued attrition until only 10,000 deer were estimated to inhabit the area in 1940.

Number

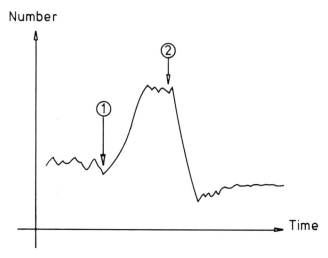

Time

Figure 3.6 Number of herbivorous animals versus time, showing increase in population at time 1 when predators are reduced, and decrease in population at time 2 as a result of overgrazing.

A phenomenon in aquatic ecosystems called the "paradox of the plankton," originally described by Hutchinson (1961), also illustrates the importance of time scales. Hutchinson argued that there is an apparent contradiction of competitive exclusion in aquatic ecosystems, where plankton develop great species diversity in a seemingly homogeneous environment. But one theory, explored in a modeling effort by Kemp and Mitsch (1979), is that "diversity among species which compete for the same niche may require an environment which fluctuates with a periodicity approaching the turnover time for those species." The model, which used turbulence as the environmental variable, illustrated that diversity was maintained when the frequency of turbulence change approached that of the turnover time of the organisms, but was less when the turbulence changed an order of magnitude faster (Figure 3.7).

6. Ecosystem components have characteristic space scales. Manipulation of ecosystems should take into account the appropriate size necessary to achieve the desired result.

The rapid growth of the human population has caused an increased need for agricultural land. Drainage of wetlands and deforestation have been widely used to meet this need. There are numerous examples of how land has been cleared on too large a scale with desertification as a result. Forests—especially tropical rain forests—maintain a high humidity of the soil. When the trees are removed, the soil is exposed to direct solar radiation and dries, causing the organic matter to be "burned" off. Similarly, a wetland slows

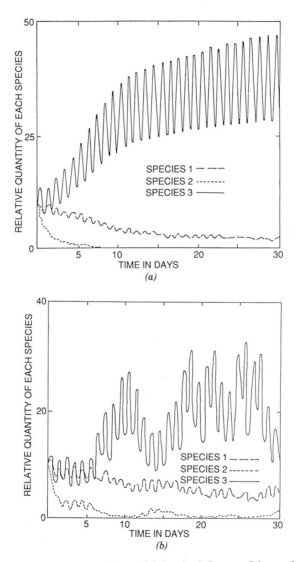

Figure 3.7 Results of plankton simulation model showing influence of time scales on ecosystem structure (Kemp and Mitsch, 1979; reprinted by permission, Elsevier Science Publishers B.V.). Relative abundance of three coexisting species of plankton are shown for (*a*) turbulent frequency of 0–0.01/day (much more rapid than turnover time of plankton) and (*b*) turbulent frequency of 1.0/day (close to turnover time of plankton organisms). Diversity appears to be greater in (*b*).

down the transport of water and at the same time organic matter is formed. When a wetland is drained, water is transported rapidly downstream, the soil dries, organic matter decomposes rapidly, and the rate of new organic matter production decreases significantly.

The wetlands of western White Russia were drained in large scale during the 1960s to meet the increasing demand for food production in the Soviet Union. Sushenya and Parfenov (1982) describe how plant pests increased on cultivated soil and the yield became lower than expected on agricultural land owing to a rapid decomposition of organic compounds in the soil. Therefore preserved bog ecosystems were created to protect valuable species of flora and fauna and to maintain a pattern of wetlands to minimize the damages of the large-scale reclamation of drained lands.

Ecologically sound management that recognizes spatial patterns of resources is needed to overcome these problems. Agro-forestry offers an important solution. Forestry is mixed with agriculture in a pattern that considers the ecosystem characteristics. The same consideration of space scale should be applied by drainage of wetlands. When wetlands are lost in a piecemeal fashion, the impact is not seen immediately but in a gradual fashion until the "cumulative loss" of the wetland leads to the loss of ecosystem function at a landscape scale.

7. Chemical and biological diversity contribute to the buffering capacities of ecosystems. When designing ecosystems, one should introduce a wide variety of parts for the ecosystem's self-designing ability to choose from.

Table 3.2 describes a hierarchy of regulation and feedback mechanisms, which attempt to meet the changes in external factors with the smallest possible internal change in function. Evidently, regulation mechanisms 3, 4, 5, and in part, 6 are related to species diversity. The more possibilities the ecosystem possesses, the higher its buffering capacity. This is especially

Table 3.2 Hierarchy of Regulation and Feedback Mechanisms in Ecosystems

1. Regulation of rates, e.g. uptake of nutrient by algae.
2. Feedback regulation of rates; e.g., by high nutrient concentration in algae the uptake rate will slow down.
3. Adaptation of process rates, e.g., by changing the dependence of nutrient uptake rate on temperature.
4. Adaptation of species to new conditions, e.g., adaptation of insects to DDT.
5. Shift in species composition. Species better fitted to new conditions will be more dominant.
6. A more pronounced shift in species composition causing a shift in the structure of the ecosystem.
7. Change in the genetic pool available for selection.

true for the buffering capacities related to the function of the system, for instance, maintaining photosynthesis at a certain level. This must not be interpreted as the widely discussed relationship between diversity and stability (see for example, McNaughton and Wolf, 1973). An ecosystem with high diversity may change its species composition radically and therefore be considered unstable in the population dynamic sense, but still have a high buffering capacity related to its function as an ecosystem. The buffering capacity is defined as the change in a state variable relative to the change in the forcing function. If the state variable is selected as a concentration or number of one species, the buffering capacity may be low when the ecosystem uses species composition as a regulation mechanism, but the buffering capacity may still be very high when the concentration or the number of organisms in a trophic level is used.

Another example is the use of diversity in crop and vegetable production. If several plant species are cultivated on the same field each becomes less susceptible to attack by insects. Experiments on the use of integrated cultures of vegetables illustrate this principle very clearly. Numerous examinations of integrated cultivation are found in the literature. One example out of many involves onions and beans cultivated together, one row onions, one row haricots, and so forth. With a small distance between the rows, pest damage was reduced by 50% compared with nonintegrated cultures.

Chemical diversity is equally important as biological diversity. All components needed for biological growth must be present in an ecosystem (see also principle 4). The biological diversity, in other words, depends on the chemical diversity.

8. Ecosystems are most vulnerable at their geographical edges. Ecological management should take advantage of ecosystems and their biota in their optimal geographical range.

Ecosystems have defined geographical ranges in which they are well tuned to the climatological (and sometimes geological) features of the landscape. When ecological engineering involves ecosystem manipulation, the stability of the system will be enhanced if the species are in the middle range of their environmental tolerance. If some or a great number of species are at their northern or southern extreme, for example, the ecosystem may be poorly suited for ecological engineering. It is often at this ecological edge that rare and endangered species exist, making the unsuitability even more apparent.

9. Ecotones are formed at the transition zones between ecosystems. The interfaces between human settlement and nature should be designed as gradual transitions, not as sharp boundaries.

Nature has developed transition zones, or *ecotones*, between ecosystems to make a soft transition. Ecotones may also be considered buffer zones between two ecosystems. They are able to absorb undesirable changes imposed on an ecosystem from neighboring ecosystems. Humans must use the same concepts when they design interfaces between human settlement and nature. It is a typical mistake, which can be seen all over the world, to construct houses, hotels, and so on close to a lake or sea shoreline. Emissions coming from the human settlement will be transferred directly to the ecosystem under such circumstances. If a buffer zone such as a wetland were maintained, the emission would be at least partly absorbed. The use of a wetland as an ecotone between an aquatic and terrestrial ecosystem is mentioned in Chapters 8 and 11. In Chapter 16, studies of Lake Glumsø in Denmark show that a wetland around the lake is the only available satisfactory solution to the eutrophication problem of the lake.

10. Ecosystems are coupled with other ecosystems. This coupling should be maintained wherever possible and ecosystems should not be isolated from their surroundings.

Ecosystems are open systems and as such exchange mass and energy with their environment. Ecosystems and their environments are therefore coupled. It is important to take that into consideration whenever ecological engineering is implemented. If a component is removed from one ecosystem, the problem is not solved if the component harms another ecosystem.

The proper use of manure as natural fertilizer in agriculture requires that the evaporation of ammonia be minimized, for the ammonia in air will dissolve in rainwater and return to other ecosystems, where it may be harmful. Ammonia in manure can be bound by the use of minerals and/or by adjustment of pH. Biogas production of manure also produces a residue, which contains less ammonia. A research project, as yet unpublished, at the Danish University of Agriculture shows that by precipitation with bentonite or activated zeolite, the loss of ammonia to the environment can be reduced by 20–30%.

11. Ecosystems with pulsing patterns are often highly productive. The importance of pulsing subsidies should be recognized and taken advantage of where possible.

Ecosystems with pulsing patterns often have greater biological activity and chemical cycling than systems with relatively constant patterns. For example, the highest productivities of forested wetlands in Florida were found in environments that had pulsing hydroperiods, with lower productivities in either drier or wetter environments (Figure 3.8). This recognizes the role that pulsing subsidies such as floods can have on ecosystem. However, as discussed by Odum (1982), the pulses, whether externally or internally generated, must be of the correct frequency and duration to allow a system to maximum production, consumption, and recycling. Otherwise little advan-

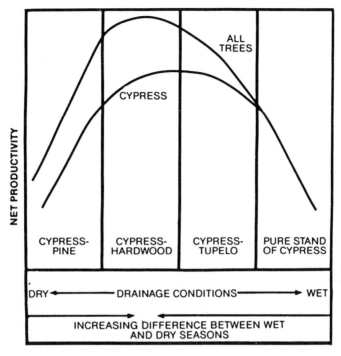

Figure 3.8 Productivity versus hydroperiod for cypress wetlands in Florida. Highest productivity was found in systems influenced by flooding pulses (Mitsch and Ewel, 1979; reprinted by permission, American Midland Naturalist.).

tage may be realized from pulses to frequent or rare relative to ecosystem structure (see also principle 5 above).

Another specific case study will illustrate the recognition of the pulsing force and how it is possible to take advantage of it in ecological engineering. Figure 3.9 shows a map of an estuary in Brazil, named Cannaneia. The shores of the islands and the coast are very productive mangrove wetlands and the

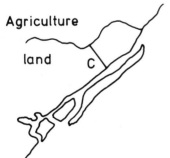

Figure 3.9 Map of Cannaneia Estuary in Brazil. Channel C was build to avoid upstream flooding of agricultural land.

entire estuary is an important nesting area for fish and shrimp. Channel C was built to avoid flooding upstream, where productive agriculture land is situated. The construction of the channel has caused a conflict between farmers, who want the channel open all year round, and fishermen, who want the channel closed. The estuary is exposed to the tide, which is important for maintenance of a good water quality with a certain minimum salinity. However, because the channel transports a significant amount of fresh water to the estuary, it was hardly possible to maintain the right salinity for the mangroves and for the nesting area. The conflict can be solved by use of an ecological engineering approach that takes advantage of the pulsing force (the tide). A sluice in the channel could be constructed to discharge the fresh water when it is most appropriate. The tide would in this case be used to transport the fresh water as rapidly as possible to the sea; the sluice should be closed when the tide is on its way into the estuary. The tidal pulse frequency is selectively "filtered" to produce an optimal management situation.

12. Everything is linked to everything else in the ecosystem. It is impossible to manage one component of an ecosystem without affecting other parts.

All components in an ecosystem are linked to each other either directly or indirectly (see also Patten, 1983). Although this statement may appear to be ecologically trivial and obvious, its message must be emphasized in all ecological engineering approaches to avoid undesired effects on components other than the one that is managed. Sometimes effects are even observed on the managed components that are opposite to what is planned, owing to indirect effects among biological components of ecosystems, for instance, by use of biomanipulation. A good model will be able to predict such indirect effects and thereby prevent the application of unacceptable management strategies. The Kaibab plateau case study refered to under principle 5 illustrates the result of a management that did not consider this proposition.

13. Ecosystems have feedback mechanisms, resilience, and buffer capacities in accordance with their preceding evolution. Existing ecosystems do not match well with man-made synthetic chemicals, although new emerging ecosystems can develop to deal with them in some cases.

As shown in Table 3.2, ecosystems have a hierarchy of feedback mechanisms. They have evolved to cope with the fluctuations and changes that nature has imposed for millions of years. Ecosystems are therefore able to meet problems related to nature's own components, but often have difficultues in coping with man-made synthetic chemicals. The examples on the use of adapted plants and microorganisms used in relation to principle 2 show, however, that it is possible to help nature to cope with synthetic chemicals. Chapter 10, for example, discusses strains of bacteria that can degrade organic pesticides such as parathion and *para*-nitrophenol.

REFERENCES

Calvin, M. 1983. New sources for fuel and material. *Science 219:*24–26.

Hutchinson, G. E. 1961. The paradox of the plankton. *Am. Nat. 95:*137–146.

Kemp, W. M. and W. J. Mitsch. 1979. Turbulence and phytoplankton: A general model of the paradox of the plankton. *Ecol. Modelling 7:*201–222.

Knight, C. 1965. *Basic Concepts of Ecology.* Macmillan, New York.

McNaughton, S. J. and L. L. Wolf. 1973. *General Ecology.* Rinehart and Winston, New York, 710 pp.

Mitsch, W. J. 1988. Productivity–hydrology–nutrient models of forested wetlands. In W. J. Mitsch, M. Straškraba, and S. E. Jørgensen, Eds., *Wetland Modelling.* Elsevier, Amsterdam, pp. 115–132.

Mitsch, W. J. and K. C. Ewel. 1979. Comparative biomass and growth of cypress in Florida wetlands. *Am. Midl. Nat. 101:*417–426.

Mitsch, W. J., C. L. Dorge, and J. W. Wiemhoff. 1979. Ecosystem dynamics and a phosphorus budget of an alluvial cypress swamp in southern Illinois. *Ecology 60:*1116–1124.

Odum, H. T. 1982. Pulsing, power and hierarchy. In W. J. Mitsch, R. K. Ragade, R. W. Bosserman, and J. A. Dillon, Jr., Eds., *Energetics and Systems.* Ann Arbor Science, Ann Arbor, MI, pp. 33–59.

Sushenya, L. M. and V. I. Parfenov. 1982. The impact of drainage and reclamation on the vegetation and animal kingdom of ecosystems on Byelo-Russian bogs. In *Proceedings of International Scientific Workshop on Ecosystem Dynamics in Freshwater Wetlands and Shallow Water Bodies,* Vol. 1. SCOPE and UNEP, Moscow, pp. 218–226.

Patten, B. C. 1983. On the quantitative dominance of indirect effects in ecosystems. In W. K. Lauenroth, G. V. Skogerboe, and M. Flug, Eds., *Analysis of Ecological Systems: State-of the-Art in Ecological Modelling.* Elsevier, Amsterdam, pp. 27–37.

4

PRINCIPLES OF ECOLOGICAL MODELING

Sven Erik Jørgensen

Department of Environmental Chemistry, Institute of Chemistry AD, The Royal Danish School of Pharmacy, Copenhagen, Denmark

4.1 MODELING AND ECOTECHNOLOGY

The topic of this book is ecological engineering or ecotechnology. Nevertheless, a chapter on modeling must be included, because the final selection of an ecotechnological method will almost always be based on model results, as it will be demonstrated in some of the case studies in Part Two.

A relation between modeling and ecotechnology already exists. Many ecotechnological methods have been inspired by application of models. The use of eutrophication models is an illustrative example. The input of nutrients to aquatic ecosystems is either from point sources (wastewater) or non-point sources (agricultural runoff). The model results may show that it is not sufficient to remove the nutrients from wastewater (see, e.g., Chapter 16), but further reduction in the nutrients input is needed. This is possible either by use of wetlands as sinks of nutrients or by agricultural management, which includes better control of fertilizer use (see Chapter 8).

This example illustrates, furthermore, that ecotechnological methods will provoke the development and application of additional models. The use of a wetland as a sink of nutrients raises the question: how can we obtain the maximum effect of the wetland as a nutrient trap? Because a wetland is a complex ecosystem, the answer can be found quantitatively only by use of a wetland model, which considers all the factors that determine the removal of nutrients from the water. Similarly, a better plan for the use of fertilizers requires that the plan take into consideration the composition of soil, the development of the plants, pH and humidity of the soil, irrigation and/or the expected precipitation—again a complex problem, which can be solved only by the use of a field model.

The relation between ecotechnology and modeling, exemplified by the above-mentioned case of eutrophication, is often realized. It is shown in a more generalized form in Figure 4.1.

Ecological models have been increasingly used in environmental management. The idea behind the use of ecological management models is demonstrated in Figure 4.2. Urbanization and technological development have had an increasing impact on the environment. Energy and pollutants are released into ecosystems, where they may cause more rapid growth of algae or bacteria, damage species, or alter the entire ecological structure. An ecosystem is extremely complex, and so it is an overwhelming task to predict the environmental effects that a given emission will have. It is here that the model comes into the picture. With sound ecological knowledge it is possible to extract the features of the ecosystem that are involved in the pollution problem under consideration; these form the basis of the ecological model (see also the discussion in Chapter 2). As indicated in Figure 4.2, the resulting model can be used to select the environmental technology best suited for the solution of specific environmental problems, or legislation reducing or eliminating the emission can be passed.

Models are, in other words, powerful tools in selecting the optimal combination of environmental technology and ecotechnology for pollution control. The model used for this problem must relate the emission with the effects on the ecosystem level. Another model is needed for the use of ecological engineering. This model must relate the management or even manipulation of the ecosystem and/or adjacent ecosystems with environmental quality in its broadest sense (see Figure 4.1). It implies that we must construct a model that considers the processes that we can regulate and/or manipulate, including their effects on the general conditions of the ecosystem and the effect of anthropogenic emissions. Of course it may sometimes be

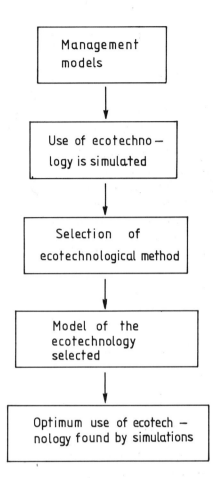

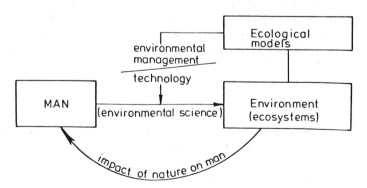

Figure 4.1 The relation between modeling and ecotechnology.

Figure 4.2 Relations among environmental science, ecology, ecological modeling, and environmental management and technology.

advantageous to aggregate the two models into one, more comprehensive model. How related the two models are will depend on many factors, including the application of computer languages.

It is not surprising that the application of ecological engineering often requires a more comprehensive ecological knowledge that is needed by the use of a solution based on environmental technology. Let us use the eutrophication case again to illustrate this. A eutrophication model attempts to capture all the essential relations between input of nutrients and the concentration of phytoplankton, but if wetlands have to be used as nutrient traps and agricultural management for control of fertilizer use, then it is necessary also to describe quantitatively the cycling of nutrients in wetlands and agricultural lands, as well as their relation with nutrient cycling in the lake.

The basic concepts of modeling are presented in this chapter, but those who want a more comprehensive treatment of ecological modeling are referred to *Fundamentals of Ecological Modelling*, second edition (Jørgensen, 1988).

4.2   ELEMENTS OF MODELING

An ecological model consists, in its mathematical formulation, of five components, as discussed below.

4.2.1   Forcing Functions or External Variables

Forcing functions or external variables are functions or variables of an external nature that influence the state of the ecosystem. In a management context the problem can often be reformulated as follows: If certain forcing functions are varied, what will be the influence on the state of the ecosystem? In other words, the model is used to predict what will change in the ecosystem when forcing functions are varied with time. Examples of forcing functions are the input of pollutants to the ecosystem, the consumption of fossil fuel, or a fishery policy, but temperature, solar radiation, and precipitation are also forcing functions (which, however, we cannot at present manipulate). Forcing functions that are controllable by humans are often named *control functions*.

4.2.2   State Variables

State variables describe, as the term indicates, the state of the ecosystem. The selection of variables is crucial for the model structure, but in most cases the choice is obvious. If, for instance, we want to model the eutrophication of a lake it is natural to include the phytoplankton concentration and the concentrations of nutrients. When the model is used in a management context the value of the state variables predicted by changing the forcing functions can be considered as the result of the model, because the model will contain relations between the forcing functions and the state variables. Most models will contain more state variables than are *directly required* for purposes of management, because the relations are so complex that they require the introduction of additional state variables. For instance, it would be sufficient in many eutrophication models to relate the input of one nutrient with the phytoplankton concentration, but because this variable is influenced by many factors (it is influenced by other nutrient concentrations, temperature, hydrology of the water body, zooplankton concentration, solar radiation, transparency of the water, etc.) a eutrophication model will most often contain a number of state variables.

4.2.3   Mathematical Equations

The biological, chemical, and physical processes in the ecosystem are represented in the model by means of mathematical equations that are the relations between forcing functions and state variables. The same type of processes can be found in many ecosystems, which implies that the same equations can be used in different models. It is, however, not possible today to have one equation that represents a given process in all ecological contexts. Most of the processes have several mathematical representations that are *equally* valid, either because the process is too complex to be understood in sufficient detail at present, or because some specified circumstances allow us to use simplifications.

4.2.4   Parameters

The mathematical representation of processes in the ecosystem contains *coefficients* or *parameters*. They can be considered constant for a specific ecosystem or part of ecosystem. In causal models the parameters will have a scientific definition, for example, the maximum growth rate of phytoplankton. Many parameter values are known within limits. In Jørgensen (1979) can be found a comprehensive collection of ecological parameters. However, only a few parameters are known exactly and so it is necessary to calibrate the others.

By *calibration* we attempt to find the best accordance between computed and observed state variables by variation of a number of parameters. The calibration may be carried out by trial and error or by use of software developed to find the parameters that give the best fit.

In many static models, where process rates are given as average values in a given time interval, and in many simple models that contain only a few well defined or directly measured parameters, calibration is not required. In models aimed at simulating the dynamics of the ecological process, the calibration is crucial for the quality of the model for the reasons summarized below:

1. The parameters are in most cases known only within limits.

2. Different species of animals and plants have different parameters, which can be found in the literature (see Jørgensen 1979). However, most ecological models do not distinguish between different species of phytoplankton, but consider them as one state variable. In this case it is possible to find limits for the phytoplankton parameters, but because the composition of the phytoplankton varies throughout the year an exact average value cannot be found.

3. The influence of the ecological processes that are of minor importance to the state variables under consideration, and therefore not included in the model, can to a certain extent be considered by the calibration, where the results of the model are compared with observations in the ecosystem. This might also explain why the parameters have different values in the same model when used for different ecosystems. The calibration can, in other words, take the site differences and the ecological processes of minor importance into account, but obviously it is essential to reduce the use of calibration for this purpose. The calibration must never be used to force the model to fit observations if unrealistic parameters are thus obtained. If a reasonable fit cannot be achieved with realistic parameters the entire model should be questioned. It is therefore extremely important to have realistic ranges at least for the most sensitive parameters. This implies that a sensitive analysis must be carried out, whereby the influence of changes in submodels, parameters, or forcing function on the most crucial state variables is determined.

4.2.5 Universal Constants

Most models also contain universal constants such as the gas constant or molecular weights. Such constants are of course not subject to calibration.

Models can be defined as formal expressions of the essential elements of a problem in either physical or mathematical terms. The first recognition of the problem is often, and most likely, expressed *verbally*. This can be recognized as an essential preliminary step in the modeling procedure, but the term formal expressions implies that a translation into physical or mathematical terms must take place before we have a model.

The verbal model is difficult to visualize and it is therefore conveniently translated into a *conceptual diagram*, which contains the state variables, the forcing functions, and how these components are interrelated by processes. The conceptual diagram can be considered as a model, and is named a conceptual model. A number of models in the ecological literature stop at this stage because of lack of knowledge about the mathematical formulation of processes. They can, however, be used to illustrate the relationships qualitatively.

Figure 3.3 (Chapter 3) illustrates a conceptual model of the nitrogen cycle in a lake. The state variables are nitrate, ammonium, nitrogen in phytoplankton, nitrogen in zooplankton, nitrogen in fish, nitrogen in sediment, and nitrogen in detritus. In the diagram are shown the following forcing functions: inflow, outflow, and the concentrations of nitrate and ammonium in the in- and outflow. Other forcing functions not shown are solar radiation and temperature. The arrows in the diagram illustrate the processes numbered 1 to 18. If we want to proceed to a quantitative model, it is necessary to formulate these processes by the use of mathematical expressions (equations).

It is of great importance to verify and validate models. *Verification* is a test of the *internal logic* of the model. Typical questions in the verification phase are: does the model react as expected? For example, will increased discharge of organic matter give a lower concentration of oxygen in a river model concerned with the oxygen balance of the system? Is the model long-term stable? Does the model follow the law of mass conservation?

Verification is therefore largely a subjective assessment of the behavior of the model. To a large extent verification will inevitably go on during the use of the model before the calibration phase, which has been mentioned above.

Validation must be distinguished from verification, but previous use of the words has not been consistent. Validation consists of an objective test of how well the model outputs fit the data.

4.3 MODELING PROCEDURE

The primary focus of all research at all times is to define the problem. Only in that way can it be assured that limited research resources can be correctly allocated and not dispersed into irrelevant activities.

The definition of the actual problem must be bound by the constituents of space, time, and subsystems. The bounding of the problem in space and time is usually easier, and consequently more explicit, than the identification of the ecological subsystems to be incorporated in the models.

Some of the projects of the International Biological Programme (IBP) assumed that it was necessary to model the whole ecosystem and that it was unnecessary to define subsystems of that ecosystem. When the final synthesis was attempted, major gaps were found in many of the projects, which could not be filled by any of the experimental or survey results, and these gaps were frequently emphasized by the absence of any preliminary synthesis (Jeffers, 1978).

The experience of IBP has led many ecologists to question the need for studies of whole ecosystems and to focus their attention on carefully designed sets of subsystems. In the synthesis of the eutrophication of lakes, for example, attention must be concentrated on algal growth and nutrient

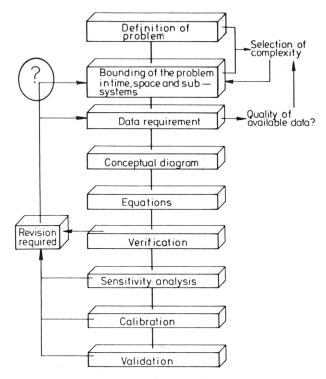

Figure 4.3 A tentative modeling procedure.

cycles as a basis for the prediction of the effect of the nutrients on the eutrophication process.

The use of models in the ecological context is relatively new, and therefore few guides are available for the construction of ecological management models. A tentative guideline is presented in Figure 4.3.

In addition to defining the problem and its parameters in space and time, it is important to emphasize that this procedure is unlikely to be correct at the first attempt, and so there is no need to aim at perfection in one step. The main requirement is to get started (Jeffers, 1978). All ecosystems have a distinctive character, and a comprehensive knowledge of the system that is going to be modeled is often needed to get a good start.

It is difficult to determine the optimal number of subsystems to be included in the model for an acceptable level of accuracy, and often it is necessary to accept a lower level than intended at the start owing to a lack of data.

It has been argued that a more complex model should be able to account more accurately for the complexity of the real system, but this is not true. Some additional factors have to be included in this consideration. As increasing numbers of parameters are added to the model there will be an increase in uncertainty. The parameters have to be estimated by observations in the field, by laboratory experiments, or by calibrations, which again are based on field measurements. Parameter estimates are therefore never errorfree. Because these errors are carried through into the model they contribute to the uncertainty of the prediction derived from the model, and there seems therefore to be great advantage in reducing the complexity of models.

Some ecologists argue that ignorance of species diversity increases the risk of neglecting important elements of their dynamics. However, comparison of models with different complexities (Jørgensen et al., 1978, 1981) demonstrates that the deviations of simpler models from alternative models, which take biological diversity into account, might be negligible for the purpose of the model. This trade-off between complexity and simplicity in the choice of model is one of the most difficult modeling problems. Attempts have been made to provide some general rules. The method published by Jørgensen et al. (1978) measures the *response* of the model to more state variables and concludes that only the major influences of importance for the problem in focus should be included in the model. The method might also be interpreted as a sensitivity test on the addition of state variables.

Once the model complexity has been selected, at least for the first attempt, it is possible to conceptualize the model (e.g., in the form of a diagram such as those shown for the nitrogen cycles in Figure 3.3). This will give information on which state variables and processes are required in the model. For most processes a mathematical description is available, and most of the parameters have, at least within limits, known values from the literature. Tables of parameters used in ecological models can be found in Jørgensen (1979).

It is possible at this stage to set up alternative equations for the same process and apply the model to test the equations against each other. However, the many ecological processes not included in the model have some influence on the processes in the model. Furthermore, the parameter values used from the literature are often not fixed numbers, but are rather indicated as intervals. Biological parameters can most often not be determined with the same accuracy as chemical or physical parameters owing to changing and uncontrolled experimental conditions. Consequently, calibration by the application of a set of measured data (see also the discussion in Section 4.2) is almost always required. However, the calibration of several parameters is not realistic. Mathematical calibration procedures for 10 or more parameters are not available for most problems. Therefore, it is recommended that sound values from the literature be used for all parameters, and that a sensitivity analysis of the parameters (see Figure 4.3) be made before the calibration. The most sensitive parameters should be selected, for an acceptable calibration of four to eight parameters is possible with the present techniques.

If it is necessary to calibrate 10 parameters or more it is advantageous to use two different series of measurements for the calibration of five parameters each, preferably by selecting measuring periods when the state variables are most sensitive to the parameters calibrated (Jørgensen et al., 1981). It is very important to make the calibration on the basis of reliable data; unfortunately, many ecological models have been calibrated against inaccurate information.

It is characteristic of most ecological models that analysis and calibration of submodels are required. If ecological models are built without knowledge of the ecosystem and its subsystems, they are often not realistic. These considerations are included in the procedure indicated above for the calibration and also in the modeling procedure presented in Figure 4.3.

After the calibration it is important to validate the model, preferably against a series of measurements from a period with changed conditions, for example, with changed external loading or climatic conditions.

The right complexity cannot, as already discussed, be selected generally. It seems that there has been a tendency to choose too complex rather than too simple a model, probably because it is very easy to add to the complexity; it is far more troublesome to obtain the data that are necessary to calibrate and validate a more complex model. As we have repeated here several times, it is necessary to select complexity on the basis of *the problem, the system, and the data available*.

The tentative modeling procedure presented in Figure 4.3 is only one among many workable procedures. However, the components of other possible procedures are approximately the same. The goals and the objectives of the model determine its nature. The steps of setting up a conceptual diagram, verification and validation are repeated in all procedures. Calibration and thereby sensitivity analysis might sometimes be redundant, when the parameters already are known with sufficient accuracy.

Modeling should, however, be considered an iterative process. When the model in the first instance has been verified, calibrated, or validated, new ideas will emerge on how to improve the model. Modelers will wish to build into the model new data, knowledge, or experience either from their own experiments or from the scientific literature; this implies that they must go through at least part of the procedure again to come up with a better model. Modelers know that they can always build a better model, which has higher accuracy, is a better prognosis tool, or contains more relevant details than the preceding model. They will approach the ideal model asymptotically, but will never reach it. However, *limited resources* will sooner or later stop the iteration and the modelers will declare their model to be good enough within the *given limitations*.

4.4   CLASSES OF ECOLOGICAL MODELS

It is useful to distinguish between various types of models and discuss the selection of model types briefly. In the introduction models were divided into two groups; research or scientific models and management models. In

Table 4.1 Classification of Models (Pairs of Model Types)

Type of Models	Characterization
Research models	Used as a research tool
Management models	Used as a management tool
Deterministic models	Predicted values are computed exactly
Stochastic models	Predicted values depend on probability distribution
Compartment models	Variables defining the system are quantified by means of time-dependent differential equations
Matrix models	Matrices are used in the mathematical formulation
Reductionistic models	As many relevant details are included as possible
Holistic models	General principles are used
Static models	Variables defining the system do not depend on time
Dynamic models	Variables defining the system are a function of time (or perhaps of space)
Distributed models	Parameters are considered functions of time and space
Lumped models	Parameters are within certain prescribed spatial locations and time, considered as constants
Linear models	First-degree equations are used consecutively
Nonlinear models	One or more of the equations are not first degree
Causal models	Inputs, states, and outputs are interrelated by use of causal relations
Black box models	Input disturbances affect only the output responses. No causality is required
Autonomous models	Derivatives do not explicitly depend on the independent variable (time)

Table 4.1 other pairs of model types are shown. A stochastic model contains *stochastic input disturbances* and *random measurement errors;* see Figure 4.4. If these are both assumed to be zero the *stochastic model* reduces to a *deterministic model*, provided that the parameters are known exactly and not estimated in terms of statistical distributions. It is worth underlining that a deterministic model is tantamount to the assumption that one has perfect knowledge of the behavior of the system. Such a model implies that the future response of the system is completely determined by a knowledge of the present state and future measured inputs.

The application of the expressions *compartment* and *matrix models* is not consistent, but some modelers distinguish between these two classes of

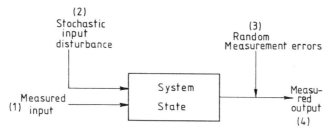

Figure 4.4 A stochastic model considers (1)–(3) whereas a deterministic model assumes that (2) and (3) are zero.

models entirely by the mathematical formulation, as indicated in Table 4.1. The classification is not widely applied.

The classification *reductionistic* and *holistic models* is based upon a difference in the scientific ideas behind the model. The reductionistic modeler will attempt to incorporate as many details of the system as possible in order to capture its behavior. This type of modeler believes that the properties of the system are the sum of all the details. The holistic modeler, on the other hand, attempts to include in the model properties of the whole ecosystem by use of general system principles. In this case it is the properties of the system, not the sum of all the details, that are considered, but the system possesses some additional properties because the subsystems are working as a unit.

Dynamic systems might have four classes of state. *The initial state* changes through *transient states* to a state where the system *oscillates around a steady state*, as shown in Figure 4.5. The transient phase can be described only by use of a dynamic model, which uses differential or difference equations to describe the system response to external factors. Differential equations are used to represent continuous changes of state with time, whereas difference equations use discrete time steps. The steady state corresponds to all derivatives equal to zero. The oscillations around the steady state are described by use of a dynamic model, and *steady state* itself can be described by use of a *static model*. Because all derivatives are equal

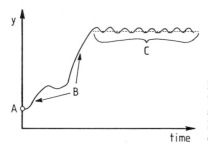

Figure 4.5 y is a state variable expressed as a function of time. A is the initial state, B transient states, and C oscillation around steady state. The dotted line corresponds to the steady state, which can be described by use of a static model.

to zero in steady states the static model is reduced to *algebraic equations*. Some dynamic systems have no *steady state*, for instance, systems that show *limit cycles*. This fourth state possibility obviously requires a dynamic model to describe the system behavior. In this case the system is always nonlinear, although there are nonlinear systems that have steady states.

A *static model* assumes consequently that all the variables and parameters are independent of time. The advantage of the static model is its potential for simplifying subsequent computational effort through the elimination of one of the independent variables in the model relationships. A typical example is a model that computes an average time-invariant set of wastewater discharge, temperature, and stream flow rate conditions. The model can be used as a management tool by comparison of various steady-state situations, but the model cannot be used to predict when these situations will occur. If forecasting systems must be applied it is necessary to use dynamic models, which are characterized by time-variant state variables.

A *distributed model* accounts for variations of variables in time and space. A typical example would be an advection–diffusion model for the transport of dissolved substances along a stream. It might include variations in the three orthogonal directions. The analyst might, however, decide on the basis of prior observations that gradients of dissolved material along one or two directions are not sufficiently large to merit inclusion in the model. The distributed model would then be reduced by that assumption to a *lumped parameter model*. A typical example of this kind of model is the continuously stirred tank reactor idealization of lake water quality dynamics. Whereas the lumped model is frequently based upon ordinary differential equations, the *distributed parameter model is usually defined by partial differential equations*.

Most distributed and lumped models are nonlinear, and a special case of this general class of nonlinear models is the *linear model*. The great advantage of the linear model is that it obeys the principle of superposition. If the input forcing function IF gives the output response OR, and likewise the forcing function IFF is related to the output ORR, then the combination of inputs (*a* IF + *b* IFF) will produce the model response (*a* OR + *b* ORR), where *a* and *b* are constants.

The causal or *internally descriptive model* characterizes how the inputs are connected to the states and how the states are connected to each other and to the outputs of the system, whereas the *black box model* reflects only what changes the input will effect in the output responses. In other words, the causal model provides a description of the internal mechanisms of process behavior. The black box model deals only with what is measurable, the input and the output.

A model that relates the input of nutrient with the phytoplankton concentration in a reservoir directly is an example of a black box model. The relationship might be found on the basis of a statistical analysis of the forcing function (nutrient input) and the phytoplankton concentration measured in

Table 4.2 Identification of Models

Type of Models	Organization	Pattern	Measurements
Biodemographic	Conservation of species or genetic information	Life cycles	Number of individuals or species
Bioenergetic	Conservation of energy	Energy flow	Energy
Biogeochemical	Conservation of mass	Element cycles	Mass or concentrations

the reservoir water. If, on the other hand, relationships of the processes are described in the model by the use of equations, the model will be causal.

The modeler might prefer to use black box descriptions in cases where knowledge about the processes is rather limited. The disadvantage of the black box model, however, is that it is limited in application to the considered ecosystem or at least to a similar ecosystem. If general applicability is required it is necessary to set up a causal model. This latter type is much more widely used in ecology than the black box model, mainly owing to the understanding that the causal model gives the model user of the function of the ecosystem.

Autonomous models do not explicitly depend on time (the independent variable):

$$\frac{dy}{dt} = ay^b + cy^d + e \qquad (4.1)$$

Nonautonomous models contain terms, $g(t)$, that make the derivatives dependent on time, for example:

$$\frac{dy}{dt} = ay^b + cy^d + e + g(t) \qquad (4.2)$$

The expressions homogeneous and nonhomogeneous models are often used to cover autonomous and nonautonomous models, respectively, when the derivatives are linear functions.

In Table 4.2 another classification of models is shown. The differences between the three types of models are the choice of components used as state variables. If the model aims for a description of a number of individuals, species, or classes of species the model is called *biodemographic*. A model that describes the energy flows is named *bioenergetic* and the state variables are typically expressed in kilowatts or kilowatts per unit of volume or area. *Biogeochemical models* consider the flow of material and the state variables are indicated in kilograms, kilograms per cubic meter, or kilograms per square meter. Often this type of models includes one or more element cycles.

Energy can, to a certain extent, replace organic matter in the model because 1 kg of biological material can be assigned a content of energy. It is therefore often quite simple to transfer a biogeochemical model to a bioenergetic model that describes energy flow. The difference between these two model types is therefore minor and is often related to the conceptual phase.

4.5 SELECTION OF MODEL COMPLEXITY AND STRUCTURE

When the modeler has clarified the scope of the model, the basic properties of the ecosystem to be modeled, and the data availability, the next step is to set up a conceptual diagram for the model. Because modeling is an iterative process it might be necessary to go back and *redefine the problem or expand the data requirements*, perhaps just after the conceptualization phase. But sometimes it may be impossible to expand the amount of available data and the modeler is then forced to simplify the model even at this stage. The model is determined by the problem, the ecosystem, and the data. And even for the most enthusiastic modeler resources are limited.

A mathematical model will therefore always be a result of *several simplifications and assumptions*, and it is a difficult task to make the right ones. An ecosystem can be modeled in several ways according to the purpose of the project. The choice of subsystems or model compartments is arbitrary. Thus several alternative models can be derived for the same environment and usually no objective method is used to select one particular model instead of another, given the modeling goals. The choice of the compartments involves a conceptualization of the system under study so that the right information can be obtained from the model. The process of conceptualization is the most fundamental, because once decisions are made at this level, all results and conclusions depend on this choice.

Various methods of constructing a conceptual model exist, but here only the considerations involved when the modeler *selects the complexity and the structure* of the model are touched on. Only a few theoretical approaches are available to solve this crucial problem, but they will be able to provide some guidelines on model selection.

A model that is capable of accounting for the complete input–output behavior of the real ecosystem and being valid in all experimental frames can never be fully known (Ziegler, 1976). This model is called the base model by Ziegler; it would be very complex and require such a great number of computational resources that it would be almost impossible to simulate. The base model of an ecosystem will never be fully known, because of the complexity of the system and the impossibility of observing all states. However, given an experimental framework of current interest, a modeler is likely to

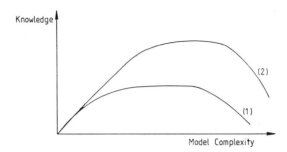

Figure 4.6 Knowledge plotted versus model complexity measured, for example, by the number of state variables. The knowledge increases up to a certain level. Increased complexity beyond this level will not add to knowledge about the modeled system. At a certain level, knowledge might decrease. (2) corresponds to an available data set that is more comprehensive or has a better quality than (1).

find it possible to construct a relatively simple model that is workable within that framework. This is a lumped model and it is the modeler's image of the ecosystem with the components lumped together and with simplified interactions (Zeigler, 1976).

It is a general assumption that a model may be made more realistic by adding more and more connections, up to a point. Addition of new parameters after that point does not improve simulations further; on the contrary more parameters imply more uncertainty, because of the possible lack of information about the flows that the parameters quantify. Given a certain amount of data, the addition of new state variables or parameters beyond a certain model complexity does not add to our ability to model the ecosystem, but only adds unaccounted uncertainty. These ideas are visualized in Figure 4.6. The relationship between knowledge gained through a model and its complexity (e.g., measured as the number of state variables or the number of connectivities) is shown for two levels of data quality and quantity. The question under discussion can be formulated with relation to this figure: how can we select the complexity and the structure of the model to assure that we reach the optimum for "knowledge gained," for the best answer to the question posed to the model? Costanza and Sklar (1985) have examined articulation, accuracy, and effectiveness of 87 different mathematical models of wetlands. The conclusion of their examination may be interpreted as support for the relationship shown in Figure 4.6.

4.6   APPLICATION OF MODELS IN ECOTECHNOLOGY

Ecological models are extensively used together with ecotechnological methods. The general questions remain:

1. which models are used in relation with which ecotechnological methods?
2. Which types of models are used? (See Section 4.4 for definitions of model types.)

Most ecotechnological methods have been used during the past two decades. Before 1970 our knowledge of quantitative ecology and modeling was too limited to consider the use of these methods in pollution control. Today we are more concerned with environmental problems, our ecological knowledge is rapidly increasing, and we steadily gain more experience in the use of ecological modeling in environmental management. We have realized that an effective reduction in the pollution problems of our environments is possible only by use of a *wide* spectrum of methods and tools. Complex problems in complex systems cannot be solved by simple methods, but require complex solutions.

It can be foreseen that more and more pollution problems will find ecotechnological solutions in the future, and these will provoke a wider and wider use of ecological models. That will give us a better understanding of our environment and thereby inspire us to develop new ecotechnological methods.

REFERENCES

Costanza, R. and F. H. Sklar. 1985. Articulation, accuracy and effectiveness of mathematical models: a review of freshwater wetland applications. *Ecol Modelling 27:*45–69.

Jeffers, N. R. J. 1978. *An Introduction to Systems Analysis with Ecological Applications.* E. Arnold, London.

Jørgensen, S. E. 1979. *Handbook of Environmental Data and Ecological Parameters.* International Society of Ecological Modelling, Copenhagen.

Jørgensen, S. E. 1988. *Fundamentals of Ecological Modelling,* 2nd ed. Elsevier, Amsterdam.

Jørgensen, S. E. and H. Mejer. 1977. Ecological buffer capacity. *Ecol. Modelling 3:*39–61.

Jørgensen, S. E., H. F. Mejer, and M. Friis. 1978. Examination of a lake model. *Ecol. Modelling 4:*253–279.

Jørgensen, S. E., L. A. Jørgensen, L. Kamp-Nielsen, and H. F. Mejer. 1981. Parameter estimation in eutrophication modelling. *Ecol. Modelling 13:*111–129.

Ziegler, B. P. 1976. *Theory of Modelling and Simulation.* Wiley, New York, 435 pp.

5

AN ECOLOGICAL ECONOMICS FOR ECOLOGICAL ENGINEERING

Judith Maxwell

School of Natural Resources, The Ohio State University, Columbus, Ohio

and

Robert Costanza*

Coastal Ecology Institute, Center for Wetland Resources, Louisiana State University, Baton Rouge, Louisiana

* Current Address: Coastal and Environmental Policy Program, Center for Environmental and Estuarine Studies, University of Maryland, Solomons, Maryland.

5.1   THE ECONOMICS OF ECOSYSTEMS

From a broad economic viewpoint, natural ecosystems are resources providing commodities and services of value to humans, and by definition, are scarce. These include raw materials, such as timber, minerals, fish, and forage, which are the basis of outputs that have well-defined commercial values (Farber and Costanza, 1987). Natural ecosystems also provide recreational activities including relaxation, hiking, nature study, and enjoyment of scenic views that may contribute to spiritual development and an experience of oneness with nature. The assimilative functions of ecosystems purify air and water and provide climate and flood control.

Natural ecosystems are reservoirs of both commodities and amenities with many as yet unrecognized potential values. These include the pharmaceutical use of plants and gene plasma to improve the productivity of domestic cultivars. In the short term, ecosystems provide habitat for both endangered and opportunistic species and a laboratory to study species interactions in response to unique problems that are relevant to our understanding of human systems. Indeed, the evolutionary economic paradigm (Nelson and Winter, 1982) applies the concept of natural selection to explain the survival characteristics of firms in a competitive environment. In the longer term, ecosystems provide the backdrop for evolution and speciation.

The past 30 years have been marked by a high and rising demand for environmental amenities, particularly in high-income countries. Concomitant with this is an increasing demand for both the commercial outputs and the assimilative capacities of ecosystems, resulting in widespread habitat modification and destruction. The rising demand for ecological services can be attributed to the lack of markets for what are the essentially free services supplied by natural ecosystems. This has favored not only excessive demand for direct consumption of many outputs but also the development of technologies that produce excessive amounts of pollution and solid waste (Ruttan, 1971; Runge, 1986). Hannon (1986) contends that the effects of resource scarcity can be observed in the substitution of high-wage for low-wage labor and high rate of return capital for low rate of return capital, reducing the returns to both wage earners and investors. He concludes that we are increasingly being forced to choose between material goods and environmental quality.

5.1.1   Basic Economic Principles and the Value of Ecosystems

In neoclassical economics, prices are relied on to provide signals regarding resource scarcity. Resource scarcity is imposed on the individual via limits on income and wealth, and technological change, in turn, is assumed to respond to market prices that signal resource scarcity (Samuelson, 1976; Varian, 1978). These are the basic tenets that underlie much traditional economic thinking.

Within this traditional framework, the subjective value of natural eco-systems arises from human preferences for their commercially valuable goods and services, for relatively free amenity values, and for such intangibles as emotional and spiritual fulfillment. Individual preferences are expressed in a voluntary exchange of endowments (income, wealth, future earning capacity) for goods and services in a marketplace. Market prices, the most widely used indicator of economic value and resource availability, are a function of preferences, endowments, and the cost of various alternatives.

However, these prices can be considered true measures of economic value only when the market is characterized by a large number of buyers and sellers and by private property rights that are enforceable and transferable. The consumption choice must be made from a basket of "rival" goods and services, that is, the choice to consume one good comes at the expense of some other good. If these conditions hold, prices can be relied on to direct the allocation of scarce resources to their highest valued use.

Unfortunately, for the many ecosystem goods and services that humans do care about, price formation is hampered owing to missing markets or distorted by subsidies and taxes. This results in the overuse and/or misal-location of resources (Baumol and Oates, 1975; Howe, 1979). The absence of efficient markets for these outputs is the major cause of what can be called "inefficient" habitat modification. This is the conversion from those land use activities that have large, real, but uncapturable benefits (i.e., natural ecosystems) to land use activities that have smaller but capturable income. In this most prevalent case, individual preferences may diverge from those of society as a whole and collective action may be warranted.

Even when preferences for the resources provided by existing natural ecosystems are strong, the inability to assign private property rights makes it impossible to exclude those who do not pay for enjoying or using them. There is no incentive to conserve such common property resources, and overuse and exhaustion result when utilization or harvest rates exceed the population growth rates of a species, as in a fishery, or forage production rates in common grazing areas. For some resource uses, such as camping sites and scenic views, utilization levels result in congestion and subsequent declines in resource quality (Cicchetti and Smith, 1973). For congestible goods, exclusion results in efficient use only if consumers pay for their use in accordance to their own personal valuation for that use. Because of obstacles to both collecting payment and eliciting true evaluations, public or collective management is usually required.

Economists tend to discount the importance of physical measures of resource availability because these do not account for the possibility of the economic feasibility of extraction or development being significantly enhanced by technological change. However, the limits of conventional technology and man-made systems to mitigate limits imposed by nature are already apparent when production is related to some measure of carrying

Table 5.1 World Production per Capita of Basic Biological Resources, 1960–1984

	Roundwood (m^3)	Fish Catch (kg)	Beef (kg)	Mutton (kg)	Wool (kg)
1960	0.63	13.2	9.3	1.9	0.84
1961	0.66	14.1	9.6	1.9	0.84
1962	0.66	14.9	9.8	1.9	0.82
1963	0.66	15.1	10.1	1.9	0.82
1964	0.67	16.1	10.0	1.9	0.79
1965	0.66	16.0	10.0	1.9	0.78
1966	0.64	16.8	10.2	1.9	0.79
1967	0.64	17.4	10.4	1.9	0.78
1968	0.71	18.0	10.7	1.9	0.79
1969	0.71	17.3	10.7	1.8	0.77
1970	0.71	18.5	10.9	1.9	0.75
1971	0.67	18.2	10.6	1.9	0.73
1972	0.66	16.7	10.5	1.9	0.71
1973	0.66	16.6	10.3	1.8	0.65
1974	0.65	17.2	10.8	1.7	0.63
1975	0.63	16.8	11.8	1.8	0.65
1976	0.64	16.8	11.5	1.8	0.62
1977	0.64	16.3	11.4	1.7	0.61
1978	0.65	16.2	11.2	1.8	0.62
1979	0.65	16.2	10.8	1.7	0.62
1980	0.66	16.2	10.4	1.7	0.63
1981	0.64	16.5	9.9	1.7	0.62
1982	0.63	16.6	9.7	1.7	0.62
1983	0.64	16.3	9.8	1.7	0.61
1984	0.64	na	9.8	1.7	0.61
1985	na	na	9.7	1.7	0.62

Source: FAO (Food and Agricultural Organization of the United Nations).

capacity such as per capita measures that relate production or consumption to the needs of an increasing population. Thus, although the production of outputs based on renewable resources has increased over time, the data in Table 5.1 show that the world production for a range of these biotic outputs, measured on a per capita basis, has peaked. In addition, much of this production is associated with a negative impact on the assimilative and rejuvenative capacity of both ecosystems and individual species owing to overuse, habitat conversion, or pollution.

Even though ecosystems are viewed as renewable resources, this renewability is contingent on resilience to perturbation and the ability to avoid the law of entropy through utilization of sunlight, water, and nutrients. Ecotechnology, the result of ecological engineering as presented in this book, offers not only the means of sustaining the current supply of environmental goods and services, but also the means to increase both the supply and the quality of natural ecosystems. This chapter focuses on the role of economics in ecosystem management in order to determine how economic factors can be used to redirect research and development to minimize negative environmental impacts and to enhance the productivity of natural ecosystems. Because economic factors determine how natural ecosystems are manipulated by humans, and this manipulation in turn affects the attributes of ecosystems that are valued by individuals, both demand and production (supply) relationships must be considered.

5.1.2   Lack of Provision for Equity and Sustainability

Another deficiency of markets, and therefore of the exclusive use of an economic approach to answer questions regarding environmental management, is that the choices and prices observed in markets are a function of types and levels of endowments. The distribution of endowments and the relative importance imparted to these endowments in the marketplace can cause significant bias in environmental valuations from individual, societal, and regional viewpoints. For example, most threatened species occur in the less developed countries, which are characterized by widespread poverty, overpopulation, and dependence on natural resource trade to secure foreign exchange. Environmental protection, to be successful in these countries, must address these socioeconomic problems. Unfortunately, when consumers and producers are geographically separated, consumers are able to internalize the benefits of natural resource trade while externalizing the cost of habitat destruction.

Markets may also fail to encourage ecologically sound management when very long time horizons are involved, resulting in intergenerational inequity and irreversible damage to ecosystems. For the individual, current consumption is preferred to future consumption so that savings or investment can be achieved only by offering positive returns. Individuals have a choice of savings options, each offering different rates of return reflecting the risk associated with receiving that return. The present value of future costs and benefits of a savings or investment program is calculated by discounting future costs and benefits with the rate of interest. If the present value of benefits minus costs is positive this indicates a viable investment opportunity. These returns can be compared with those of alternative investments to determine an optimal mix of investments.

The choice of interest rate is critical in determining the optimal harvest rate for renewable resources and in turn determines their renewability. Clark (1973) has shown that the high discount rates used by individuals to assure short-term profit maximization may cause overexploitation and exhaustion of species characterized by slow growth rates. Even if the discount rate were zero, that is, if future values equal current values, and even if such practices have a potential to cause high societal costs to future generations, the short life-span of humans will result in management practices that favor current values and incomes. For example, flourocarbons permit packaging under pressure and a finer spray than that provided by the alternative pump spray. These benefits are directly reflected in market prices and are quite modest when compared to the catastrophic cost that would result from ozone depletion. Page (1978) contends that the inability to estimate the probability of an ecocatastrophy fuels our indifference to the huge asymmetry between current benefits and the potential for future catastrophic costs making even political solutions difficult.

Where the allocation of scarce ecological resources is concerned there will invariably be a clash between individual preferences and social benefits. Markets are just one of many arenas for resolution of these disputes. Public resource management, legislation, and regulatory agencies offer alternative methods or forums for directing management decisions. In spite of the absence of market prices, environmental decision making increasingly emphasizes the quantification of benefits and costs. This has resulted in proliferation of new resource value concepts and valuation methods.

5.2   BROADENING THE ECONOMIC VALUATION OF ECOSYSTEMS

5.2.1   Willingness to Pay and Extending Existing Markets

For the individual, the economic value of an increment in any good or service is the maximum amount that he or she is willing to pay (WTP) for it. Conversely, the value of a decrement is the minimum amount that the individual is willing to accept (WTA) for it. The prices formed in well functioning markets are one source of WTP and WTA estimates. Where markets fail to provide appropriate measures of environmental values the WTP and WTA concepts of economic value are not invalidated but alternative "pseudo-markets" must be used to estimate these values.

The notion that an alternative chosen will be at the expense of some opportunity foregone is central to economic decision making. For example, the cost of providing a scenic view can be directly derived from the net value of outputs and uses foregone, such as timber harvest and dispersed grazing. This is referred to as an *opportunity cost*. To be economically efficient, the scenic view will be preferred over other uses if its value exceeds that of the cost of providing it, including opportunity costs. Ecological goods and amenities are valued for a variety of reasons. *Utilitarian* or *use value* refers to the value of using an ecosystem to derive both current and future benefits. These include commercial outputs such as timber, outdoor activities and experiences, wildlife, and aesthetics. For examples of raw material evaluation see Hyde (1984) and Bartlett (1984), who discuss valuation assumptions and methods for timber and range forage, respectively.

Option value is associated with any use if its future availability is questionable. In such cases consumers may be willing to pay an amount in addition to the present net value of the future use to secure is availability. This additional amount is termed a risk premium. Therefore, whenever uncertainty attends future use, utilitarian values will underestimate the true value of expected future use. See Bishop (1982) for a summary and review of research on option value. For empirical studies of option value see Brookshire et al. (1983) and Greenley et al. (1981).

When development, either of a new effluent-producing technology or a land use requiring habitat modification, will have an impact on future production possibilities, some still unknown, consumers and producers may be willing to forego this development to preserve future options (Arrow and Fisher, 1974). This value, termed *quasi-option value*, is frequently associated with development that has the potential to produce irreversible environmental damage or extinction of species (for example, offshore drilling, strip mining, or large scale deforestation.) It is the value of avoiding that risk. If quasi-option value is included in traditional project analysis it results in a more conservative development policy and is particularly relevant when there is concern regarding inter- and intragenerational equity (Conrad 1980).

Moral or *existence value* rises from an altruistic desire to assure availability not only for one's own generation but also for future generations. These are sometimes referred to as *philanthropic* and *bequest values*, respectively. The value of nonhuman biota for their own sake and not as sources of human satisfaction is referred to as *intrinsic value*. Randall and Stoll (1980) provide a more detailed discussion of these various types of existence values.

In practice, the measurement of these value concepts has remained difficult and largely limited to the valuation of environmental commodities and amenities that directly benefit humans. Norton's (1986) concept of *contributory value* assigns value to species not for their direct value to humans, but according to their role in maintaining and accentuating the ecosystem processes that support these direct benefits. These include the maintenance of atmospheric and aquatic quality, the amelioration and control of climate, flood control, the maintenance of a genetic library, and the supportive role of food webs and nutrient cycling. Contributory value recognizes both the long time horizons involved in many ecosystem processes and the synergism that can result from the interaction of two or more species creating benefits of which none is individually capable.

Though empirically elusive, contributory value does provide a useful framework for conceptualizing how natural ecosystems might be evaluated. However, as Randall (1986) contends, human preferences are focused more on life forms than on life processes. This bias is further distorted by the assignment of humans, in general, of higher preferences for species with commercial value, for wild relatives of domesticated species, and for those that are most familiar and/or easy to empathize with, such a large mammals. Lovejoy (1986) refers to this bias against invertebrates as "vertebrate chauvinism," and others point to interspecies inequity (Costanza and Daly, 1987).

Fortunately, species perceived as having high value are hierarchically organized in food chains and therefore depend on lower life forms. If it is accepted that each species, no matter how uninteresting or lacking in direct usefulness, has a role in natural ecosystems that do provide many direct benefits to humans, it is possible to shift the focus from preservation of individual biota and habitats to the question of whether or not a specific environmental modification will be, on balance, favorable or unfavorable to humans. Production relationships then become the main method of assessing the importance of interspecies dependencies and the effects of perturbation on these relationships.

5.2.2   Ecosystem Function and Economic Value

Assessing the contributory value of ecosystems involves the ability to understand and model the ecosystem's role in an integrated ecologic–economic system and its response to perturbation. The models must be at a level of detail and resolution that allows the assessment of impacts (marginal products) on economically important ecosystem commodities and amenities. Several types of ecological modeling can be used for this purpose which we define under the general heading of "ecological–economic" models. They range from relatively simple, static, linear input–output models (Hannon, 1973, 1979; Isard, 1972; Costanza and Neill, 1984; Costanza and Hannon, 1987), to multiple regression models (Farber and Costanza, 1986), to more sophisticated nonlinear, dynamic spatial simulation models (Costanza et al., 1986). Braat and van Lierop (1985) provide a summary of ecological–economic models currently in use.

Other chapters in this book detail the knowledge and models of ecosystems that must be incorporated in such an assessment. The point that must be stressed is that the economic value of ecosystems is directly connected to their physical, chemical, and biological role in the overall system, *whether humans recognize that role or not*. Standard economics has operated on the fundamental assumption that the *only* measures of value are humans' subjective preferences. However, humans are often woefully uninformed about an ecosystem's true contribution to their own well being, and their preference-based valuations cannot always be trusted (Costanza, 1984). In addition, we cannot assume that local optimizing of independent individuals will produce optimal results for the society in the absence of perfect markets. Perfect markets are the exception rather than the rule in the natural resource area. A better model for the behavior of humans in imperfect markets may be the concept of a "social trap."

5.2.3   Social Traps and Ecological Engineering

A *social trap* is any situation in which the short-run, local reinforcements guiding individual behavior are inconsistent with the long-run, global best interest of the individual and society (Cross and Guyer, 1980; Platt, 1973; Teger, 1980). This state of affairs frequently holds for decisions concerning ecosystem management, and the study of how to recognize, avoid, and escape from social traps can help us manage ecosystems more effectively (Costanza, 1987).

We go through life making decisions about which path to take based largely on "road signs," the short-run, local reinforcements that we perceive most directly. These short-run reinforcements can include monetary incentives, social acceptance or admonishment, and physical pleasure or pain.

Traps can also arise out of simple ignorance of the relevant reinforcements, from the change of reinforcements with time (*sliding reinforcer traps*), from the externalization of some important reinforcements from the accounting system (*externality traps*), from the actions of some individuals affecting the group in adverse ways (*collective traps*), or from a combination of these causes (*hybrid traps*).

The tragedy of the commons is a well-known social trap used to study overexploitation of natural resources (Hardin, 1968). The classic commons trap goes something like this. There is a common property resource (say, grazing land). Each individual user (rancher) sees the individual cost for consuming an additional unit of the resource (adding one more animal) as small and constant, and much less than the private benefits (from selling an animal). However, the overall cost to all the users of each additional resource unit consumed (animal added) increases exponentially as the resource is stressed. Eventually, one additional animal (which costs its owner no more than the first) leads to the destruction of the resource (resulting in great cost to both the animal's owner and the rest of the ranchers). The tragedy of the commons is a collective trap that occurs because the costs and benefits apparent to the individual are inconsistent with the costs and benefits to the collective society.

Edney and Harper (1978) experimented with a simple game designed to test people's behavior in a commons game. In this game a pool of resources is represented by poker chips. The resource pool is renewable; it is replenished after each round in proportion to the number of chips left in the common pool. The objective for each player is to accumulate as many chips as possible from the common pool. At each round players can take either one, two, or three chips. If all players take three chips per round, the resource pool is quickly depleted, and the players end up with far fewer chips than if they had all taken only one chip per round, since doing so would have allowed the resource pool to replenish itself. This game is a trap (and a good analogy for many real-world common-property resource problems) because the short-term, narrow incentive (to take as many chips as possible each round) is inconsistent with the long-term incentive (to accumulate as many chips as possible by the end of the game).

Cross and Guyer (1980) list four broad methods by which traps can be avoided or escaped from. These are education (about the long-term, distributed impacts); insurance; superordinate authority (i.e., legal systems, government, religion); and converting the trap to a trade-off, that is, correcting the road signs.

Many trap theorists believe that converting the trap to a trade-off is the most effective method for avoiding and escaping from social traps because it does not run counter to our normal tendency to follow road signs; it merely corrects the signs' inaccuracies by adding compensatory positive or negative reinforcements. A simple example illustrates how effective this method can be. Playing slot machines is a social trap because the long-term costs and benefits are inconsistent with the short-term costs and benefits (Cross and Guyer, 1980). People play the machines because they expect a large short-term jackpot, even though the machines are in fact programmed to pay off, say, $0.80 on the dollar in the long term. People may "win" hundreds of dollars playing the slots (in the short run), but if they play long enough they will certainly lose $0.20 for every dollar played. To change this trap to a trade-off, one could simply reprogram the machines so that every time a dollar was put in $0.80 would come out. In this way the short-term reinforcements ($0.80 on the dollar) are made consistent with the long-term reinforcements ($0.80 on the dollar), and only the dedicated aficionados of spinning wheels with fruit painted on them would continue to play.

In terms of Edney and Harper's common property resource consumption game, one could turn the trap into a trade-off by taxing any consumption above the optimal level for resource stability. For example, if players took two or three chips they could be "taxed" one or two chips, respectively, so that the short-term benefits of taking more than one chip were offset by short-term costs. This would remove the short-term incentive to take more than one chip and make the long- and short-term incentives in the game consistent, thereby eliminating the trap.

Social traps are one way of generalizing those situations in which the local optimizing of independent agents goes afoul. In this sense they indicate imperfections in the free market approach to resource allocation, which relies on local optimizing of independent agents. It can be argued that the proper role of a democratic government is to eliminate social traps (no more and no less) while maintaining as much individual freedom as possible. This can be accomplished most effectively by turning the traps into trade-offs that can be handled within the current market system as modifications to the cost of potentially entrapping activities.

Social traps abound in the environmental policy area because of the abundance of imperfectly owned and common property resources. To turn these traps into trade-offs, we must calculate the long-term social cost of activities with environmental impacts and charge those costs to the responsible parties in the short run. Environmental traps exist in large part because the producers of ecological damage do not bear the risk, but pass it on to the general population and future generations.

5.2.4 Dealing with Risk and Uncertainty

When there is considerable uncertainty regarding the magnitude and distribution of the costs of ecological damage, one approach to solving this problem involves changing the burden of proof. This reverses the tacit presumption of innocence until proven guilty when chronic effects are difficult to quantify. The effect of smoking on lung cancer is a case in point. Cigarettes had been presumed innocent until the overwhelming weight of evidence proved them guilty. Although such presumption of innocence is appropriate for individual users, it is not appropriate when there exists a potential to impair the health of others. This is true for hazardous or potentially hazardous materials or potential ecological damages that could cause collective harm. We should instead presume the opposite, or guilty until proven innocent, when it comes to these impacts.

Thus, to turn environmental traps into trade-offs, we could charge the responsible parties the worst-case costs. By worst case we mean our current best estimate of the largest potential damages. This does not mean that the worst case cannot be exceeded at some point in the future as new information is accumulated. If it is, the new information can be used to increase the worst-case costs. This money (essentially a tax) could be put into a trust fund that would be returned to the producer (with interest) if and when the effect is proved to be innocuous, or could be used to compensate for damages caused by the effect if and when damage becomes apparent. This would change the short-term incentive structure for ecological damages from one that discourages studies to determine the effects to one that encourages these studies so that some of the trust fund monies could be returned. The economic incentives to eliminate ignorance about ecological damages would increase and by discouraging activities whose negative effects are unknown, the worst-case tax attributes a quasi-option value to the undeveloped ecosystem. The following section provides a detailed discussion on the use of a variety of mechanisms to correct the incentive structure.

5.3 MODIFYING THE ECONOMIC INCENTIVE STRUCTURE

5.3.1 Perfecting Imperfect Markets

As we have seen, even when society accepts the biophysical production constraints imposed by nature, it does not follow that individualistic market processes will derive a set of prices that accurately reflect the marginal social value of conservative or sustainable management. Markets are no more able to remedy unjust income distributions than they are to achieve sustainability, even though better knowledge of these constraints may lead to a change in consumption patterns, via changes in consumer tastes and preferences, even when relative prices do not change. However, the undervaluation of the social costs of habitat destruction and waste disposal, as well as related biases in input prices, indicates the necessity of redirecting economic activity to limit negative impacts on natural ecosystems.

This redirection has historically been achieved by institutional change. For example, the proliferation of environmental legislation in the 1970s can be directly attributed to the interaction between economic agents (individuals, households, firms) who sought to internalize the benefits and externalize the costs of natural resource and environmental uses.

The most widely used government regulation is direct controls that rely on the establishment of an acceptable standard of environmental quality or sustainable level of biological harvest. Permits are then allocated to users specifying the amount of discharge or harvest allowed. The sum of these permitted uses equals the total allowed for the environment or ecosystem.

This regulatory approach has been widely criticized. A long-standing criticism is that regulatory agencies have more discretion over interpretation of the law than they have over enforcement. This provides an incentive to polluters and other natural resource users to "bargain" with the agency, with the end result that the agency becomes staffed from the ranks of the regulated (Kolhmeier, 1969; Stigler, 1971) and vice versa.

Pointing to the failure of direct government regulation, economists have almost universally supported the use of institutional innovations that use market mechanisms, such as price controls to provide for more accurate evaluation (e.g., subsidies, taxes, or user fees such as entrance fees, licenses, or emission fees), or private property rights that provide incentives to manage environmental or ecological systems rationally.

The preference for market-like mechanisms is based on efficiency arguments. For example, inefficiencies arise when a standard for compliance treats all sources of pollution or congestion as homogeneous, as in the direct regulation approach. Under a pricing approach, polluters are charged according to the amount and toxicity of the effluents emitted. From an efficiency standpoint an optimal level of pollution abatement is achieved because each polluter individually chooses the level of pollution abatement that equates their own marginal costs and benefits of pollution control. Polluters facing relatively low pollution abatement costs will cut back more than will polluters with higher costs. If the overall level of pollution abatement is shown to be inadequate, the effluent (price) charge can be raised to achieve the desired reduction, allowing more timely adaptation to differing and changing conditions.

In contrast, under the direct control system, optimal pollution abatement would require that the government distinguish between polluters both in terms of toxicity and amount of waste, and in terms of abatement costs. In most cases, the cost of obtaining this information is prohibitive. In the absence of this information, the government resorts to requiring that all polluters cut emissions by the same amount, resulting in a greater total cost of pollution control than would occur under the pricing approach. The inefficiency of direct regulation has been demonstrated in a number of empirical studies. In three studies of the St. Louis metropolitan area, Atkinson and Lewis (1974, 1976) and Atkinson and Tietenberg (1982) found that current management for particulate pollution cost three to five times as much as the theoretically efficient program providing the same level of ambient air quality. Seskin et al. (1983), in a study of nitrogen dioxide in Chicago, show a 14-fold increase in costs for direct regulation over the least cost strategy.

Other advantages of price controls, as discussed by Haveman (1976), include (1) the negative effects of price controls on the production costs of high pollution goods, which filter through to final prices and in turn reduce demand, production, and pollution; and (2) the lower relative costs of administrating price controls that rely on voluntary compliance reinforced by random audits (much like income tax) as opposed to the costly monitoring of individual dischargers needed to achieve similar levels of compliance under direct regulation.

In spite of this, direct regulation has not been entirely dismissed by economists. Consider, for example, the many problems encountered in an open access fishery. Optimal rates of harvest over time involve both economically efficient harvest of existing stocks and reservation of a portion of stock for future production. However, with no secure property rights, individual firms do not take the longer view and must seek to maximize profits on a year to years basis. A positive income elasticity of demand for fish products will lead to an excessive inflow of both human and technological capital as new firms enter and existing firms increase fishing effort, resulting in dissipation of any economic surplus. A collective trap occurs because short-term economic optimization leads to depletion of fish stocks and unintended extinction if personal discount rates exceed biological growth rates (Clark, 1973). Crutchfield and Pontecorvo (1969) estimated that the United States and Canada could maintain the same catch of Pacific salmon at a savings of $50 million annually if overcapitalization were eliminated. These problems indicate the need for a mixture of fisheries management techniques including (1) regulations related to the characteristics of the resource itself, such as closed seasons or limits on total catch (direct regulation), (2) limits on inputs, such as boat and engine size and constraints on trawling technology (direct regulation), and (3) limits to entry into the fishery by means of licensing (private property rights) or through taxes and user fees (pricing mechanisms) that lower incentives to investment.

Although it is always possible to formulate an optimal regulatory policy with complete information, in reality regulations will be determined on the basis of incomplete information on economic, political, distributional, and scientific factors. And no matter what regulatory device has been chosen, it is likely to be impervious to new information, resulting in social traps as discussed above. The probability of regulatory failure is higher if there exist problems related to intertemporal equity, latency, and the potential for individual actions to lead to high collective risk. Therefore we need to direct our efforts toward defining what types of institutional innovations are the most responsive to changing conditions and uncertainty, and most able to reform technical effort so as to reduce environmental stress and/or enhance the productive and assimilative capacities of ecosystems.

5.3.2   A Flexible Ecological Cost Charging Scheme

As noted earlier, the most widely used approaches in the United States for avoiding and escaping environmental traps have been education and governmental regulation. Although these methods are essential elements in the overall picture, they may not be the most effective means available. As has been pointed out, converting traps to trade-offs seems to be a more effective method in many experimental trap situations, but it has been little used in the environmental area. In terms of environmental management, converting traps to trade-offs implies determining the long-run, distributed costs of environmentally hazardous activities and charging those costs to the responsible parties in the short run. Pollution taxes are the best-known example of this approach (and they have been quite effective in the few cases in which they have been tried), but it is possible to extend this approach to a much broader range of environmental problems. Below we give two examples of how this approach might be applied to improve the economics of ecological management and engineering. The specific examples deal with hazardous wastes and controlling coastal erosion, but the approach is general and can be applied (with some modifications) to many ecological engineering problems.

The hazardous waste management problem can be viewed as a hybrid trap containing elements of time delay, ignorance, externality, and collective traps. The negative environmental effects of hazardous waste do not become evident until long after they are produced (time delay); their ultimate effects are largely unknown at the time of their production and release into the environment (ignorance); the negative effects are borne by parties other than the producer without sufficient compensation (externality); and common property resources are consumed (i.e., groundwater contamination) by individual agents who do not bear the true cost of that consumption. The current regulatory approaches to hazardous waste management are well known and are not adequately addressing the problem.

To turn this trap into a trade-off one must charge the producers of hazardous waste for the ultimate long-run environmental and health costs of these wastes, and the charges must be imposed at the time of the waste's production. The Superfund can be seen as a small step in the right direction even though it relates only to the cleanup of abandoned hazardous waste sites with damage being assessed long after the wastes are produced. It is also severely underfunded for the task.

The ignorance component of the hazardous waste management trap is the most difficult to deal with. How can we charge producers of hazardous wastes for the ultimate long-run costs of their waste if we have no idea what those costs will actually be (if any)? Part of the problem is that this ignorance and uncertainty about future costs is itself a cost, or more precisely a risk of unknown magnitude. The trap exists in large part because the producers of the hazardous waste do not bear this risk in the short run, but pass it on to the general population and future generations.

The problem of coastal wetland management is another example of a complex hybrid trap, with time delay, ignorance, and collective elements. Coastal erosion in Louisiana is a particularly severe example. Canal dredging and other hydrologically disruptive activities have resulted in a current land loss rate of more than 100 km^2/yr (Craig et al., 1979; Gagliano et al., 1981; Scaife et al., 1983). It may already be too late to arrest or reverse this trend, particularly because many of the responsible parties have already left the state. The situation is a trap because the narrow, short-term incentives of those damaging the wetlands are or were inconsistent with the long-term good of the system.

To turn this trap into a trade-off, one should charge the responsible parties the full cost of the ultimate environmental damage, at the time the damage is done. To do this one needs to know the economic value to society of coastal marshes and the amount of marsh destroyed by each activity. As with the hazardous waste issue, there is much uncertainty involved in these estimates, but the worst-case costs should be assumed and the burden of proof that the damages are in fact less than the worst case shifted to the parties who benefited.

For example, a recent study concluded that each acre of coastal wetlands in Louisiana has a present value to society of roughly $5,000–$10,000/ha ($2000–$4000/acre) (Farber and Costanza, 1986). This range of values was due to uncertainties in the valuation procedure. Increasing the accuracy of the valuation estimates is an expensive proposition, and one that would stress the research budget of the state.

To eliminate this trap effectively, one could charge the responsible parties for marsh destruction (i.e., oil companies for dredging access canals through wetlands) the $10,000/ha ($4000/acre) worst-case cost. These fees would go into a trust fund to be used for mitigating environmental damages by purchasing marshland elsewhere, backfilling canals, diverting fresh water, and so on. The responsible parties could lower the fee by proving that the damages are actually less than the worst-case assumption (by funding independent studies) or by minimizing the amount of wetlands they damage in the process of accomplishing their goal (by directional drilling, immediate backfilling, etc.). In either case the cause of wetland conservation would be served without completely prohibiting the search for oil and gas.

5.3.3 Inducing Innovation in Ecological Engineering

Beyond policies designed to protect and conserve natural ecosystems, opportunities also exist to improve and expand these systems via the development of ecotechnologies. However, research resources are also limited and ecological engineering research competes with all other forms of research for support. Given the difficulty of valuing the benefits of this research, it is important to assess the potential for and impediments to ecotechnological research. If we view research funding as a resource allocation problem, we are concerned with the following questions posed by Binswanger and Ruttan (1978) and adapted here to differentiate between conventional and environmentally enhancing technologies. First, what quantity of resources should be allocated to technological innovation with long-term payoffs relative to social programs with immediate benefits in order to maximize social benefits? Second, how should these resources be allocated between conventional and ecotechnological research and adoption? Third, for ecotechnologies, how should resources be allocated among basic and applied research, and between development and diffusion activities? Finally, if the market economy cannot be relied on to induce an optimal allocation of resources to this spectrum of research categories, what institutional mechanisms or economic policies will induce the optimal reallocation?

As discussed earlier, technological change is endogenous to the economic system; that is, economic variables, such as resource endowments, relative factor prices, and final product prices, determine the rate and direction of technical innovation. As relative prices change, firms and individuals can choose to reallocate resources among known production inputs and consumption choices or to allocate some resources in order to capture new production or consumption opportunities. For example, a firm or government can allocate some of its budget to research with the objective of developing more efficient production processes or alternatively may choose to pay higher dividends or increase funds for social programs.

There is considerable controversy in determining the amount of resources to be allocated to research relative to nonresearch activities. A 1987 article in the Wall Street Journal, reporting that the White House had called for a doubling of the National Science Foundation's (NSF) budget over the next five years, provides a classic example of this dilemma (Murray, 1987). An NSF spokesman says that the bulk of this money will be used to fund university-based science, technology, and engineering centers, citing the government's responsibility to build up domestic research capacity in order to boost competitiveness. However, the NSF is funded in the same appropriations bill as the Department of Housing and Urban Development and the level of funding realized will depend on the willingness of legislators to trade housing programs for research. Although it can be argued that the long-term returns to research may be very large, it is difficult to weigh such risky ventures against immediate hardship.

Once a research budget is determined, achieving an optimal balance between the funding of ecotechnological and conventional technological research will be biased by the lack of markets for many environmental goods and services. This usually results in (1) technological innovation that permits the achievement of harvest levels that can cause extinction or overload the assimilative capacities of ecosystems through excessive production of wastes, and (2) underinvestment in ecotechnological innovation owing to the public (free) goods nature of its outputs.

In a generalized model, ecotechnology can be viewed as information in the form of scientific data that cannot be embodied in a proprietary product. That is, information can be shared among many users without being depleted, and even though research and development costs may be substantial, users can usually avoid all but the most inconsequential transfer costs. Arrow (1962) identified this as the classic "free rider" problem associated with public goods. Because innovators are not able to internalize any of the payoffs derived from ecotechnologies there is no incentive to produce them. This is true of much biological and ecological research and explains the need for government support. Ideally, the innovator's compensation should reflect the total social net benefits that arise from the innovation.

Another impediment to ecotechnological research is the uncertainty surrounding its potential usefulness, particularly in the case of basic research. In many cases it is impossible to predict the returns to basic research because future applications are yet unknown; that is, a quasi-option value is involved. Binswanger and Ruttan (1978) argue that the state of basic research determines and creates the potential payoffs from applied research. However, the uncertainty regarding payoffs and the inability to capture payoffs means that such research is not likely to be undertaken by a profit-maximizing firm. Thus the government or other public institutions must usually be involved in ecotechnological research and development.

Binswanger and Ruttan also point out that agricultural (and forest) experiment stations are examples of institutional innovations designed to overcome this lack of incentive for private development due to both the risk involved and the lack of direct compensation. However, a bias toward underinvestment in ecotechnological research persists in a way that does not affect forestry or agriculture. That is, farmers and forest growers have a strong demand for research when they are able to internalize the benefits of that research. Potential adopters of ecotechnologies, on the other hand, are usually unable to exclude other users from the benefits generated, and therefore are unlikely to adopt such practices even when societal benefits are high.

For example, it has been shown that wetlands in strip-mined areas reduce the amount of potentially harmful chemicals in downstream water, many of which are not yet recognized by clean water standards nor treated for by current water purification technologies (see Chapter 12). Thus the mining company, by using wetlands, not only internalizes the costs of its own pollution, but also provides a level of water quality that is higher than that required to meet health standards. If the mining company cannot internalize the benefits of raising water quality it is unlikely to do so.

This lack of immediate user demand, coupled with the shortsightedness imposed on the political system by the relatively short length of elective office and the annual budgetary process, results in a significant impediment to the funding of ecotechnological research from government sources. For example, to receive NSF funding for university-based research centers, it is necessary to demonstrate the ability to generate support from private industry first. Profit-maximizing firms, by their very nature, will not demand ecotechnological research unless adoption of ecotechnologies is encouraged via legislation and regulation. Therefore, the government must also take an active role in fostering the diffusion and adoption of these technologies. This usually includes the use of ecotechnologies in the management of public lands and waters, economic mechanisms to encourage adoption by the private sector (subsidies, taxes and credits), and extension activities to demonstrate to the public the appropriate use of ecotechnologies in practical land and water management applications.

The flexible ecological cost charging scheme suggested above would eliminate many of these problems and induce ecotechnological innovation. By presuming guilt until proven innocent but allowing reduction of taxes for improved performance the economic incentives to reduce costs are brought to bear on the problem. Under this system, *not* reducing pollution becomes expensive to the individual firm (because of the high default tax) and the incentives are for the firm to reduce pollution in order to reduce costs. Firms are left to innovate their own technological means to this end and the net effect is to stimulate ecotechnological innovation.

5.4 SUMMARY AND CONCLUSIONS

Markets for ecological goods and services are far from perfect and we cannot rely on the free market to allocate these resources efficiently. The current system, which misallocates these resources, is better described as a social trap. Escaping from the trap involves turning it into an economic trade-off by making long-run social costs and risks incumbent on individuals and firms in the short run using taxes and subsidies. Quantifying the proper level of taxes and subsidies should be tied to models of the ecological impact of activities, but the burden of proof as to the magnitude of these damages should fall on the parties that stand to profit from them, not the general public. A flexible ecological cost charging scheme can be designed that induces ecotechnological innovation by making it the most economically attractive option in the short run (as well as the long run). Implementation of such a scheme could go a long way toward allowing the development of an effective ecological engineering.

REFERENCES

Arrow, K. J. 1962. Economic welfare and the allocation of resources for invention. In R. R. Nelson, ed. *The Rate and Direction of Inventive Activity*. Princeton University Press, Princeton, NJ.

Arrow, K. J. and A. C. Fisher. 1974. Environmental preservation, uncertainty, and reversibility. *Quart. J. Econ. 55*:313–319.

Atkinson, S. E. and D. H. Lewis. 1974. A cost effectiveness analysis of alternative air quality control strategies. *J. Environ. Econ. Management 1*:237–250.

Atkinson, S. E. and D. H. Lewis. 1976. Determination and implementation of optimal air quality standards. *J. Environ. Econ. Management 3*:363–380.

Atkinson, S. E. and T. H. Tietenberg. 1982. The empirical properties of two classes of designs for transferrable discharge permit markets. *J. Environ. Econ. Management 9*:101–121.

Bartlett, E. T. 1984. Estimating benefits of range for wildland management and planning. In G. L. Peterson and A. Randall, Eds., *Valuation of Wildland Benefits*. Westview Press, Boulder, CO, pp. 143–156.

Baumol, W. J. and W. E. Oates. 1975. *The Theory of Environmental Policy*. Prentice Hall, Englewood Cliffs, NJ.

Binswanger, H. P. and V. W. Ruttan. 1978. *Induced Innovation: Technology, Institutions, and Development*. The Johns Hopkins University Press, Baltimore, MD.

Bishop, R. 1982. Option value: an exposition and extension. *Land Econ. 58*:1–15.

Braat, L. C. and W. F. J. van Lierop. 1985. *A Survey of Economic-Ecological Models*. International Institute for Applied Systems Analysis, Laxenburg, Austria.

Brookshire, D. S., L. S. Eubanks, and A. Randall. 1983. Estimating option prices and existence values for wildlife resources. *Land Econ. 59*:1–15.

Brown, Gardner M., Jr. and J. Swierzbinski. 1985. Endangered species, genetic capital, and cost reducing R&D. In D. O. Hall, N. Myers, and N. S. Margaris, Eds., *Economics of Ecosystem Management*. Dr. W. Junk, Dordrecht, The Netherlands, pp. 111–127.

Cicchetti, C. J. and V. K. Smith. 1973. Congestion, quality, deterioration, and optimal use: Wilderness Recreation in the Spanish Peaks Primitive Area. *Social Sci. Res. 2*:18–30.

Clark, C. W. 1973. The economics of overexploitation. *Science 181*:630–634.

Conrad, J. M. 1980. Quasi option value and the expected value of information. *Quart. J. Econ. 94*:813–820.

Costanza, R. 1987. Social traps and environmental policy. *BioScience 37*:407–412.

Costanza, R. and H. E. Daly. 1987. Toward an ecological economics. *Ecol. Modeling 38*:1–7.

Costanza, R. and B. Hannon. 1988. Multicommodity ecosystem analysis: dealing with the "mixed units" problem in flow and compartmental analysis. In B. C. Patten and S. E. Jorgensen, Eds., *Systems Analysis and Simulation in Ecology* (in press).

Costanza, R. and C. Neill, 1984. Energy intensities, interdependence, and value in ecological systems: a linear programming approach. *J. Theoret. Biol. 106*:41–57.

Costanza, R., F. H. Sklar, and J. W. Day, Jr. 1986. Modeling spatial and temporal succession in the Atchafalaya/Terrebonne marsh/estuarine Complex in South Louisiana. In D. A. Wolfe, Ed., *Estuarine Variability*. Academic, New York, pp. 387–404.

Craig, N. J., R. E. Turner, and J. W. Day, Jr. 1979. Land loss in coastal Louisiana (U.S.A.). *Environ. Management 3*:133–144.

Cross, J. G. and M. J. Guyer. 1980. *Social Traps*. University of Michigan Press, Ann Arbor, MI.

Crutchfield, J. A. and G. Pontecorvo. 1969. *The Pacific Salmon Fisheries: A Study of Irrational Conservation*. The Johns Hopkins University Press, Baltimore, MD.

Deegan, L. A., H. M. Kennedy, and C. Neill. 1984. Natural factors and human modifications contributing to marsh loss in Louisiana's Mississippi River Deltaic Plain. *Environ. Management 8*:519–528.

Edney, J. J. and C. Harper. 1978. The effects of information in a resource management problem: a social trap analog. *Human Ecol. 6*:387–395.

FAO (Food and Agricultural Organization of the United Nations). 1986. *Production Yearbook*. Rome (and previous issues).

FAO (Food and Agricultural Organization of the United Nations). 1986. *Yearbook of Forest Products*. Rome (and previous issues).

Farber, S. and R. Costanza. 1987. The economic value of wetlands systems. *J. Environ. Management 24*:41–51.

Gagliano, S. M., K. J. Meyer-Arendt, and K. M. Wicker. 1981. Land loss in the Mississippi River Deltaic Plain. *Trans. Gulf Coast Assoc. Geol. Soc. 31*:295–299.

Greenley, D. A., R. G. Walsh, and R. A. Young. 1981. Option value: empirical evidence from a case study of recreation and water quality. *Quart. J. Econ. 95*:657–673.

Hannon, B. 1973. The Structure of ecosystems. *J. Theoret. Biol. 41*:535–546.

Hannon, B. 1979. Total Energy Costs in Ecosystems. *J. Theoret. Biol. 80*:271–293.

Hannon, B. 1986. Foreword. In C. A. S. Hall, C. J. Cleveland, and R. Kaufman. *Energy and Resource Quality: The Ecology of the Economic Process*. Wiley, New York, p. ix.

Hardin, G. 1968. The tragedy of the commons. *Science 162*:1243–1248.

Haveman, R. H. 1976. *The Economics of the Public Sector*. Wiley, New York.

Howe, C. W. 1979. *Natural Resource Economics*. Wiley, New York, 350 pp.

Hyde, W. F. 1984. Timber valuation. In G. L. Peterson and A. Randall, Eds., *Valuation of Wildland Resource Benefits*. Westview Press, Boulder, CO, pp. 131–142.

Isard, W. 1972. *Ecologic-Economic Analysis for Regional Development*. The Free Press, New York.

Kohlmeir, L. M., Jr. 1969. *The Regulators: Watchdog Agencies and the Public Interest*. Harper & Row, New York.

Lovejoy, T. E. 1986. The species leave the ark. In B. G. Norton, Ed., *The Preservation of Species*. Princeton University Press. Princeton, NJ, pp. 13–27.

Murray, A. 1987. National Science Agency, bucking tide of cuts in budgets, wins White House show of largesse. *Wall Street Journal*, April 27.

Nelson, R. R. and S. G. Winter. 1982. *An Evolutionary Theory of Economic Change*. Harvard University Press, Cambridge, MA, 437 pp.

Norton, B. G. 1986. On the inherent danger of undervaluing species. In B. G. Norton,

Ed., *The Preservation of Species*. Princeton University Press, Princeton, NJ, pp. 110–137.

Page, T. 1978. A generic view of toxic chemicals and similar risks. *Ecol. Law Quart.* 7:207–244.

Platt, J. 1973. Social traps. *Am. Psychol. 28:*642–651.

Randall, A. and J. Stoll. 1980. Consumer's surplus in commodity space. *Am. Econ. Rev. 70:*449–455.

Randall, A. 1986. Human preferences, economics, and the preservation of species. In B. G. Norton, Ed., *The Preservation of Species*. Princeton University Press. Princeton, NJ, pp. 79–109.

Runge, C. F. 1986. Induced innovation in agriculture and environmental quality. Department of Agricultural and Applied Economics, University of Minnesota, Staff Paper 86-16. St. Paul.

Ruttan, V. W. 1971. Technology and the environment. *Am. J. Agric. Econ. 53:*707–717.

Samuelson, P. A. 1976. *Economics,* 10th ed. McGraw-Hill, New York, 917 pp.

Scaife, W. W., R. E. Turner, and R. Costanza. 1983. Coastal Louisiana recent land loss and canal impacts. *Environ. Management 7:*433–442.

Seskin, E. P., R. J. Anderson, Jr., and R. O. Reid. 1983. An empirical analysis of economic strategies for controlling air pollution. *J. Environ. Econ. Management 10:*112–124.

Stigler, G. J. 1971. The theory of economic regulation. *Bell J. Econ. Management Sci. 2*(Spring):321.

6

ECOLOGICAL ENGINEERING AND SELF-ORGANIZATION

Howard T. Odum

Department of Environmental Engineering Sciences and Center for Wetlands, University of Florida, Gainesville, Florida

6.1   INTRODUCTION

As the resources of the world become increasingly limiting to the expansion of the human economy, the management of the planet must turn more and more to a cooperative role with the planetary life support system, sometimes called *stewardship of nature*. The pattern of humanity and nature that prevails is symbiotic because two coupled systems have higher performance than two separate systems that are not mutually reinforcing.

The study of the system of humanity and environment is sometimes called human ecology, economic geography, landscape ecology, or ecological economics, but study is not enough. The techniques of designing and operating the economy with nature are called *ecological engineering*. Ecological engineering develops designs that can compete and survive so that humans become partners with their environment. This chapter describes some principles of ecological engineering with examples.

The essence of ecological engineering is managing *self-organization*. Traditional engineering replaces nature with new structure and process, but ecological engineering provides designs that use environmental structures and processes. Self-organization designs a mix of man-made and ecological components. It develops a pattern that maximizes performance, because it reinforces the strongest of alternative pathways that are provided by the variety of species and human initiatives. Patterns are retained that utilize all resources and by-products in a competitively efficient manner.

By lightly managing self-organizational processes, human management causes ecological engineering designs to supersede the wasteful displacement of nature's useful work. Self-organization also occurs with the trial and error of human efforts in finding management that works. A better pattern of humanity and nature results from fitting humanity to encourage the useful services of ecosystems. Ecological engineering designs make an economy competitive.

Ecological engineering, like other kinds of engineering, develops designs to solve problems. By reinforcing what works, the self-design process provides solutions to problems involving the economic–environmental sectors. The process identifies the management alternatives for public policies that maximize the mutual reinforcement of the economies of humanity and nature.

Environmental engineering that responds to unbridled human wishes wastes the environmental resource contributions. Ecological engineering reduces costs by fostering nature's inputs. Instead of eliminating a former system in order to replace it with new construction, ecological engineering fits the human interface to use the work contributions of the environment.

Whereas engineers once stated to society, "You decide what you want and we will design and build it to specification," ecological engineers should state, "You give us the problem situation and we will manage for a high environmental and economic performance." Just as an engineer is asked to make a bridge that works and lasts, the ecological engineer should provide a pattern with nature that works and lasts. Such designs are those that are self-organizing for maximum performance.

Part of ecological engineering is acceptance of humans as a part of nature and having faith in the self-organizing by humans and by nature's processes as a reality of life. Some people are trained in an environmental protectionist mode that sees human-induced changes as bad, to be fought.

Sometimes environmental activism develops an adversary position to protect nature against the equally extreme viewpoints of economic development to eliminate the natural processes and substitute human settlement. Ecological engineering seeks the middle ground, requiring a symbiosis of both the life support of nature and the flexibility and intelligent work of humans. Development of ecological engineering, which is half science and half engineering, can substitute knowledge and procedure for the wasteful adversarial methods that often struggle in court.

6.2 CHARACTERISTICS OF ECOLOGICAL ENGINEERING

Ecological engineering differs from traditional engineering and from most economic approaches to environmental management. Some of the characteristics and differences are given next.

6.2.1 Control by Large Scale Mechanisms

An important component of ecological engineering is looking to larger-scale systems than the one in which the problem or interface is located. Success depends on the way the landscape works as it interacts with the controlling interactions of the human economy. Thus ecological engineering designs involve a larger scale than most engineering.

In the hierarchy of components of the environment and the economy, control is exerted by the larger components on the small because the larger ones have longer time constants, longer turnover times, larger territories of actions, and more information storages. The fluctuations in the smaller components are absorbed and filtered out as they converge in support of the elements higher in the hierarchy. Thus larger animals control plants and microbes. The human programs generally can control the cycles and timing of ecosystems of the landscape, and public policies of human society control the personal and market fluctuations of human economy.

Ecological engineering contrasts with the economic concept that unlimited free human competition at a level of individual human action will automatically generate successful patterns on a large scale involving nature. The ecological engineering view is that coordination of the work of environment and people is necessary for maximum performance of the economy. Designing for individual profits ignores and therefore does not reinforce the environmental resource basis for the economy. Maximizing individual incomes often detracts from the vitality of the whole economy by knocking out such environmental works as water purification, soil development, forest production, fishery production, and waste disposal capacities of wetlands.

Many sciences are analytic, breaking down parts for experimental testing and relating. Ecological engineering is synthetic, putting data together with such techniques as systems modeling, benefit evaluations, and long-term prognoses. Its models and performance criteria are middle and long range.

6.2.2   Self-Organizational Principles from Microcosms

Starting in 1954 we studied experimental microcosms in the laboratory, small ecosystems in containers with conditions arranged to resemble a larger ecosystem. In 1962 and 1963 we defined ecological engineering as environmental manipulations by humans using small amounts of supplementary energy to control systems in which the main energy drives are still coming from natural sources. Formulas for ecological engineering may begin with natural ecosystems as a point of departure, but the new ecosystems that develop may differ somewhat, because boundary conditions established in the engineering process are different (Odum, 1962; Odum et al., 1963). These studies convinced us that the process of self-design observed in these experimental ecosystems was the mechanism for developing functional ecosystems with high metabolism rapidly. Then we were able to see the processes in the larger out-of-doors as well. Over the years evidence has gradually increased, showing that after the first period of competitive colonization, the species prevailing are those that reinforce other species through nutrient cycles, aids to reproduction, control of spatial diversity, population regulation, and other means.

The self-organizing process is readily followed with microcosms. These partially closed ecosystems in containers are given the same inputs as the larger ecosystems they resemble such as light and stirring energy. They are seeded with the same material components and as many species as possible from the larger reference ecosystem. Standard ecosystem patterns form rapidly with production, consumption, recycle, diversity, hierarchies, organic storages, and so on.

Different combinations of species develop as fine tuning for the different conditions applied to different experiments. The rapid flexible self-organization during succession in microcosms shows the power of letting self-organization generate adapted systems. Setting up microcosms is ecological engineering in miniature.

Microcosms may be replicated and made similar by intermixing components at regular intervals. Because the information and other inputs from the larger surroundings are not available in the partially sealed container, microcosms don't achieve as high levels of production as the larger reference realms. They do help people see the way the tendencies of individual species populations become constrained and organized as systems reinforcement develops.

Now there is a large literature especially concerned with the responses of these microecosystems to chemical substances. Because of the cooperative mechanisms that develop in self-organization, adapted ecosystems are often more resistant to chemical toxicity than individual organisms alone. In this they are like the larger ecosystems where the environment often protects species by being a buffer for toxicities.

Microcosms developing with special conditions such as with toxic chemical substances usually involve unusual species combinations. Other examples where distinctive patterns result from microcosm self-organization are the microecosystems developing in hot water microcosms, microcosms adapted to brines, microcosms adapted to sewage, and microcosms adapted to cave conditions. Often self-organization produces similar results, even when done in different parts of the world, provided specialized species adapted to the special conditions are available.

6.2.3 Self-Organization in the Larger Ecosystems

The self-organization of new ecosystems forming adaptations to new conditions is about us everywhere that humans are changing the landscape. Islands provide good examples because special insular conditions and limited availability of seeding and immigration helps generate odd combinations. A few distinctive examples are given in Table 6.1.

Seeding from an ecosystem that is already well organized may accelerate organization in another area where conditions are similar. There may be an order of seeding also that maximizes development. The cluster of genetic

Table 6.1 Examples of Ecological Engineering

Salt Manufacture in Evaporite Basins. An ancient ecological engineering technology is the brine saltern, in which seawater is passed through a series of stages in which solar energy eventually creates salt. Essential aspects are the blue-green algae that absorb the heat, the photorespiring plants, and bacteria that consume organic matter. Human participation is in controlling the water flow, sometimes adding fertilizer to the blue-green mat stages, and scraping salt out of the final lagoon.

Fishponds. Many fishponds and aquacultures use a self-organizing pond ecosystem for much of the support of desirable yield species of fish and other organisms. The human role concerns the initial seeding of the yield species, supplementary direct feeding of yield species, fertilization with inorganic or organic substances, control of predators, harvest schedules, and so on.

Grassy Lawns. The grassy plots around housing and public buildings in western society are used to fit a cultural concept of neatness, to protect against fire, and to eliminate cover for criminals. The human role includes management for a predominant grass species using fertilization, cutting, pesticide, planting, and replanting.

Fallow Stage in Land Rotation. Ancient agricultural practices rotate land from crops back to a successional stage where many species usually seeded naturally start to develop the characteristic ecosystems of the region, but self-organize differently because of the different conditions left by agriculture such as residual chemicals, residual crop seeds, and depleted soils. Effective restoration depends on there being plots of the main climax ecosystems in the vicinity to make seeding rapid. The human role may be simply abandoning cultivation.

Ecosystems in Hot Waters. In the hot springs of volcanic regions and in the thermal wastewaters from industrial plants, hot water ecosystems develop dominated by blue-green algae and bacteria, with a few higher animals appearing as temperatures decrease. Such systems found after long periods of self-organization have very high photosynthetic efficiencies. The thermal waters from industry are often interrupted, so that the ecosystem's self-organizing processes operate only for a short time before there is change. Under intermittent conditions, gross production may be less than in undisturbed adaptation (Odum et al., 1974–1979).

Grazing-Induced Thorn Forests. In dry, tropical areas where goats and other grazing animals are turned loose in the environment, self-organization develops vegetation with thorns, which may prevent overgrazing thereafter. In a national park on Bonaire, Dutch West Indies, a goat thorn forest has developed with considerable structure and beauty. The human role is mainly the regulation of goat populations.

Table 6.1 Examples of Ecological Engineering (*continued*)

Ornamental Interior Plantings. Because the shade plants of rain forests are adapted to very low light intensities, developing very high chlorophyll concentrations, and very broad leaf surfaces, they are adapted to an entirely different habitat, the interiors of houses and public buildings, so long as frost is prevented. They provide a bit of the lush aesthetics of nature's magnificent ecosystem as part of human structures. They require only available watering, because their net growths are small and nutrient requirements small after the initial seeding. Increasingly, cultivation of single house plants is being replaced with small ecosystems with self-organization of species allowed to develop green mosaics, with some animals included.

information involved in seeding a self-organization process is collectively a message from the past that facilitates future production.

Ecological engineering here may consist of setting the conditions and inputs, sometimes called boundary conditions, and then developing packages of seeding suitable for accelerating ecosystem development for these conditions.

6.2.4   The Product of Self-Organization—Domestic Ecosystems

After self-organization has operated for a time, responding to exchanges through interface with the human economy, changes slow down as considerable complexity, hierarchy, and other properties of more mature ecosystems appear. The adapted ecosystems that result may be called *domestic* ecosystems, because they now provide services to humans. To use such an ecosystem in another location with similar interface conditions, one has only to transport the information, which includes the seeding of species and the information about the appropriate management.

Some new useful ecosystems have been developed from conscious efforts to develop them, but so far most self-organization has been happenstance, often in spite of management efforts in some other direction. Examples of useful ecosystems which also serve the human economy are given in Table 6.1. In the next section we discuss techniques of ecological engineering.

6.3   TECHNIQUES OF ECOLOGICAL ENGINEERING

6.3.1   Interface Ecosystems for Recycling Materials to the Environment

The most famous domesticated ecosystems are those used for the treatment of wastes, such as the trickling filter and anaerobic digester systems for receiving sewage that is partly consumed with carbon dioxide and inorganic nutrient by-products released to the environment. All over the world these ecosystems are copied and transplanted. More recently, it has been realized that almost any ecosystem can be a waste-treatment ecosystem. It can be in a container such as the present sewage treatment plants or in a suitable terrain where inputs and outputs can be controlled. Already, wetlands are extensively used as interfaces between wastewaters and the public domain.

We proposed wastewater treatment as a conscious ecological engineering self-organization methodology in 1962 (Odum et al., 1963; Odum, 1967; 1971), and implemented it for marine lagoons in 1967 (Odum, 1970; 1985) and for cypress swamps in 1973 (Odum et al., 1975; 1977; Ewel and Odum, 1984). However, the use of self-organizing ecosystems to receive wastes is as old as human cultures, developed by nature automatically wherever wastes were released to the environment. Many of our best long-range data come from studies of wastewater systems that just happened (Ewel and Odum, 1984), including studies of municipal wastewater flows that had been going into wetlands for many years at Morehead City, North Carolina and Wildwood, Waldo, Jasper, and Naples, Florida.

6.3.2 Introduction of Exotic Species

The self-organization that is observed taking place where exotic animals and plants have been introduced is controversial because it is often resisted, yet is part of the continual fitting of ecosystems to the landscape. To accelerate the contribution of self-organization to the landscape, it may be argued that exotic species introduction should be maximized. When added to a well adapted, high-diversity ecosystem, exotics don't take over, but become additional minor members of the gene pool.

One place where invading exotics may dominate, displacing previous systems, is where conditions have changed, the old system is no longer well adapted, and self-organization is in process. For example, the giant toad, *Bufo marinus*, an exotic in Puerto Rico, moved into the rain forest as a rare member of the gene pool, but in the new environment of cut-grass lawns around public buildings, it became a dominant insect consumer adapted to the cutting and fertilizing management of the new ecosystem.

6.3.3 Exotics on Islands

Another place where an invading exotic may dominate is in islands or land-locked lakes that have not had access to many species. Although the species that did arrive formed typical ecosystems with production, consumption, mineral cycles, and hierarchies, the resource use and performance is less than that possible with a better adapted set of species for self-organization to draw from. The situation is like the "gnotobiotic" microcosms where pure cultures of species are combined to form ecosystems (Nixon, 1969; Taub, 1969; 1974). If it were desired to maximize production and ultimate ability to compete, then one might recommend adding more exotic species. Even though the species are not normally associated, self-organization does take place, although the ecosystem design would not compete if exposed to other invasions. An example of invading exotics becoming dominant is the nitrogen-fixing shrub-tree, *Myrica*, invading the new lava areas in the Hawaiian Islands (Vitousek, 1987). Formerly, algae and other microbes were fixing nitrogen, but not as effectively as the new exotics that now prevail.

However, introductions to islands or isolated lakes may cause interesting ecosystems and species to be lost. The species that developed on isolated islands are as much a part of the world's needed gene pool as those on the continents. Sometimes, species threatened by change on their island of origin, if transported to a new situation, become dominants and thus major contributors to productivity in other situations.

Also, the agricultural and forestry basis of the human economy depends on keeping agroecosystems simple so that more production goes into economic use. Agricultural varieties are bred for yield often at the expense of ability to compete against wild varietites or defend against diseases. Thus our public policies are against purposeful introductions that may add competitors or diseases. Exotic seeding accelerates self-organization that works to increase complexity. This is not the purpose of agriculture, which is to redirect energies from complexity to yield.

To summarize, ecological engineering, to accelerate ecosystem development for maximum performance, may use multiple seeding of new species, but should keep introductions from within the same region to avoid introducing entirely new problems for agriculture in the short run. Perhaps this method should not be used on islands or isolated lakes if the diversity there has been restricted by isolation. The gene pool and system integrity there may be more important than increased performances that might be obtained if mainland ecosystems were transferred there.

6.3.4   Accelerating Arrested Succession

Species colonizing early succession usually have abundant reproduction with means for widespread dispersal of seeds, larvae, and so forth. Seeding from species that develop later in succession is slow because reproductive individuals are usually fewer and larger, and often require animal transport. Therefore in new conditions, succession stops after the initial weedy species have become established. There is then an *arrested succession* that remains with less than full performance until suitable species that can generate a higher diversity climax can arrive. The weedy, arrested successions that sometimes form with exotics in disturbed areas may be improved by accelerating further introduction of species so that regular organization and controls can redevelop.

Examples are the miles and miles of gorse scrub in New Zealand and herbaceous cover in Iceland where many generations of sheep eliminated the later stages of tree succession so that present vegetation is an arrested stage with low diversity and probably with lower gross production. The net accumulation of organic matter may be greater because a complete ecosystem of consumers is lacking.

Another example is the ponds that develop willows in disturbed lands in Florida but do not continue succession to cypress and gum for lack of seeding. The willows may remain for 60 years as an arrested succession (Rushton, 1987).

The phosphate mining in Florida has been creating new kinds of geological substrates to which older succession is not adapted. Consequently, exotics dominate succession. Lack of available seed trees and the birds and animals to transplant seeds has caused the new self-organization to stop in an arrested state.

Experiments in the field by Kangas (1983) and Rushton (1987) show that the arrested succession can be broken and further succession released by planting of seedlings. Ecological engineering here consists of supplying the missing ingredients in self-organization.

6.3.5   Appropriate Environmental Technology

As pollution problems developed, various technological solutions were developed for waste processing. Many of these solutions were expensive and intensive in their use of energy and other resources. Many people thought of these technologies as a contribution to the economy, providing jobs.

However, some of this technological processing is an unnecessary displacement of nature's role, and thus an unnecessary cost, diverting resources from more productive purposes. Such technology makes industries compete poorly. Appropriate environmental technology minimizes human services and costs while better utilizirg solar energy and other environmental methods.

6.3.6   Finding Appropriate Concentrations

Because of savings in construction, some technologies were concentrated into large installations serving larger areas such as power plants, waste disposal plants, and manufacturing plants. Such concentrations make it more difficult to use the work of local environments, to provide life support services, absorb wastes, and reduce costs. Appropriate use of environment as a partner may require reduction in size and dispersal of industrial plants.

6.3.7   Improving Impact Evaluations

Many evaluations of impact consider parts and processes separately. Often the catalog of items and possible responses to environmental change are encyclopedic without generating much understanding. A better impact study results if the parts and processes are related with an energy systems diagram that includes the environmental systems involved and the new economic development whose impact is of concern. Then an *emergy* analysis table can be developed that evaluates the changes in overall economic impact.

6.4   DESIGN CHARACTERISTICS

Usually designs for functional systems have purpose. Sometimes engineering designs are said to have an object function that measures the success of the design, such as profit or efficiency. Theoretical reasoning suggests that self-organization designs also have goals and an object function.

6.4.1   Maximum Power Design Criterion

The designs that emerge from self-organization are those that cause the parts and processes of the system to contribute to the system's function. Such parts and processes are reinforced by large-scale feedback mechanisms. For example, larger animals may reinforce the plants on which they depend by their transport of seeds and recycle of mineral nutrients. On a larger scale, the efforts of humans to control nature by trial and error are self-organizing because the patterns that reinforce productive processes are copied and continued. Depending on the circumstance, self-organization may reinforce different abilities such as biomass, diversity, efficiency, ability to process a chemical, ability to concentrate a nutrient, or organic reserves. But all the reinforced components have in common the property of contributing to the overall useful energy flow. In other words, systems that draw more resources and use them to generate more resources and be more efficient in this use can overcome limitations and displace alternative designs.

This maximum performance criterion for system success was called *the maximum power principle* by Lotka in the 1920s (Lotka, 1922; 1925), with earlier roots attributed to Boltzmann and others.

6.4.2   Maximum Empower

Recognizing that energy flows of different type are not equal in their contribution to useful work, we have tried to clarify the principle as the "maximum empower principle." We defined *emergy* as the energy of one type required for a process or a stored product (Odum, 1986). The rate of flow of *emergy* is *empower*. For example, the inputs of sunlight, rain, fuels, fertilizers, pesticides, and human service to forestry production are each accompanied by energy, but these energies are of different type.

The concept of energy equivalents of one kind was loosely used in *Environment, Power, and Society* (Odum, 1971), expressing the energy required in organic matter equivalents. Names used at first were "energy cost" and later "embodied energy" and finally "emergy," a name suggested by David Scienceman as short for "energy memory." In one paper an extra "n" was added by the volume editor to decrease possible confusion between energy and emergy (Odum, 1986; see also Scienceman, 1987).

The *transformity* is the energy of one type required to make one unit of another. This was first defined as "energy quality factor" at the award ceremony of the Prize of the Institute de la Vie, Paris, in 1975 (Odum, 1976) and later as energy transformation ratio and finally transformity (Odum, 1986). By multiplying each energy flow by its solar transformity, each is expressed in solar emergy units (solar emjoules). The rate of flow of solar emjoules is solar empower. By using emergy and empower, high-quality inputs such as information, human service, and scarce chemical requirements are recognized for the high energy requirements in the trajectory that makes them available. To maximize empower is to maximize the convergence of the products of power use.

6.4.3   Emergy Use and Effect on Production

The theory relating emergy to performance is stated as follows. The ability to meet contingencies and compete is proportional to the empower (rate of emergy use), because self-organization retains only structures and pathways whose emergy use is commensurate with the amplifier effects of the transformed product used. In other words, after self-organization what is left is circular pathways that feed back useful effects in proportion to what was required to generate them. Components that contribute to the rest of the system are reinforced by control feedbacks from the higher levels of system hierarchy. For example, animal–plant associations that develop patterns of seed planting during self-organization are mutually reinforcing. The complexes of microbes and soil animals that develop effective recycling of minerals reinforce those kinds of plant producers whose organic matter is compatible with the decomposition and become mutually reinforcing.

6.4.4   Teleological Mechanisms in Self-Organization

If the mechanisms of higher-level reinforcement provide goals that the mechanisms at lower levels must conform to to be reinforced, they are teleological mechanisms (Hutchinson, 1948). Those who were taught that teleology was a sin do not allow themselves to think of mechanistic performance goals for survival in self-organization. This becomes a mental obstacle to understanding what controls ecosystems or how ecological engineering works. Understanding ecosystems and their management requires recognition of the control by higher levels of the components of a smaller level. Ecosystems cannot be understood by studies at one level alone (i.e., population ecology) because control comes from the next level.

Maximizing empower usually maximizes power as well. In other words, bringing to bear all the high-quality (high-transformity) inputs ensures maximum use of available energies of high quantity but low quality as well. For example, adding human services and seeds (high transformity and high emergy, but low energy) may help a vegetational area maximize its use of sunlight (low transformity, but high energy).

6.4.5   Mathematical Emulation of Systems

If mathematical models can be discovered that rigorously represent the self-organizational process including the constraints of energy laws and the maximum power principle, then simulation of these models will show what patterns of energy, matter, and information will be reinforced. The same models will apply to many different kinds of systems. Already, for example, similar pulsing oscillator models are found in many sciences and on many different scales of size.

Simulations will show the performances that will be reinforced from among the trial and error variations always present at each size level. Although we have only parts of the general models identified so far, it appears possible that simulation of systems performance can be achieved by models at a larger level without much reference to the smaller, faster components. That real systems have commonalities in process and design when appropriately generalized is an explanation of the top-down modeling advocated by many. System designs that prevail follow the energy-based mathematics of self-organization.

When one type of computer operation faithfully represents another but with different detailed mechanisms, it is said to emulate. Perhaps this word is appropriate for the technique of representing many systems with general mathematical models that generate correct results from energy principles without reference to the mechanisms that are different in each kind of system. This principle suggests that simple models can incorporate the principles and energy constraints of self-organization and simulate the important aggregate properties of more complex systems. It may be appropriate to call this kind of simulation, *emulation*.

By showing the mathematical properties of the self-organizational process, models clarify some of the reasons why such configurations as autocatalytic loops, feedbacks of high-quality to lower-quality production, material recycle, hierarchical structure, and diversified branches become reinforced and prevail. Configurations observed in self-organizing systems that also generate reinforcements in computer simulation are given below.

6.4.6   Generic Characteristics of Self-Organized Design

Generally occurring self-organized designs are described in paragraphs that follow (Odum, 1975; 1979; 1983a; 1983b, 1988).

6.4.6.1   Power-Maximizing Reinforcement by Autocatalysis

A typical autocatalytic loop design has an energy transformation process that is aided by the feedback of the product to help with more transformations. The product has higher quality than the original input so that a small amount has a multiplier effect on the input. Between the production and the feedback, the product is stored and some is lost by depreciation (second law losses). This design prevails because it is an energy reinforcement loop. The products that continue to be made are those that do reinforce by their feedback amplifier actions, thus maximizing inputs and commensurate efficiency.

Because considerable energy is used in a transformation process, in storage of the product and feedback, autocatalytic loops do not prevail at very low energy levels. Only simple linear, diffusion-type processes prevail when available energy is small. Examples are laminar flow, molecular diffusion without eddies, and ecosystems without enough energy to support life.

6.4.6.2   Energy Availability Controlling the Exponent of Autocatalysis

With higher levels of available energy, a quadratically autocatalytic pathway draws more resources and converts them more efficiently, taking energy away from linear or autocatalytic pathways. See simulations by Odum and Richardson (Richardson and Odum, 1981; Odum, 1982). In the choice of available conditions and organisms in a well seeded system, means for quadratic interactions are available to take over when energy levels rise, reinforcing such species and behaviors. Cooperative interactions that have quadratic mathematics are general in ecosystems (Allee, 1951; Odum and Allee, 1954).

6.4.6.3   Paradigm of Alternating Production and Consumption

In theoretical studies and in nature, evidence is accumulating that maximum power in the long run comes from systems that alternate periods of excess production with a short pulse of rapid consumption. For example, a farmer may keep his cows off his pasture while the grass is still sparse so as not to interfere with getting a heavy green cover as soon as possible. Trees grow up; when they fall there is a pulse of consumption followed by slow regrowth. Patterns in mountain forests have alternating patches of regrowth and consumption triggered by storms and landslides. Patterns of grassland growth alternate with waves of consumption by fire or herbivore herds.

6.4.6.4   Surges of Consumption Ahead of Production

In regimes of day and night and winter and summer, rising and falling rates of primary production are expected because of variation in the energy inputs. It might be expected and is sometimes observed that production increases first, followed by increase in the consumers. In such systems the return of nutrient materials and services from the consumer could lag, delaying potential production.

However, in many systems what emerges is a surge of consumption that occurs at the same time or before the surge of production. For example, synchrony of gross production and total respiration was observed in Texas bays (Odum, 1967). Consumer species use energy storages, migration, or seasonal temperature increases to accelerate consumption. Species strategies develop with the greatest energy demands timed with maximum food availability. For example, reproduction is timed so that the period of most rapid growth of the new generation occurs when food is most available. For another example, the timing of emerging leaves, insects, and bird migrations help keep production and consumption coupled in deciduous forest cycles. Sometimes we say that the species adapted to a system of pulsing production are those that are at the dinner table when dinner is served.

Thus the timing of the consumption by coupling with the production can reinforce the system under some conditions, causing those species to prevail over those with a lag. Design causes reinforcements organizing with the congruent coupling of production and consumption.

Is in-phase pulsing a contradiction of the alternating production–consumption paradigm in item Section 6.4.6.3 above? Possibly not if the nutrient raw materials are coming from the consumers; then the surge of consumption is required for the surge of production. The pattern represents a consumer frenzy delayed by winter and necessary for a net production period to follow.

If higher-level consumers (carnivores) surge in phase with production, they may keep the intermediate consumers (herbivores) out of phase, protecting the surge of production necessary to maximize use of available sunlight and nutrient resources.

Perhaps nearly synchronous production and consumption result from self-organizing when the consumption and the production surges have a time duration of the same order of magnitude as the rhythms of the incoming resource availabilities. The adaptation to diurnal and seasonal variations in solar energy may have more effect on overall performance than the criterion of keeping consumption out of the way of production. Considerable work may be needed to determine what timing of production–consumption coupling maximizes power in different conditions.

6.4.6.5 Storage of Recycling Materials

When consumption is well coupled to production with material recycle, no bottlenecks develop owing to material shortages. Then neither recycling materials nor input requirements are the more limiting to production. Whatever storages of recycling materials are necessary to keep materials from limiting will reinforce the production process. Comparative studies of many ecosystems show storages of critical materials to be commonly observed after self-organization. For example, open-sea plankton store iron and phosphorus; desert plants store water; and ecosystems receiving inorganic materials store organic matter reserves.

6.4.6.6 Nonautocatalytic Support of Consumers

Because services cannot be stored, different designs are required to maintain the consumer services to producers. Examples of services are seed planting, genetic controls, spatial controls, and mechanical cultivation. When stocks of service-coupled producers and consumers are both small, neither can start normal autocatalytic growth for lack of the other. In other words, a system of production requiring consumer actions cannot grow, as simulation shows.

However, if there is a second source for the consumers, either an immigration from outside, or a linear pathway of growth from producers, the consumers can reach a population size so as to augment production. Thus designs that supply consumers by nonautocatalytic means are necessary designs to take advantage of consumer services.

6.4.6.7 Competitive and Cooperative Components in Sequence

A design found in succession is the sequence of two stages. First, components prevail by faster individual-acting, competitive, and inefficient start-up of resource use. Second, more complex, organized, cooperative designs can take over by contributing efficiency. For example, first come wind-distributed weeds; whichever ones make a fast start accelerate succession. The later members are aided by animals and other interrelationships, have longer time constants, lower depreciation rates, and other adaptations to last longer.

Models to represent these designs have two different producer configurations in parallel. The simulation shows rapid transformation from the successional state based on the one pathway to the more efficient one based on better storages and organization later.

6.4.6.8 Power-Maximizing Frequency of Long-period Pulses

When there are surges of activity of components with longer turnover times and larger realms of support and influence, all of the smaller components in the energy support territory are entrained. The timing of the large-scale surge is impressed on the small-scale members. The reverse is not true, for the surges of the small ones have short periods, small realms, and unless they are synchronous are absorbed by the larger components as a filter absorbs noise.

As described in Section 6.4.6.3 above, a surge of function of a large member may be reinforcing to the long-range power supplies of the system. By conducting a surge of consumption that is quickly over, all the resources stored in the smaller members are rapidly consumed, after which the large-member surge has to shut down. After that there is a long period of gradual buildup of storages in the smaller members before the level reaches that which makes an autocatalytic or quadratic autocatalytic curve. Where the input energy is fixed, this pattern allows maintenance of production at a higher level. Postponed consumption develops a large pulse after which production is quickly reestablished.

Simulation experiments by Richardson (1988), working with Odum, show an optimal frequency for the consumptive pulse to maximize power utilization in the long run, including both net production and net consumption parts of the oscillatory regime.

6.4.6.9 Self-Regulating Time Constants to Maximize Useful Filtration

Where there are fluctuations and oscillations in availability of input resources, receiving pathways are reinforced if they develop time constants that absorb the energies. The more frequencies there are in the input resources, the more kinds of pathway niches there are for adapting absorption frequency. In other words, self-organization develops a set of frequency domain niches to fit a set of resource frequencies in the inputs. The many simulation studies by Campbell (1984), working with Odum, show the performance of various basic ecosystem chains and webs as resource absorbers. There are similarities between an ecosystem and an FM (frequency-modulated) radio.

6.4.6.10 Impedance Characteristics for Control of Timing

Another type of design for maximizing the energy absorption from externally varying resources uses "inductance"-like elements that exert back force to acceleration. In electronics, the back-force property is delivered by coils that generate magnetic field storages in response to input acceleration. In ecosystems this kind of impedance may be provided by the behavior of some organisms. Stubborn characteristics of human behavior may have this property.

Some of the configurations form internal oscillations, which can be tuned to absorb pulsing inputs of available energy or pass them without absorption. The many simulations by Zwick (1985), working with J. Alexander, show that there are optimal impedance characteristics for maximum power utilization.

6.5 ECOLOGICAL ENGINEERING EDUCATION

Ecological engineering requires a different kind of background from traditional engineering, one that few engineers or ecologists have received in their formal training. After the potentials of ecological engineering were realized and it was defined as a field in 1962, courses were tentatively offered on an experimental basis thereafter at three universities. Based on this experience, suggestions follow for appropriate courses and curricula.

6.5.1   A Course in Ecological Engineering

The following is a statement for the course catalog: *"Principles and practices in design and management of environment with society, including systems concepts for organization of humanity, technology, and nature."* The course starts with the concept of a unified system. The general patterns of ecosystems with and without the economic interfaces are given, using synthetic diagrams, models, and microcomputer simulation to relate structure and process.

As part of the explanation for the generic patterns, the maximum power principle is introduced. Emergy, transformity, and other indexes are used to evaluate alternatives. Designs are sought that maximize emergy contribution.

Using microcosms and case histories, self-organizational processes, succession, and their management are described. These show how successful designs are achieved whether the design is intentional or not.

Main designs for symbiotic environmental–economic systems are examined. Main categories of environmental impact are considered as well as the way improved management generates working interfaces between the economy and environment. Techniques are given for ecosystem control with minimal effort. Table 6.2 lists the main topics with examples.

6.5.2   Appropriate Education for Career-Level Ecological Engineering

Although there has been a long enough tradition in many kinds of engineering to set standards for training and performance, ecological engineering is only now being recognized, and people with the necessary training have been few. Customary engineering procedures eliminating nature, simplifying structures, and utilizing economies of larger scale are inappropriate for ecological engineering. Knowledge of environmental systems, self-organizational processes, and evolutionary adaptation is required as well as engineering subjects such as mathematics, systems, concepts of design, and problem solving.

The environmental realm is just as much an area of basic science as other sized realms of the universe, but education has traditionally divided the area into specialties (soils, forests, geology, meteorology, environmental chemistry, economics, etc). There is a wide range of basic courses needed from these fields, not the superficial cultural versions, but the rigorous introductory versions required for majors in these specialties. Sometimes we use the phrase "the environmental generalist" because many sciences are combined. In addition, the core includes career-level introductory courses to

Table 6.2 Topics for a Course in Ecological Engineering

- Self-organizing process and its representation with systems diagrams and models.
- Concept of maximum power and its role in self-organization.
- Environmental design corollaries from the maximum power principle.
- Review of multiple-seeded microcosms to illustrate the self-organizing process, principles, and management.
- Developing a new ecosystem to fit new conditions by arranging multiple seeding and allowing self-organization.
- Diagramming the coupled systems of human economy and environmental systems, identifying the reward feedback linkages.
- Review of types and examples of interface coupling that contribute environmental services.
- Conversion of wastes into by-products by arranging for their use by environmental systems.
- Use of systems diagrams of economic–environmental interfaces to organize impact statements.
- Concept of emergy and transformity and their use to evaluate environmental contributions, impacts, and alternatives.
- Equating emergy of exchanges between two systems for equity and reinforcement.
- Identify economic potentials from the emergy signature of available resources.
- Determining macroeconomic value of an item as the proportion that emergy is of the total economy, expressing the contribution in dollars of gross economic product.
- Increasing environmental roles by reducing the ratio of investment inputs to environmental inputs.
- Microcomputer simulation of ecological engineering designs.
- Determining risk from the pulsing frequency obtained by identifying position in hierarchies.
- Adapting systems to potentially catastrophic events.
- Adapting a system to use a former stress as a resource.
- Review of domesticated ecosystems available for solving problems.
- Using diversity for stable maintenance of landscapes, lakes, and estuaries.
- Use of wetlands in landscape organization for water conservation and quality control; connecting human settlements and wetlands for effective recycling.
- Controlling time, place, and frequency of waste release to fit ecosystem patterns.
- Locating development spatially according to high levels of emergy availability.
- Developing symbiotic interfaces between thermal heat releases from power plant thermal heat and environment.

Table 6.2 Topics for a Course in Ecological Engineering (*continued*)

- Controlling vegetation type for water conservation, heating, and air conditioning.
- Management of succession to accelerate post-mining reclamation.
- Arresting succession by restricting seeding; releasing arrested succession by seeding.
- Identifying appropriate agroecosystems for regional productivity.
- Managing forests for sustained yield by control of cutting area and frequencies.
- Managing fisheries by adjusting capital investment according to position in energy hierarchies.
- Managing ponds for sustained, low-energy aquaculture.
- Review of progress in development of closed ecosystems containing humans.
- Locating highways and bridges to maintain environmental systems and wildlife corridors.
- Organizing corridors of city transportation to converge and diverge between levels of energy hierarchy.
- Determining sustainable loading of parks by human users.
- Controlling housing densities to maximize environmental support.
- Recycling biodegradable solid wastes to environmental processes in place of landfills.

environmental sciences and engineering. Few programs now exist with the needed balance. More time is required than for most degrees.

Specifically, the general education should include typing and programming, math through calculus, English composition and other languages, world history and geography, economics, physics, biology, and chemistry. The main, rigorous introductory courses to the environmental sciences should follow, such as, limnology, soils, oceanography, geography, forestry, agronomy, geology, hydrology, meterorology, environmental chemistry, and systems engineering.

Because of the spread of tough basic courses, the curriculum is a good base for many other careers. It is a good program for students who are not yet sure of their ambitions. It might be called "liberal science."

6.5.3   Accreditation of Ecological Engineering

Training of ecologists in biology departments rarely includes the essentials given above. Increasing emphasis on molecular biology makes it unlikely that this will change. Not only are the environmental science courses missing, but the perspectives and ways of thinking are on mechanisms two orders of magnitude too small. Some people from this training often deny that there is an ecological system or an environmental system.

Engineering organizations apply accreditation criteria to environmental engineers, with ambiguous results. Criteria used are those for the technology, construction, and operation of installations, not for determining what is appropriate to build in the first place nor for design of environmental pattern. As a result, the students don't get adequate environmental science, ecology, or ecological engineering as defined here.

Attempts by those with only biological education to write accreditation specifications for applied ecology are just as inappropriate as efforts to write accreditation specifications for environmental engineering by those with only engineering training.

The solution may be to recognize the need under the appropriate name "ecological engineering" and if accreditation is necessary, possible, or desirable, to constitute broad committees including public policy representatives to define requirements.

Once the discipline is defined by a substantial clientele, curricula and departments of universities may be restructured and courses condensed so that the design of humanity and nature becomes part of the mainstream of higher education.

REFERENCES

Allee, W. C. 1951. *Cooperation among Animals with Human Implications*. Schuman, New York, 233 pp.

Campbell, D. E. 1984. Energy filters and ecosystem design. Ph.D. Dissertation. Department of Environmental Engineering Sciences, University of Florida, Gainesville, FL, 450 pp.

Ewel, K. C. and H. T. Odum, eds. 1984. *Cypress Swamps*. University of Florida Press, Gainesville, FL, 472 pp.

Hutchinson, G. E. 1948. Circular causal systems in ecology. *Ann. N.Y. Acad. Sci. 50:*221–246.

Kangas, P. C. 1983. Landforms, succession, and reclamation. Ph.D. Dissertation. Department of Environmental Engineering Sciences, University of Florida, Gainesville, FL, 187 pp.

Lotka, A. J. 1922. A contribution to the energetics of evolution. *Proc. Natl. Acad. Sci., 8:*147–155.

Lotka, A. J. 1925. *Physical Biology*. Williams and Wilkins, Baltimore.

Nixon, S. W. 1969. Synthetic microcosm. *Limnol. Oceanogr. 14:*142–145.

Odum, H. T. 1962. Man in the ecosystem. Proc. Lockwood Conference on the Suburban Forest and Ecology. *Bull. Conn. Agric. Station 652:*57–75.

Odum, H. T. 1967. Biological circuits and the marine systems of Texas. In T. A. Olson and F. J. Burgess, Eds., *Pollution and Marine Ecology*. Wiley-Interscience, New York, pp. 99–157.

Odum, H. T. 1971. *Environment, Power and Society,* Wiley, New York, 336 pp.

Odum, H. T. 1975. Combining energy laws and corollaries of the maximum power principle with visual systems mathematics. In *Ecosystem Analysis and Prediction.* Proceedings of conference on ecosystems at Alta, Utah. SIAM Institute for Mathematics and Society, pp. 239–263.

Odum, H. T. 1976. Energy quality and carrying capacity of the earth. *Trop. Ecol.* *16*(1):1–8.

Odum, H. T. 1979. Energy quality control of ecosystem design. In R. F. Dame, Ed., *Marsh Estuarine Systems Simulation.* Belle Baruch Library in Marine Science No. 8. University of South Carolina Press, pp. 221–235.

Odum, H. T. 1982. Pulsing, power, and hierarchy. In W. J. Mitsch, et al., Eds., *Energetics and Systems.* Ann Arbor Science, Ann Arbor, MI, pp. 33–54.

Odum, H. T. 1983a. *Systems Ecology*, Wiley, New York, 644 pp.

Odum, H. T. 1983b. Maximum power and efficiency, a rebuttal. *Ecol. Modelling* *20*:71–82.

Odum, H. T. 1985. Self-organization of Ecosystems in Marine Ponds Receiving Treated Sewage. University of North Carolina Sea Grant Publication #UNC-SG-85-04. North Carolina State University, Raleigh, NC, 250 pp.

Odum, H. T. 1986. Enmergy in ecosystems. In N. Polunin, Ed., *Ecosystem Theory and Application.* Wiley, New York, pp. 337–369.

Odum, H. T. 1988. Self-organization, transformity, and information. *Science* *242*:1132–1139.

Odum, H. T. and W. C. Allee. 1954. A note on the stable point of populations showing both intraspecific cooperation and disoperation. *Ecology 35*:95–97.

Odum, H. T., John W. Caldwell, W. Smith, M. Kemp, H. McKellar, M. Lehman, K. Benkert, M. J. Lucas, R. Knight, D. Hornbeck, D. Young, M. L. Homer, and F. Ramsey, et al. 1974, 1975, 1978, 1979. Power Plants and Estuaries at Crystal River Florida; Annual Record of Metabolism of Estuarine Ecosystems and Salt Marsh at Crystal River, Fl. Reports to Florida Power Corporation from Systems Ecology and Energy Analysis Group, Department of Environmental Engineering Sciences, University of Florida, Gainesville, FL.

Odum, H. T. and A. F. Chestnut. 1970. Studies of Marine Estuarine Ecosystems Developing with Treated Sewage Wastes. Annual report for 1969–1970. Institute of Marine Sciences, University of North Carolina, Morehead City, NC, 364 pp.

Odum, H. T., K. C. Ewel, W. J. Mitsch, and J. W. Ordway. 1975. Recycling treated sewage through cypress wetlands in Florida. Occasional Publication 1975-1, Center for Wetlands, University of Florida, Gainesville. Also in F. D'Itri, Ed., 1977. *Wastewater Renovation and Reuse. Marcel Dekker,* New York pp. 35–68.

Odum, H. T., W. L. Siler, R. J. Beyers, and N. Armstrong. 1963. Experiments with engineering of marine ecosystems. *Publ. Insti. Marine Sci. Univ. Texas 9*:374–403.

Richardson, J. F. 1988. Spatial designs for maximum power. Ph.D. Dissertation. Department of Environmental Engineering Sciences, University of Florida, Gainesville, FL (in committee).

Richardson, J. F. and H. T. Odum. 1981. Power and a pulsing production model.

In W. J. Mitsch, R. W. Bosserman, and J. M. Klopatek, Eds., *Energy and Ecological Modelling*. Elsevier, Amsterdam, pp. 641–647.

Rushton, B. T. 1987. Wetland reclamation by accelerating succession. Ph.D. Dissertation. Department of Environmental Engineering Sciences, University of Florida, Gainesville, FL, 199 pp.

Scienceman, D. M. 1987. Energy and emergy. In G. Pillet and T. Murota, Eds., *Environmental Economics*. Roland Leimgruber, Geneva, Switzerland, pp. 257–276.

Taub, F. 1969. A biological model of a freshwater community. A gnotobiotic ecosystem. *Limnol. Oceanogr. 14:*136–141.

Taub, R. B. 1974. Closed ecological systems. *Ann. Rev. Ecol. Syst. 5:*139.

Vitousek, P. 1987. Crafoord Day Symposium, Royal Swedish Academy of Sciences, Stockholm, Sweden.

Zwick, P. 1985. Energy systems and intertial oscillators. Ph.d. Dissertation. Department of Environmental Engineering Sciences, University of Florida, Gainesville, FL, 239 pp.

7

AGRICULTURE AND ECOTECHNOLOGY

David Pimentel

*Department of Entomology,
Cornell University, Ithaca, New York*

7.1 INTRODUCTION

There is unfortunately no accurate historical account of the development of agriculture in the world; however, we can logically reconstruct what might have happened in the evolution of agriculture. It appears that agriculture evolved slowly from the less-structured societies of food gatherers. We know that gatherers brought grains, vegetables, and fruits to camp for cooking and consumption. Probably some seeds were dropped on the soil in the clearing of the camp, and had the opportunity to grow there. One can imagine the discovery of the concentration of grains, vegetables, and fruits upon the return of the gatherers to the campsite some time later. This probably encouraged some of the more venturesome people to associate seeds with plants and to begin to plant some spare seeds themselves.

The relative ease of finding and harvesting such crops compared to random gathering would encourage more plantings. The trend to produce food is thought to have been slow at first; perhaps only a small percentage of the food supply was produced from the gardens, but gradually this percentage increased.

One important step in helping the new seeds to get a good start was the deliberate removal of the existing natural vegetation, including shrubs and trees, that would interfere and compete with crop growth. Burning was the easiest and most common means of clearing the land. Following burning, the plots were generally clear except for a few large trees and charred stumps.

The seeds were planted by poking holes in the soil with digging sticks and dropping the seeds into the holes. Placing seeds in the cleared ground increased germination and subsequent growth so the plants could compete successfully with weeds.

Little or no care was given to the early crop plantings. A few months to a year later the early farmers would return to harvest their crop, or what was left of it. Insects, diseases, birds, and mammals shared the harvest, and weed competition reduced yields just as many of these same pests reduce crop yields today.

The final step in the development of early agriculture was to expand the crop plantings sufficiently to produce most of the food supply. With time, as the camps became relatively permanent because food supply was nearby and ample, men and women no longer had to travel to find food. Then, too, living close to the plantings allowed the group to claim ownership and protect the plantings from other people as well as from birds, mammals, weeds, and other pests.

These early agriculturalists and hunter/gatherers expended approximately three-quarters of their time and energy in meeting their food needs. In fact, obtaining food and collecting fuel wood for food preparation usually dominated the activities of these societies.

In addition to having to move relatively frequently, hunter/gatherers had to search wide areas for food and then transport it back to camp. Under the most favorable environmental conditions a family of five would require about 750 ha of terrestrial ecosystem, whereas under more marginal ecosystem conditions the family might search 50,000 ha for food (Clark and Haswell, 1970). In both situations, but especially in the marginal ecosystem, the hunter/gatherers had to search for and transport food long distances to supply the family with food.

Whether hunter/gatherers or agriculturalists, humans used land, water, energy, and biological resources to obtain their food supplies. Particularly in agriculture, ecotechnology or the management of natural resources plays a vital role in food production. In this chapter, I assess the role of land, water, energy, and biological resources in agricultural production and what ecotechnology might do to improve the efficiency of food production while sustaining the productivity of natural resources.

7.2 LAND RESOURCES

Next to sunlight, land is the most vital resource for agriculture. In the United States, about 160 million ha of cropland is planted and cultivated to provide food for 240 million people, or 0.7 ha/person (USDA, 1985). Excluding the land devoted to food exports, the arable land planted to feed each person is about 0.6 ha.

Worldwide, with arable land resources estimated to be about 1.5 billion ha (Buringh, 1979) and a world population of 5 billion (PRB, 1986), the available arable land per person amounts to only 0.3 ha. According to FAO predictions (FAO, 1982), the possible net expansion in total world cropland is 3.9 million ha/yr. At this rate world cropland will expand to 2 billion ha by the year 2110.

Undoubtedly, it will be possible to bring some additional land, much of which is considered marginal, into production. Optimistic estimates propose that worldwide cropland can be expanded to 3.4 billion ha (Buringh, 1979). However, others conservatively project that worldwide cropland can be expanded to 2 billion ha, but only with the use of large amounts of energy to make the marginal land productive (NAS, 1977). Assuming that the world population reaches 12 billion and expands the land base to 2 billion ha, only 0.2 ha or less of cropland would be available per person (Figure 7.1), and most people would have to consume essentially a vegetarian or plant protein diet.

With excellent soils, favorable temperatures and rainfall, heavy fertilization, and effective pest control, it is theoretically possible to provide 20 people with adequate calories and protein from 1 ha of land. However, with average land, climate, and other factors, it would be optimistic to hope to feed half that number. This is especially true considering the current worldwide soil erosion crisis (Pimentel et al., 1986). In the United States, for example, several million hectares of marginal land, which is highly susceptible to severe soil erosion, is already being cultivated (OTA, 1982; Naegeli, 1986).

The available arable land, water, and energy resources, as well as the kind of food a particular society desires or can afford, determine the crops produced in a given region. At present, two-thirds of the people in the world consume a primarily vegetarian diet in contrast to industrialized countries where diets are characteristically high calorie/high animal protein. This latter diet requires large amounts of land and energy to produce. Based on all available data concerning future availability of arable land and energy, it will not be possible to provide all people of the world with such a high calorie/

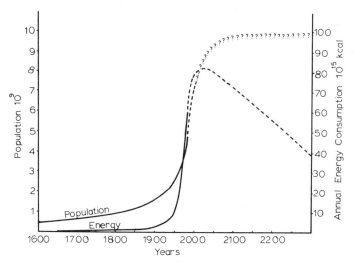

Figure 7.1 Estimated world population numbers (——) from 1600 to 1975 and projected numbers (– – – –) (?????) to the year 2250. Estimated fossil fuel consumption (——) from 1650 to 1975 and projected (– – – –) to the year 2250 (Pimentel and Hall, 1984).

high animal protein diet. In fact, if U.S. population growth continues for less than 200 years (Pimentel et al., 1988), it will be similar to that of China, and diets will have to be modified to include larger amounts of plant protein and less animal protein.

The quality of land is as important as the number of hectares available. As mentioned, to bring land of marginal quality into production, water and energy inputs for fertilizers and/or irrigation are essential. The extent of expansion will depend on the availability of all these resources.

Meanwhile, farmers are using soils that are eroding at an alarming rate (Pimentel et al., 1976; Holdgate et al., 1982; Eckholm, 1983). Soil erosion reduces the natural productivity of soils by removing nutrients and organic matter, and by reducing topsoil depth and water availability (Lal, 1984).

Global dimensions of land destruction are a major concern. About 35% of the earth's land surface is affected (Mabbutt, 1984). The natural productivity of many soils has been reduced 25–100% because of erosion (Langdale et al., 1979; Lal, 1984; Mabbutt, 1984). In the past, several civilizations, including those in Mesopotamia and Greece, failed in part because of the degradation of their agricultural lands (Jacks and Whyte, 1939; Lowdermilk, 1953; Troeh et al., 1980).

Worldwide, an estimated 6 million ha of agricultural land is irretrievably lost each year because of water and soil erosion, salinization from irrigation, and other factors (UNEP, 1980; Dudal, 1981; Kovda, 1983). Based on current worldwide soil loss, a recent study for the period 1975–2000 projected that rain-fed land degradation will depress food production another 15–30% (Shah et al., 1985).

Recent surveys in Iowa, which has some of the richest agricultural lands in the world, indicate that about half of the original topsoil has been lost (Risser, 1981). Under normal agricultural conditions, the formation of 2.5 cm (1 in.) of soil requires 200–1000 years (Larson, 1981; Hudson, 1981; McCormack et al., 1982; Sampson, 1983; Lal, 1984). Fortunately, so far, increased fertilizers and other fossil energy inputs have been available to offset the reduced productivity of some world cropland caused by soil erosion. However, the loss of about 20 cm of topsoil from a topsoil base of 30 cm requires double the usual energy inputs just to maintain crop yields (Pimentel et al., 1981). In future decades these large energy inputs may not be affordable or even available.

In developing countries, the rate of cropland erosion is nearly twice that experienced in the United States (Ingraham, 1975). For example, reports indicate that each year an average of 38 t/ha of soil is lost from over half of India's total land area because of serious erosion (CSE, 1982). Erosion is also a problem in China, where it is reported to average 43 t/ha-yr (Brown, 1976; AAC, 1980).

Resources to offset the productivity losses are not always readily available or affordable. In some regions, to obtain more land for crop production, forests are being cut (Pimentel et al., 1986) and steep slopes are being used to grow crops. As vegetation has been removed, erosion has intensified. Often flooding in the lowlands has become a serious environmental problem to the extent that, in India for example, losses from flooding have doubled during the past 10 years (USDA, 1965; Sharma, 1981). Certainly, as more marginal land is put into agricultural production, land degradation and flooding will increase (CEQ, 1980).

7.3 WATER RESOURCES

Along with sunlight and land, water is a vital resource for agricultural production. Sufficient rain falls upon most arable agricultural land, but periodic droughts continue to limit yields in some areas of the world. All crops require and transpire massive amounts of water. For example, a corn crop that produces 7000 kg of grain per hectare will take up and transpire about 4.2 million L of water per hectare during the growing season (Leyton, 1983). To supply this much water each year, about 10 million L (100 cm) of rain must fall per hectare, and furthermore it must be evenly distributed during the year and growing season.

A decrease in rainfall of only 5 cm during the growing season reduces corn production by about 15% (Finkel, 1983). Decreasing rainfall to 30 cm per season reduces yields to about one-fifth of those from areas receiving 100 cm of rainfall. Sorghum and wheat require less water than corn and could be grown instead, but these crops produce about one-third less grain per hectare than corn (USDA, 1984a).

Irrigation is essential if rainfall cannot be relied upon to supply the moisture needed for crop production. Irrigated crop production requires the movement of large quantities of water. For example, using irrigation to produce 1 kg of the following food or fiber products requires 1400 L of water for corn, 4700 L for rice, and 17,000 L for cotton (Ritschard and Tsao, 1978). In the United States, agricultural irrigation presently consumes 83% of the total 360 bld (billion liters per day) that is consumed by all sectors of society (Murray and Reeves, 1977).

Irrigation water is either piped long distances from rain-fed reservoirs or pumped from wells supplied by aquifers. These large underground storage areas are slowly filled by rainfall. Intense and prolonged irrigation stresses aquifers to such an extent that, in some areas of the world, water is being "mined" and used more quickly than it can be replaced by rainfall (CEQ, 1980). Even now in the United States, water overdraft exceeds replenishment by at least 25% (USWRC, 1979).

In addition to its large water usage, irrigated crop production is costly in terms of energy consumption. Nearly one-fifth of all energy expended in U.S. agricultural production is used to move irrigation water (USDA, 1974). For example, in Nebraska, rain-fed corn production requires about 630 L/ha of oil equivalents, whereas irrigated corn requires an expenditure of about three times more energy (Pimentel and Burgess, 1980).

Periods of drought, whether for a season or extending over many years, also influence crop production. All countries experience water shortages at times, which influence agricultural production (Ambroggi, 1980). The current severe drought in the Sahel, which extends through much of Africa south of the Sahara, clearly illustrates the complexity of the problem. A substantial decrease in the normal rainfall results in reduced crop production (Hare, 1977). The effect, however, has been magnified by the large human population per available arable land as well as the long-time mismanagement of all land resources, including forest areas (Biswas, 1984).

Water resources are becoming a global problem. From 1940 to 1980 worldwide water use more than doubled. The anticipated growth in world population and agricultural production can be expected to double water needs again in the coming two decades (Ambroggi, 1980).

By the year 2000, it is estimated that the world agricultural production will consume nearly 80% of the total water withdrawn (Biswas and Biswas, 1985). Because more water will be needed to support agricultural production, both the extent and location of water supplies will become major constraints on increased crop production. Considering that about one-third of the major world river basins are shared by three or more countries (CEQ, 1980), water availability is certain to cause competition and conflicts between countries and even within countries (Biswas, 1983).

In contrast to limited water supplies is the problem of too much water, accompanied by rapid water runoff. Both may cause waterlogging of soils and flooding. Fast water runoff decreases crop yield per hectare by reducing the water available to the crop, by removing soil nutrients, and even by washing away the soil and crops themselves (OTA, 1982).

When forests on slopes are replaced with crops to augment food supplies, runoff and soil erosion increase and serious flood damage occurs to valuable crops and pasture (Beasley, 1972). Economic costs of these losses can be significant. In the United States, damage from sediments and flooding to surrounding areas costs an estimated $6 billion each year (Clark, 1985). In Bangladesh in 1974, a severe flood diminished the productivity of the soil significantly and reduced the rice harvest, which ultimately led to severe food shortages and famine (Brown, 1976).

In general, most water runoff, which brings with it loose soil, has negative effects on agriculture. But where annual foods can be managed such as those characteristic of the Nile River basin, the water and rich soil sediment benefit the crops growing in the flood plains (Biswas, 1984).

7.4 BIOLOGICAL RESOURCES

Almost 90% of human food comes from just 15 species of plants and 8 species of livestock, all of which were domesticated from the wild (Mangelsdorf, 1966; Myers, 1979). However, high agricultural productivity as well as human health depend upon the activity of a myriad of natural systems composed of an estimated 10 million species of plants and animals that inhabit the world ecosystem (Pimentel et al., 1980a; Ehrlich and Ehrlich, 1981; Myers, 1983). We know that humans cannot survive with the presence of only their crop and livestock species, but exactly how many and which plant and animal species are essential is unknown.

Natural species are eliminated in many ways. The most serious causes include clearing land of natural vegetation for agriculture, urbanization, and chemical pollution (pesticides, etc.) of the environment.

Of great concern is the growing rate of species elimination and the subsequent loss of genetic diversity that human encroachment is causing (Biswas and Biswas, 1985). Predictions are that an estimated 1 million species of plants and animals will be exterminated by the end of this century (Eckholm, 1978; Ehrlich and Ehrlich, 1981). Even now in the Indian subcontinent, 10% of the plant species are threatened or endangered (Sharma, 1981). This high rate of extinction is alarming because it is not known how many of these organisms may be necessary to maintain food production and other vital human activities (Myers, 1984, 1985).

Natural biota perform many functions essential for agriculture, forestry, and other sectors of the environment. Some of these include providing genetic diversity basic to successful crop breeding; recycling vital chemical elements such as carbon and nitrogen within the ecosystem; moderating climates; conserving soil and water; serving as sources of certain medicines, pigments, and species; and supplying fish and other wildlife (Pimentel et al.,

1980a; Ehrlich and Ehrlich, 1981; Myers, 1979, 1984). In addition, some natural biota such as bacteria and protozoans help prevent human diseases. Some prey on and destroy human pathogens, others degrade wastes, and still others remove toxic pollutants from water and soil and help buffer the impact of air pollutants (Pimentel et al., 1980a).

The relative impact of a species in the environment can be judged in part by its biomass per unit area. Groups such as insects, earthworms, protozoans, bacteria, fungi, and algae average 6500 kg/ha, more than 350 times the average human biomass of 18 kg/ha in the United States. Reducing or exterminating some species groups could seriously disturb the normal or natural functioning of important environmental systems in nature. For example, soil quality and agricultural productivity jointly depend upon soil biota, organic matter, and the presence of certain inorganic elements (Brady, 1984). Each year various soil organisms break down and degrade about 20 tonnes of organic matter per hectare (Alexander, 1977). This degradation is essential for the release of bound nutrients and their subsequent recycling for further use in the ecosystem (Golley, 1983). Some soil biota help control pests in crops, and others "fix" atmospheric nitrogen for use by plants (Pimentel et al., 1980a). Earthworms, insects, and other biota open holes, improve water percolation, and loosen the soil to enhance new soil formation. Because the complex roles of these biota are not clearly understood, great care should be taken to prevent the extinction of any species.

7.5 ENERGY RESOURCES

Most recent increases in crop yields have been achieved by using enormous amounts of fossil energy to supply fertilizers, pesticides, irrigation, and fuel for machinery (Leach, 1976). For example, in the United States crop yields have increased about 3.5 times during the last 60 years, while fossil energy inputs rose 40-fold (Pimentel and Wen, 1988). A significant amount of this energy was used to compensate for degraded land (Pimentel et al., 1988).

How can food supply and energy expenditures be balanced against the needs of the growing world population? Doubling the food supply during the next 25 years would help offset the serious malnourishment that 1 billion humans presently endure (Latham, 1984), as well as help feed the additional people. However, such an increase, assuming current land degradation and minimal substitution of labor with mechanization, would require about a five- to tenfold increase in the total amount of energy expended for food production. Such a large energy input for food production would be necessary to balance the decreasing return of crop output per fertilizer energy input.

Throughout the world, energy use for food production continues to grow dramatically. Chinese agriculture provides a striking example of the increased reliance on fossil energy in food production. During the past three

Table 7.1 Energy Inputs and Outputs in Corn (Maize) Production in Mexico Using Only Manpower

	Quantity/ha	kcal/ha
INPUTS		
Labor	1,144 h[a]	624,000
Ax and hoe	16,570 kcal[b]	16,570
Seeds	10.4 kg[b]	36,608
Total		677,178
OUTPUTS		
Corn yield	1,944 kg[a]	6,901,200
kcal output/kcal input		10.19

[a] Lewis, 1951.
[b] Estimated.

decades, fossil fuel inputs in Chinese agriculture rose 100-fold, and crop yields tripled (AAC, 1980; Taylor, 1981; Lu et al., 1982).

The United States, like most developed countries, is a heavy energy user. For example, 17% of the total energy supply, or about 1500 L per person of oil equivalents, is expended on production, processing, distribution, and preparation of food. This contrasts with most developing nations, which use less than one-tenth this amount of energy for all their food.

In many developing nations, crops are still produced by hand, requiring about 1200 h of labor (Table 7.1). In contrast, in the highly mechanized U.S. agriculture, crops like corn require only about 10 h of labor (Table 7.2). The energy output to input ratio for the hand-produced corn is about 1:10, whereas in the U.S. mechanized system the ratio is 1:2.2.

Using U.S. agricultural technology to feed the current world population of 5 billion a high protein/calorie diet would require 7.6×10^{12} L of fuel per year (Pimentel and Hall, 1984). At this rate, assuming petroleum was the only source of energy for food production, and all known reserves were used solely for this purpose, world oil reserves would last only 11 years. Of course, not all nations desire the diet typical of the United States.

One practical way to increase food supplies with minimal fossil energy inputs and expenditures would be for everyone—especially those living in the industrialized nations—to consume less animal protein (Pimentel et al., 1980b). This diet modification would reduce energy expenditures and increase food supplies because less edible grain would be fed to livestock to produce costly animal protein.

The average yield from 10 kg of plant protein fed to animals is only 1 kg of animal protein. If the 130 million tons of grain that is fed yearly to U.S. livestock were consumed directly as human food, about 400 million people—1.7 times more than the U.S. population—could be sustained for 1 year

Table 7.2 Energy Inputs for Corn Production in the United States

	Quantity/ha	kcal/ha
INPUTS		
Labor	10 h	5,000
Machinery	55 kg	1,018,000
Gasoline	40 L	400,000
Diesel	75 L	855,000
Irrigation	2.25×10^6 kcal	2,250,000
Electricity	35 kwh	100,000
Nitrogen	152 kg	3,192,000
Phosphorus	75 kg	473,000
Potassium	96 kg	240,000
Lime	426 kg	134,000
Seeds	21 kg	520,000
Insecticides	3 kg	300,000
Herbicides	8 kg	800,000
Drying	3,300 kg	660,000
Transportation	300 kg	90,000
Total		11,037,000
OUTPUTS		
Total yield	7,000 kg	24,500,000
Kcal output/kcal input		2.22

Source: After Pimentel and Wen, 1988.

(Pimentel et al., 1980b). However, dietary patterns and favorite foods are based not only on availability and economics, but on social, religious, and other personal factors. For these reasons major diet modifications of any kind are difficult to achieve, especially within a short time span.

In view of the heavy drain on fossil fuel supplies, biomass (grain, sugar bagasse, other crop residues, and fuel wood) has been suggested as a major substitute fuel. To some extent biomass has always been used for fuel, but to use it in the amounts needed to spare fossil fuels requires careful analysis.

One constraint operating against the increased use of biomass for fuel is the amount of land required to grow it. When the amount of cropland needed to feed one person and that required to provide biomass to fuel one average U.S. automobile with ethanol for one year are compared, nine times more cropland is used to fuel the automobile than to feed a person (Pimentel et al., 1984a).

Biomass in the form of wood has long been a major fuel for humans. Presently, more than half the world's population depends on firewood as its primary fuel source (Pimentel et al., 1986). But supplies of firewood are diminishing as an estimated 12 million ha of forests are cut and cleared each year primarily to provide more lands for agricultural production (Spears and Ayensu, 1984). As a result, the total amount of wood biomass available in the world per person has declined 10% during the past 12 years (Brown et al., 1985). Because more food must be produced to maintain the ever-increasing world population, the supply of fuel wood can be expected to diminish.

If biomass usage is increased, we must consider the effect this would have on available quantities of arable land for food/fiber production. Furthermore, the removal of trees and crop residues is known to decrease soil fertility and facilitate soil erosion. From this one can conclude that biomass resources are limited in their usefulness as fossil fuel substitutes.

Solar energy technology shows some potential for the future. Photovoltaics and solar thermal energy should supply limited amounts of renewable energy (ERAB, 1982); however, these techniques do require land and also may affect some wildlife (Pimentel et al., 1984b).

7.6 ECONOMIC AND ENVIRONMENTAL BENEFITS OF ECOTECHNOLOGICAL AGRICULTURAL MANAGEMENT

The economic and environmental benefits of several ecotechnological conservation methods for cultivating corn are compared to conventional corn production in this section. Today, for instance, it costs $523 to produce a hectare of conventional corn (Table 7.3). The ecotechnological practices used include "no-till," "ridge planting," and "low-input system." Included with no-till is the alternative practice of rotating corn with another appropriate crop. Both reduce erosion, and the rotation eliminates the need to use an insecticide treatment to control the corn rootworm complex, a typical pest problem in continuously grown, conventional corn (Pimentel et al., 1977). Selecting the appropriate crops for rotation with corn reduces corn diseases (Pearson, 1967; Mora and Moreno, 1984) and weed problems (NAS, 1968; Mulvaney and Paul, 1984). When appropriately integrated into the agricultural system, crop rotations can provide major benefits of soil and water conservation as well as insect, disease, and weed control (Richey et al., 1977). Although rotations offer many advantages, some disadvantages include inconvenience of producing multi-crops and sometimes less profit if the alternate crop produces less net return than corn.

For the ridge-planting system developed in China (Wan et al., 1959), several low-input alternative practices are added (Table 7.3). These include livestock manure and use of cover crops with continuous corn. The use of legume cover crops is of value in reducing soil erosion and water runoff, reducing weed problems, and conserving soil nutrients—soil nutrients are picked up and stored by the cover crop. However, ridge planting is not suitable for all soils, rainfall levels, and crops (Lal, 1977, 1985), thus em-

Table 7.3 Energy and Economic Inputs per Hectare for Conventional and Alternative Corn Production Systems

	Conventional			No-Till and Rotations			Ridge Planting and Livestock Manure			Low-Input Ecological System		
	Qty.	10³ kcal	Econ.	Qty.	10³ kcal	Econ.	Qty.	10³ kcal	Econ.	Qty.	10³ kcal	Econ.
Labor (h)	10[a]	7[f]	50[r]	7[cc]	6[f]	35[r]	12[jj]	9[f]	60[r]	12[jj]	9[f]	60[r]
Machinery (kg)	55[b]	1,485[g]	91[s]	45[dd]	1,215[g]	75[s]	45[dd]	1,215[g]	75[s]	45[dd]	1,215[g]	75[s]
Fuel (L)	115[b]	1,255[h]	38[t]	70[ee]	764[h]	23[t]	70[ee]	764[h]	23[t]	70[ee]	764[h]	23[t]
N (kg)	152[b]	3,192[i]	81[u]	152[ff]	3,192[i]	81[u]	(27t)[kk]	559[qq]	17[rr]	(27 t)[kk]	559[qq]	17[rr]
P (kg)	75[b]	473[j]	53[v]	75[ff]	473[j]	53[v]	34[ll]	214[j]	17[u]	34[ll]	214[j]	17[u]
K (kg)	96[b]	240[k]	26[w]	96[k]	240[k]	26[w]	15[mm]	38[k]	4[w]	15[mm]	38[k]	4[w]
Limestone (kg)	426[b]	134[l]	64[x]	426[ff]	134[l]	64[x]	426[ff]	134[l]	64[x]	426[ff]	134[l]	64[x]
Corn seeds (kg)	21[b]	520[m]	45[y]	24[gg]	594[m]	51[y]	21[b]	520[m]	45[y]	21[b]	520[m]	45[y]
Cover crop seeds (kg)	—	—	—	—	—	—	10[oo]	120[oo]	10[ss]	10[oo]	120[oo]	10[ss]
Insecticides (kg)	1.5[c]	150[n]	15[z]	0[hh]	0	0	1.5[c]	150[n]	15[z]	0[hh]	0	0
Herbicides (kg)	2[c]	200[n]	20[z]	4[ii]	400[n]	40[z]	0[pp]	0	0	0[pp]	0	0
Electricity (10³ kcal)	100[b]	100[o]	8[aa]	100[b]	100[o]	8[aa]	100[b]	100[o]	8[aa]	100[b]	100[o]	8[aa]
Transport (kg)	322[d]	89[p]	32[bb]	196[d]	54[p]	20[bb]	140[d]	39[p]	14[bb]	138[d]	38[p]	14[bb]
Total		7,845	$523		7,172	$476		3,862	$352		3,712	$337
Yield (kg)	6,500[e]	26,000[q]		6,500	26,000[q]		6,500	26,000[q]		6,500	26,000	
Output/input ratio		3.31			3.63			6.73			7.00	

[a] Labor input was estimated to be 10 h because of the extra time required for tillage and cultivation compared with no-till, which required 7 h (USDA, 1984b).
[b] Pimentel and Wen, 1988.
[c] Mueller et al., 1985.
[d] Transport of machinery, fuel, and nitrogen fertilizer (Pimentel and Wen, 1988).
[e] Three-year running average yield (USDA, 1982).
[f] Food energy consumed per laborer per day was assumed to be 3500 kcal.
[g] The energy input per kilogram of steel in tools and other machinery was 18,500 kcal (Doering, 1980) plus 46% added input (Fluck and Baird, 1980) for repairs.
[h] Fuel includes a combination of gasoline and diesel. A liter of gasoline and diesel fuel was calculated to contain 10,000 and 11,400 kcal, respectively (Cervinka, 1980). Weighted average value of 10,900 used in calculations. These values include the energy input for mining and refining.
[i] Nitrogen = 21,000 kcal/kg (Dovring and McDowell, 1980).
[j] Phosphorus = 6300 kcal/kg (Dovring and McDowell, 1980).
[k] Potassium = 2500 kcal/kg (Dovring and McDowell, 1980).
[l] Limestone = 315 kcal/kg (Terhune, 1980).

114

[m] Hybrid seed = 24,750 kcal/kg (Heichel, 1980).

[n] Energy input for insecticides and herbicides was calculated to be 100,000 kcal/kg (Pimentel, 1980).

[o] Includes energy input required to produce the electricity.

[p] For the goods transported to the farm, an input of 275 kcal/kg was included (Pimentel, 1980).

[q] A kilogram of corn was calculated to have 4000 kcal.

[r] Labor = $5/h.

[s] USDA, 1984b.

[t] Liter = $0.33.

[u] N = $0.53.

[v] P = $0.51.

[w] K = $0.27.

[x] Limestone = $0.15.

[y] (USDA, 1984b).

[z] Insecticide and herbicide treatments = $10/kg for both the material and application costs.

[aa] kwh = 7¢.

[bb] Transport = 10¢/kg.

[cc] No-till requires less labor than conventional because tillage and cultivation are reduced (Colvin et al., 1982; Mueller et al., 1985).

[dd] 20% smaller machinery was used because less power is needed in no-till and ridge planting (Colvin et al., 1982; Allen and Hollingsworth, 1983; Muhtar and Rotz, 1982; Hamlett et al., 1983; USDA, 1984c).

[ee] Nearly 40% less fuel is required compared with conventional because the soil was not tilled, only lightly cultivated (Colvin et al., 1983; Mueller et al., 1985).

[ff] Assumed that same amount of N, P, K, and Ca required in no-till.

[gg] About 15% more seed was planted to offset poor germination in no-till (USDA, 1984c).

[hh] No insecticide was used because the corn was planted in rotation after soybeans.

[ii] Twice as much herbicide was used compared with conventional tillage to control weeds.

[jj] Five additional hours were necessary for collecting and spreading 27 t of manure (Pimentel et al., 1984c).

[kk] A total of 27 t of cattle manure was applied to provide 152 kg of N.

[ll] A total of 41 kg of P was provided by the manure.

[mm] A total of 81 kg of K was provided by the manure.

[nn] Cultivation and cover crop used for weed control.

[oo] About 10 kg of cover crop seeds was used (Heichel, 1980).

[pp] No herbicide used, weed control carried out by cultivation and cover crop.

[qq] About 1.9 L of fuel was required to collect and apply 1 t of manure (Pimentel et al., 1984c).

[rr] The value of manure was given for the fuel required to transport and spread.

[ss] 1 kg of cover crop seed = $1.

phasizing the need for care in selecting appropriate technologies for ecological resource management.

The low-input ecological system employs ridge planting plus manure, cover crop, and rotation techniques (Table 7.3). This system incorporates many of the ecotechnological practices mentioned.

Numerous other alternative ecotechnologies could have been considered for this example, including other cropping systems, green manures, and pest control practices, but the technologies we selected illustrate the potential of an alternative system to conserve soil and water resources, reduce the need for pesticides, and improve the sustainability of the agroecosystem.

Average input data for conventional corn production are listed in Table 7.2. It is assumed that this crop is grown in a region where rainfall averages 1000 mm, and on land with a slope of 3–5% and erosion rate of 18 t/ha-yr. Average U.S. corn yield is 6500 kg/ha, and the energy input is calculated to be 8.0 million kcal with 12 h of human labor. The energy production ratio, that is, the ratio of kilocalories output per kilocalories input, is 2.2 (Table 7.2). Total production costs are calculated to be $523/ha.

The no-till system is assumed to be planted in an environment similar to that of conventional corn. The major differences between no-till and conventional are: (1) erosion is reduced from 18 t/ha-yr to about 1 t/ha-yr; (2) labor is reduced from 12 to 10 h; (3) smaller tractors are employed; (4) less tractor fuel is used; (5) about twice as much herbicide is used to control weeds; and (6) no insecticide is used because the corn is planted in rotation after a nonhost crop such as a legume (Table 7.3). The total energy inputs and costs are about 10% less than those for conventional. Also, the yield of corn in no-till is assumed to be similar to that of conventional (Van Doren et al., 1977; Taylor et al., 1984; Hargrove, 1985).

As mentioned, several alternative practices are integrated in the ridge-planting system (Table 7.3). For this system the assumptions are: (1) ridge planting is carried out on the contour and crop residues are left on the surface, reducing erosion from 18 t/ha-yr to a tolerable level of less than 1 t/ha-yr; (2) available livestock manure is substituted for all the nitrogen needs and most of the phosphorus and potassium needs; (3) labor input is raised to 15 h/ha to include the time required for manure spreading compared with 12 h/ha for the conventional system; (4) because of the cover crop and well-designed tillage system, no herbicide is included; and (5) smaller tractors are used and less fuel is consumed (Thompson, 1985). The total energy inputs for the ridge-planting system are reduced by nearly half, and production costs are reduced to one-third those of the conventional system (Table 7.3). Results similar to those calculated for this low-input system have been obtained by farmers who have used a like low-input system for crop production (Thompson, 1985). The production costs of these were about $100/ha less than those of conventional systems (Thompson, 1985).

The low-input ecological system includes all the beneficial practices listed for the ridge-planting system plus crop rotations. This would eliminate the need for insecticides and would add other benefits mentioned earlier.

Although the corn yield for the low-input system is assumed to be equal to that of the conventional system, long-term yields would probably be much higher. Using sound soil and water conservation measures will slow the loss of soil and decline of productivity. Over a 20-year period about 2.6 cm of soil can be expected to be lost in the conventional corn system, with a soil loss of 18 t/ha-yr. About 500 years would be required to replace this 2.6 cm of lost soil. If this soil degradation were offset with increased energy inputs such as fertilizer and irrigation, a nonrenewable resource (fossil energy) would be substituted for a renewable resource. With the cost of fuel projected to rise in the coming decades, the substitution of a nonrenewable resource for a renewable resource will become very costly to farmers and society. Thus ecotechnological practices can pay major dividends in the long term.

This analysis suggests that the use of ecologically sound practices will maintain high yields while reducing production costs and protecting the environment—especially soil, water, energy, and biological resources. For example, with the ridge-planting system, soil erosion and water runoff are controlled and pesticide use is reduced. All of these reduce costs by decreasing fertilizer, pesticide, and machinery costs. Of major importance is the maintenance of the productivity of the soil and integrity of the entire agroecosystem for the future. Fortunately, numerous alternative practices for soil and water conservation and pest control are readily available for use in productive agriculture (PSAC, 1965; Troeh et al., 1980). Each set of ecotechnologies, however, has to be selected and adapted to the particular environment of the region.

7.7 SITE-SPECIFIC AGROECOSYSTEM

A perspective of resource use in an agroecosystem can be gained by examining energy flow at a specific site. For this example, the organic agricultural system in Hailun County, China is used (Wen and Pimentel, 1984).

The Hailun agroecosystem in the 1950s was a self-sustaining system in terms of energy, food, and feed production and household supplies for the people and nearly so in soil nutrients. The energy flow of the total agroecosystem is shown diagrammatically in Figure 7.2.

The total output from the agroecosystem in terms of crops was 5.3 million kcal/ha (Wen and Pimentel, 1984). To produce these crops, the total input was 2.7 million kcal/ha. The tools, representing the smallest (1.8%) input, were the only fossil fuel input in the system; 98.2% energy input was renewable energy. The energy input/output ratio for this system as a whole was 1:1.96. This energy input/output ratio is not large, but it compares favorably with the input/output ratios for corn, wheat, and sorghum produced in Mexico, Guatemala, India, Nigeria, and the United States (Pimentel and

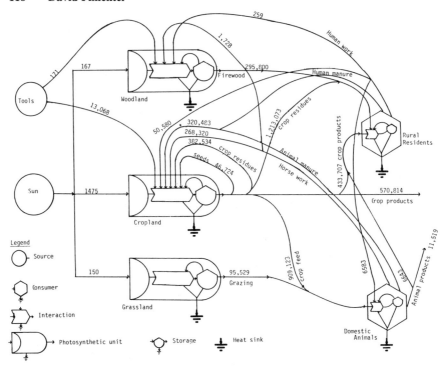

Figure 7.2 Energy flow in Hailun agroecosystem in China during 1952–1954 (Mkcal/yr). Note: energy flow after Odum (1971, 1972) (From Wen and Pimentel, 1984).

Pimentel, 1979). Based on an input/output ratio for fossil energy only (1:107), it is most efficient.

7.8   ECOTECHNOLOGY AND AGRICULTURE

Soil, water, air, energy, and biological resources are essential for agricultural production. The complex ecological interaction among these resources must be understood so that they can be managed as an integrated system for a productive, sustainable agriculture.

The ecotechnological principles that underlie a productive, sustainable agricultural system are as follows (Pimentel et al., 1988). (1) Adapting and designing the agricultural system to the environment of the region. This means, for example, culturing crops and/or forages (livestock) that are eco-logically adapted to the soil, water, climate, and biota present at the site. (2) Optimizing the use of biological resources in the agroecosystem. This includes making effective use of biological pest control, green manures, cover crops, rotations, agricultural wastes, and other biological resources. (3) Developing strategies that induce minimal changes in the natural ecosystem to protect the environment and minimize the use of fossil energy in manipulating the agroecosystem.

The ecotechnological approach is complex, but it can be simplified by focusing primarily on four factors that are commonly manipulated in an agroecosystem—soil nutrients, water, energy, and pests. The goal is to conserve soil nutrients and water, while at the same time encouraging beneficial organisms and discouraging pests. Soil nutrients (nitrogen, phosphorus, potassium, etc.) and water are essential to a productive agriculture. Conserving soil and water resources reduces the inputs of commercial fertilizers and irrigation needed and thus decreases costs. Similarly, manipulations of the agroecosystem that encourage biological pest control and make the environment unfavorable for pests reduce the use of pesticides. Combined, these strategies will reduce input costs and help maintain a highly productive, ecologically sound agriculture (Pimentel et al., 1988).

REFERENCES

AAC *(Zhonggue Nongye Nianjian)*. 1980. *Agricultural Almanac of China*. Agriculture Press, Beijing.

Alexander, M. 1977. *Introduction to Soil Microbiology,* 2nd ed. Wiley, New York.

Allen, R. R. and L. D. Hollingsworth. 1983. Limited tillage sorghum on wide beds. ASAE Paper 83-1517.

Ambroggi, R. P. 1980. Water. *Sci. Am. 243(3)*:101–106, 111–114, 116.

Beasley, R. P. 1972. Erosion and sediment pollution control. Iowa State University Press, Ames.

Biswas, A. K. 1983. 2. Shared natural resources: Future conflicts or peaceful development? In R.-J. Dupuy, Ed., *The Settlement of Disputes on the New Natural Resources*. Martinus Nijhoff, The Hague, pp. 197–215.

Biswas, A. K. 1984. *Climate and Development*. Tycooly, Dublin.

Biswas, M. R. and A. K. Biswas. 1985. The global environment. Past, present and future. *Resources Policy 3*:25–42.

Brady, N. C. 1984. *The Nature and Properties of Soils,* 9th ed. Macmillan, New York.

Brown, L. R. 1976. World population trends: signs of hope, signs of stress. Worldwatch Paper 8. Worldwatch Institute, Washington, DC.

Brown, L. R., W. U. Chandler, C. Flavin, C. Pollock, S. Postel, L. Starke, and E. C. Wolf. 1985. *State of the World 1985*. W. W. Norton, New York.

Buringh, P. 1979. Introduction to the Study of Soils in Tropical and Subtropical Regions. Centre for Agricultural Publishing and Documentation, Pudoc, Wageningen, The Netherlands.

CEQ. 1980. *The Global 2000 Report to the President,* Vol. 2. Council on Environmental Quality and the Department of State. U.S. Government Printing Office, Washington, DC.

Cervinka, V. 1980. Fuel and energy efficiency. In D. Pimentel, Ed., *Handbook of Energy Utilization in Agriculture*. CRC Press, Boca Raton, FL, pp. 15–24.

Clark, C. and M. Haswell. 1970. *The Economics of Subsistence Agriculture*. Macmillan, London.

Clark, E. H. II. 1985. The off-site costs of soil erosion. *J. Soil Water Conserv. 40:*19–22.

Colvin, T. S., C. A. Hamlett, and A. Rodriguez. 1982. Effect of tillage system on farm machinery selection. ASAE Paper 82-1029.

Colvin, T., D. Erbach, S. Marley, and H. Erickson. 1983. Large-scale evaluation of a till plant system. ASAE Paper 83-1027.

CSE. 1982. The State of India's Environment. 1982. A Citizen's Report. Centre for Science and Environment, New Delhi.

Doering, O. C. 1980. Accounting for energy in farm machinery and buildings. In D. Pimentel, Ed., *Handbook of Energy Utilization in Agriculture*. CRC Press, Boca Raton, FL, pp. 9–14.

Dovring, F. and D. R. McDowell. 1980. Energy use for fertilizers. Dept. Agr. Econ. Staff Paper 80 E-102, University of Illinois, Urbana.

Dudal, R. 1981. An evaluation of conservation needs. In R. P. C. Morgan, Ed., *Soil Conservation. Problems and Prospects*. Wiley, New York, pp. 3–12.

Eckholm, E. P. 1978. Disappearing species: The social challenge. Worldwatch Paper 22. Worldwatch Institute, Washington, DC.

Eckholm, E. P. 1983. *Down to Earth. Environment and Human Needs*. W. W. Norton, New York.

Ehrlich, P. R. and A. H. Ehrlich. 1981. *Extinction: the Causes and Consequences of the Disappearance of Species*. Random House, New York.

ERAB. 1982. Solar energy research and development: Federal and private sector roles. A report of the Energy Research Advisory Board to the United States Department of Energy. Department of Energy, Washington, D.C.

FAO. 1982. *1981 Production Yearbook*. Food and Agricultural Organization of the United Nations, Rome.

Finkel, H. J. 1983. Irrigation of cereal crops. In H. J. Finkel, Ed., *CRC Handbook of Irrigation Technology,* Vol. 2. Boca Raton, FL, pp. 159–189.

Fluck, R. C. and C. D. Baird. 1980. *Agricultural Energetics*. AVI, Westport, CT.

Golley, F. B., Ed. 1983. *Tropical Rainforest Ecosystems: Structure and Function*. Elsevier, Amsterdam.

Hamlett, C. A., T. S. Colvin, and A. Musselman. 1983. Economic potential of conservation tillage in Iowa. *Trans. ASAE 26(3):*719–722.

Hare, F. K. 1977. Climate and desertification. In *Desertification: Its Causes and Consequences*. Pergamon Press, Oxford, pp. 63–129.

Hargrove, W. L. 1985. Influence of tillage on nutrient uptake and yield of corn. *Agron. J. 77:*763–768.

Heichel, G. H. 1980. Assessing the fossil energy costs of propagating agricultural crops. In D. Pimentel, Ed., *Handbook of Energy Utilization in Agriculture*. CRC Press, Boca Raton, FL, pp. 27–33.

Holdgate, M. W., M. Kassas, and G. F. White. 1982. *The World Environment. 1972–1982.* A Report by the United Nations Environment Programme. Tycooly, Dublin.

Hudson, N. 1981. *Soil Conservation,* 2nd ed. Cornell University Press, Ithaca, NY.

Ingraham, E. W. 1975. A query into the quarter century. On the interrelationships of food, people, environment, land and climate. Wright-Ingraham Institute, Colorado Springs, Colo.

Jacks, G. V. and R. O. Whyte. 1939. *Vanishing Lands. A World Survey of Soil Erosion.* Doubleday, Doran, New York.

Kovda, V. A. 1983. Loss of productive land due to salinization. *Ambio 12:*91–93.

Lal, R. 1977. Soil-conserving versus soil-degrading crops and soil management for erosion control. In D. Greenland and R. Lal, Eds., *Soil Conservation in the Tropics.* Wiley, London, pp. 81–86.

Lal, R. 1984. Productivity assessment of tropical soils and the effects of erosion. In F. R. Rijsberman and M. G. Wolman, Eds., *Quantification of the Effect of Erosion on Soil Productivity in an International Context.* Delft Hydraulics Laboratory, Delft, Netherlands, pp. 70–94.

Lal, R. 1985. A soil suitability guide for different tillage systems in the tropics. *Soil Till. Res. 5:*179–196.

Langdale, G. W., R. A. Leonard, W. G. Fleming, and W. A. Jackson. 1979. Nitrogen and chloride movement in small upland Piedmont watersheds: II. Nitrogen and chloride transport in runoff. *J. Environ. Qual. 8:*57–63.

Larson, W. E. 1981. Protecting the soil resource base. *J. Soil Water Conserv. 36:*13–16.

Latham, M. C. 1984. International nutrition and problems and policies. In *World Food Issues.* Center for the Analysis of World Food Issues, International Agriculture, Cornell University, Ithaca, NY, pp. 55–64.

Leach, G. 1976. *Energy and Food Production.* IPC Science and Technology Press Limited, Guilford, Surrey.

Lewis, O. 1951. *Life in a Mexican Village: Tepoztlan Restudied.* University of Illinois Press, Urbana.

Leyton, L. 1983. Crop water use: principles and some considerations for agroforestry. In P. A. Huxley, Ed., *Plant Research and Agroforestry.* International Council for Research in Agroforestry, Nairobi, Kenya, pp. 379–400.

Lowdermilk, W. C. 1953. Conquest of the Land Through Seven Thousand Years. USDA Agric, Info. Bull. No. 99.

Lu, M., J. Ysheng, and S. Chenyueng. 1982. Typical analysis of rural energy consumption in China. (Zhonggue Nongchun Nengliang Xiaofei diansing fenshi). *Nongye Jingji Luenchung 4:*216–223.

Mabbutt, J. A. 1984. A new global assessment of the status and trends of desertification. *Environ. Conserv. 11:*103–113.

Mangelsdorf, P. C. 1966. Genetic potentials for increasing yields of food crops and animals. In *Prospects of the World Food Supply.* Symp. Proc. Natl. Acad. Sci., Washington, DC.

McCormack, D. E., K. K. Young, and L. W. Kimberlim. 1982. Current Criteria for Determining Soil Loss Tolerance. ASA Special Publication Number 45, American Society of Agronomy, Madison, WI.

122   David Pimentel

Mora, L. E. and R. A. Moreno. 1984. Cropping pattern and soil management influence on plant diseases: I. *Diplodia macrospora* leaf spot of maize. *Turrialbo 34(1):*35–40.

Mueller, D. H., R. M. Klemme, and T. C. Daniel. 1985. Short- and long-term cost comparisons of conventional and conservation tillage systems in corn production. *J. Soil Water Conserv. 40:*466–470.

Muhtar, H. A. and C. A. Rotz. 1982. A multi-crop machinery selection algorithm for different tillage systems. ASAE Paper 82-1031.

Mulvaney, D. L. and L. Paul. 1984. Rotating crops and tillage. Both sometimes better than just one. *Crops Soil Mag. 36(7):*18–19.

Murray, R. C. and E. B. Reeves. 1977. Estimated Use of Water in the United States in 1975. U.S. Geol. Surv. Circ. 765.

Myers, N. 1979. *The Sinking Ark.* Pergamon Press, New York.

Myers, N. 1983. *A Wealth of Wild Species.* Westview Press, Boulder, Colo.

Myers, N. 1984. Genetic Resources in Jeopardy. *Ambio 13:*171–174.

Myers, N. 1985. The end of the lines. *Natural History 94:*2, 6, 12.

Naegeli, W. N. 1986. Interpreting the national resources inventory for regional planners and decision makers: A case study for the Tennessee Valley Region. Ph.D. Thesis, Cornell University, Ithaca, NY.

NAS. 1968. Principles of Plant and Animal Pest Control II. Weed Control Publication 1597. National Academy of Sciences, Washington, DC.

NAS. 1975. *Population and Food: Crucial Issues.* National Academy of Sciences, Washington, DC.

NAS. 1977. *Supporting Papers: World Food and Nutrition Study,* Vol. 2. National Academy of Sciences, Washington, DC

Odum, H. T. 1971. *Environment, Power and Society.* Wiley, New York. 331 pp.

Odum, H. T. 1972. An energy circuit language for ecological and social systems: its physical basis. In B. C. Patten, Ed., *Systems Analysis and Simulation in Ecology.* Academic, New York, pp. 139–211.

OTA. 1982. Impacts of technology on productivity of the croplands and rangelands of the United States. Office of Technology Assessment, Washington, DC.

Pearson, L. C. 1967. *Principles of Agronomy.* Reinhold, New York.

Pimentel, D., Ed. 1980. *Handbook of Energy Utilization in Agriculture.* CRC Press, Boca Raton, FL.

Pimentel, D. and M. Burgess. 1980. Energy inputs in corn production. In D. Pimentel, Ed., *Handbook of Energy Utilization in Agriculture.* CRC Press, Boca Raton, FL, pp. 67–84.

Pimentel, D. and C. W. Hall, Eds. 1984. *Food and Energy Resources.* Academic, New York.

Pimentel, D. and M. Pimentel. 1979. *Food, Energy and Society.* Edward Arnold, London.

Pimentel, D. and D. Wen. 1988. Technological changes in energy use in U.S. agricultural production. In C. R. Carroll, J. H. Vandermeer, and P. M. Rosset, Eds., *The Ecology of Agricultural Systems.* Macmillan, New York (in press).

Pimentel, D., E. C. Terhune, R. Dyson-Hudson, S. Rochereau, R. Samis, E. Smith,

D. Denman, D. Reifschneider, and M. Shepard. 1976. Land degradation: effects on food and energy resources. *Science 194:*149–155.

Pimentel, D., C. Shoemaker, E. L. LaDue, R. B. Rovinsky, and N. P. Russell. 1977. Alternatives for reducing insecticides on cotton and corn: Economic and environmental impact. Environ. Res. Lab., Off. Res. Dev., EPA, Athens, GA (issued in 1979).

Pimentel, D., E. Garnick, A. Berkowitz, S. Jacobson, S. Napolitano, P. Black, S. Valdes-Cogliano, B. Vinzant, E. Hudes, and S. Littman. 1980a. Environmental quality and natural biota. *BioScience 30:*750–755.

Pimentel, D., P. A. Oltenacu, M. C. Nesheim, J. Krummel, M. S. Allen, and S. Chick. 1980b. Grass-fed livestock potential: energy and land constraints. *Science 207:*843–848.

Pimentel, D., M. A. Moran, S. Fast, G. Weber, R. Bukantis, L. Balliett, P. Boveng, C. Cleveland, S. Hindman, and M. Young. 1981. Biomass energy from crop and forest residues. *Science 212:*1110–1115.

Pimentel, D., C. Fried, L. Olson, S. Schmidt, K. Wagner-Johnson, A. Westman, A. Whelan, K. Foglia, P. Poole, T. Klein, R. Sobin, and A. Bochner. 1984a. Environmental and social costs of biomass energy. *BioScience 34:*89–94.

Pimentel, D., L. Levitan, J. Heinze, M. Loehr, W. Naegeli, J. Bakker, J. Eder, B. Modelski, and M. Morrow. 1984b. Solar energy, land and biota. *SunWorld 8:*70–73, 93–95.

Pimentel, D., G. Berardi, and S. Fast. 1984c. Energy efficiencies of farming wheat, corn, and potatoes organically. In Organic Farming: Current Technology and Its Role in a Sustainable Agriculture. ASA Spec. Publ. No. 46. American Society of Agronomy, Madison, WI, pp. 151–161.

Pimentel, D., W. Dazhong, S. Eigenbrode, H. Lang, D. Emerson, and M. Karasik. 1986. Deforestation: interdependency of fuelwood and agriculture. *Oikos 46:*404–412.

Pimentel, D., T. Culliney, I. Buttler, D. Reinemann, and K. Beckman. 1988. Ecological resource management for a productive, sustainable agriculture. In D. Pimentel and C. W. Hall, Eds., *Food and Natural Resources*. Academic, New York (in press).

PRB. 1986. World population data sheet. Population Reference Bureau, Inc., Washington, DC.

PSAC. 1965. Restoring the Quality of our Environment. Report Environmental Pollution Panel, Pres. Sci. Adv. Comm., The White House, Washington, DC.

Richey, G. B., D. R. Griffith, and S. D. Parsons. 1977. Yields and cultural energy requirements for corn and soybeans with various tillage-planting systems. *Adv. Agron. 28:*141–182.

Risser, J. 1981. A renewed threat of soil erosion: It's worse than the dust bowl. *Smithsonian 11:*121–131.

Ritschard, R. L. and K. Tsao. 1978. Energy and water use in irrigated agriculture during drought conditions. US DOE LBL-7866. Lawrence Berkeley Lab., University of California, Berkeley.

Sampson, R. N. 1983. Soil conservation. *Sierra Club Bull. 68(6):*40–44.

Shah, M. M., G. Fischer, G. M. Higgins, A. H. Kassam, and L. Naiken. 1985.

People, land and food production—potentials in the developing world. International Institute for Applied Systems Analysis. CP-85-11. Laxenburg, Austria.

Sharma, A. K. 1981. Impact of the Development of Science and Technology on Environment. Proceedings of the 68th Indian Science Congress, Varanasi, India, pp. 1–43.

Spears, J. and E. S. Ayensu. 1984. Resources development and the new century: sectoral paper on forestry. Global Possible Conference, World Resources Institute, Washington, DC.

Taylor, F., G. S. V. Raghaven, S. C. Negi, E. McKyes, B. Vigier, and A. K. Watson. 1984. Corn grown in a Ste. Rosalie clay under zero and traditional tillage. *Can. Agric. Eng. 26(2):*91–95.

Taylor, R. P. 1981. Rural energy development in China. Resources for the Future, Washington, DC.

Terhune, E. C. 1980. Energy used in the United States for agricultural liming materials. In D. Pimentel, Ed., *Handbook of Energy Utilization in Agriculture.* CRC Press, Boca Raton, FL, pp. 25–26.

Thompson, R. 1985. Personal communication. New Farm, Boone, IA.

Troeh, F. R., J. A. Hobbs, and R. L. Donahue. 1980. *Soil and Water Conservation for Productivity and Environmental Protection.* Prentice-Hall, Englewood Cliffs, NJ.

UNEP. 1980. Annual review. United Nations Environment Programme, Nairobi, Kenya.

USDA. 1965. *Losses in Agriculture.* Agricultural Handbook No. 291, Agric. Res. Serv., U.S. Government Printing Office, Washington, DC.

USDA. 1974. Energy and U.S. agriculture: 1974 data base. Vols. 1 and 2. Federal Energy Administration. Office of Energy Conservation and Environment, State Energy Conservation Programs, Washington, DC.

USDA. 1982. Fertilizer: outlook and situation. USDA Econ. Res. Serv. FS-13.

USDA. 1984a. *Agricultural Statistics 1984.* U.S. Government Printing Office, Washington, DC.

USDA. 1984b. Economic indicators of the farm sector, costs of production. U.S. Dept. of Agr., Econ. Res. Ser., ECIFS 4-1.

USDA. 1984c. Returns to corn and soybean tillage practices. USDA, Econ. Res. Ser., Agr. Econ. Rept. #508.

USDA. 1985. *Agricultural Statistics 1985.* U.S. Govt. Printing Office, Washington, DC.

USWRC. 1979. The nation's water resources. 1975–2000. Vols. 1–2. Second National Water Assessment. United States Water Resources Council. U.S. Government Printing Office, Washington, DC.

Van Doren, D. M., G. B. Triplett, Jr., and J. E. Henry. 1977. Influence of long-term tillage and crop rotation combinations on crop yields and selected soil parameters for an Aeric Ochraqualf soil [maize]. Res. Bull. Ohio Agr. Res. Dev. Cent. 1091. Map. Ref. Sept. 1977.

Wan, G., Y. Gu, and C. Li. 1959. The cultivating principles in the book, "Lu's Chinqiu." In *Agronomy History in China* (Zhonggue Nongxueshi) (First Draft), Vol. 1. Kexue Press, Beijing, pp. 77–102.

Wen, D. and D. Pimentel. 1984. Energy flow through an organic agroecosystem in China. *Agric. Ecosyst. Environ.* *11*:145–160.

PART TWO

CASE STUDIES OF ECOLOGICAL ENGINEERING

8

THE ROLE OF WETLANDS IN THE CONTROL OF NUTRIENTS WITH A CASE STUDY OF WESTERN LAKE ERIE

William J. Mitsch and Brian C. Reeder

School of Natural Resources, The Ohio State University, Columbus, Ohio

and

David M. Klarer

Old Woman Creek State Nature Preserve and National Estuarine Research Reserve, Huron, Ohio

8.1   INTRODUCTION

Eutrophication and related problems caused by excessive nutrients and other chemicals reaching bodies of water remain significant in many parts of the United States. Although there has been some progress on the control of oxygen-demanding point sources of water pollution such as municipal waste-water and industrial discharges, nutrients and sediments from agricultural areas are considered to be among the most difficult to control for achieving acceptable water quality in the nation (ASWIPCA, 1984). Furthermore, recent analysis of water quality data suggests that there are trends of increased loadings of nitrogen and phosphorus to the nation's waters in most parts of the country, mostly due to increased uses of fertilizers (Smith et al., 1987). The problem of controlling nutrients is particularly significant because we have found that technological solutions such as tertiary treatment for point sources and certain technologies for non-point sources from urban and agricultural regions are costly. Furthermore, it is expensive to do nothing to control the problem; one estimate suggested a national cost of erosion of sediments and nutrients from farmland to be $35 billion annually in the United States on downstream aquatic systems (Clark et al., 1985).

   Wetlands, both in their natural state and as created parts of the landscape, offer excellent ecological engineering possibilities for the control of nutrients and sediments from point and non-point sources. This chapter describes ecological principles and aspects of ecological engineering design that are involved in the use of wetlands for the control of nutrients. A specific wetland case study site along western Lake Erie of the Laurentian Great Lakes in North America where wetlands are being investigated for their role as buffer zones between an upland agricultural watersheds and the Great Lakes is then presented.

8.2   ECOLOGICAL PRINCIPLES

Wetlands are ecosystems generally found at the ecotone between terrestrial systems and aquatic systems and therefore often have characteristics of both systems (Mitsch and Gosselink, 1986) (Figure 8.1). Wetlands have been formally defined in the United States as "lands transitional between terrestrial and aquatic systems where the water table is usually at or near the surface or the land is covered by shallow water" (Cowardin et al., 1979). They are further defined as having at least one of the following attributes: (1) hydrophytic vegetation, (2) hydric soil, and/or (3) standing water or saturated soils during the growing season. Wetlands, for the purposes of this chapter, include freshwater marshes, tidal freshwater and salt marshes, mangrove

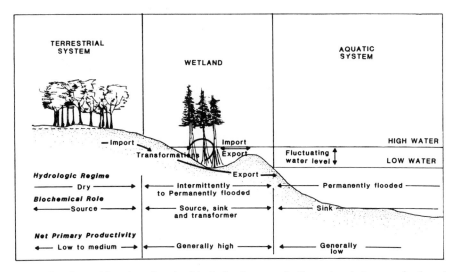

Figure 8.1 General location of wetland in the landscape as buffer system between uplands and deepwater ecosystems. (From Mitsch and Gosselink, 1986; reprinted by permission of Van Nostrand Reinhold Co.)

swamps, northern peatlands, deepwater swamps, and bottomland hardwood forests.

8.2.1   The Wetland Mass Balance

A general mass balance for nutrients and other chemicals in wetlands is shown in Figure 8.2. Nutrients and other chemicals brought into the system are called *inputs* or *inflows*. For wetlands, these inputs come primarily through hydrologic pathways such as precipitation, surface and groundwater flows, and tidal exchange. For nitrogen, atmospheric inputs such as nitrogen fixation are also possible. Hydrologic *exports* or *losses*, or *outflows* include surface water, groundwater, and tidal exchanges. Again, with the nitrogen cycle there are biologically mediated nitrogen losses to the atmosphere (via dentrification). The significance of other losses to the atmosphere, such as ammonia volatilization and methane and sulfide releases, is not well understood, although they are potentially important pathways for individual wetlands and for the global cycling of minerals. *Intrasystem cycling*, or the cycling of chemicals such as nitrogen and phosphorus within the wetland itself, includes pathways such as litter production, remineralization, nutrient uptake by the plants themselves, and nutrient translocation. These are the pathways that enable a wetland to be a chemical transformer as discussed below.

The role of wetlands in the processing of nutrients and chemicals for improving water quality is not a simple one nor is it easy to generalize for all wetlands. Nevertheless, there are a few principles that can be given:

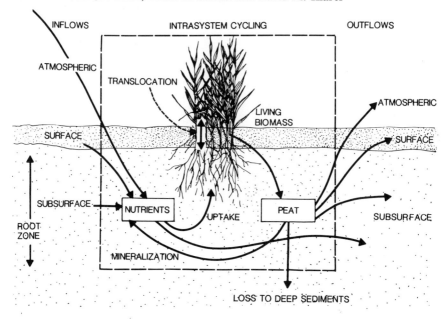

Figure 8.2 Generalized nutrient budget for wetland. (From Mitsch and Gosselink, 1986; reprinted by permission of Van Nostrand Reinhold Co.)

1. Wetlands serve as sinks, sources, or transformers of chemicals, depending on the wetland type, hydrologic conditions, and year in question.

A wetland functions as a *chemical sink* if it retains more nutrients or sediments than it releases over a given period of time (Figure 8.3*a*). This is often considered one of its most significant roles on the landscape. This ability to retain materials that enter but do not leave the system is due to an number of properties of wetlands: (1) wetlands are often peat-building (organic-accumulating) systems that retain nutrients and other chemicals in buried organic sediments; (2) wetlands are often very productive autotrophic systems, leading them to have significant amounts of inorganic nutrients converted to organic biomass; (3) wetlands are usually hydrologically isolated from high stream velocities, waves, and currents, and hence are often excellent sedimentation basins; and (4) the shallow water environment of wetlands allows maximum water–soil contact, with an array of associated biochemical processes in an aerobic–anaerobic system. A wetland can also be a *chemical transformer* when the net amount of an element entering and leaving the wetland is the same, but it is changed in form, for example, from inorganic to organic form (Figure 8.3*b*). A third possibility, that of a wetland serving as a chemical source, can occur when a wetland exports more of a certain chemical or chemicals than it takes in. This last condition can occur during unusual hydrologic events, for example, floods and seiches, and usually does

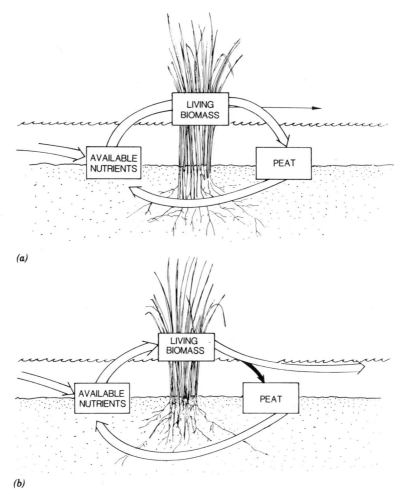

(a)

(b)

Figure 8.3 Examples of nutrient flows in wetlands where wetland is (*a*) a sink of nutrients, and (*b*) a transformer of nutrients. (From Mitsch and Gosselink, 1986; reprinted by permission of Van Nostrand Reinhold Co.)

not last for a long period. It has also been suggested that because wetlands artificially receive nutrients for a long period, they can eventually become nutrient sources to downstream systems (Richardson, 1985).

2. Seasonal patterns of nutrient uptake and release are characteristic of many wetlands.

The fact that a wetland is a sink, source, or transformer of nutrients or chemicals on a year by year basis implies nothing about the seasonal differences that may occur. During the height of the growing season, uptake of chemicals by plants and immobilization of elements by microflora and microfauna lead to the retention of nutrients. As the higher plants die, they have already translocated a significant portion of their nutrients to roots and rhizomes, but a significant portion is left to decompose and leach the chemicals back into the water. This often leads to a net release of chemicals in the fall and early spring. This seasonality of wetland nutrient retention is synchronized with the time in which most nutrient problems in lakes and other bodies of water. Thus even seasonal uptake of nutrients is desirable in many cases.

3. Wetlands are frequently coupled to adjacent ecosystems through chemical exchanges that significantly affected both systems.

Wetlands can be described as hydrologically *open* or *closed*. Open systems, with exchange of much material with surrounding ecosystems, include bottomland hardwood forests and tidal salt marshes, with exchanges due to flooding and tides, respectively. A number of wetlands, notably ombrotrophic bogs and cypress domes, have very little exchange with surrounding ecosystems and are called closed. The chemicals that pass through open wetlands are transported to downstream aquatic systems such as rivers, lakes, reservoirs, and estuaries. It is therefore not appropriate to view a wetland designed for chemical treatment without considering the downstream aquatic system.

4. Wetlands can be either highly productive ecosystems rich in nutrients or low-productivity systems with scarce nutrients.

There is a misconception in many textbooks that all wetlands are highly productive systems. There are many examples of low-productivity wetlands, the most common being the ombrotrophic bog and the cypress dome, both described above as ecosystems closed to significant hydrologic inflows. It is often the case, however, that these low-productivity wetlands are suitable systems for retention of nutrients.

8.3 PREVIOUS STUDIES OF WETLANDS AS NUTRIENT SINKS

8.3.1 Natural Wetlands as Nutrient Sinks

Several studies have demonstrated that wetlands, without any major modification of existing hydrologic conditions, act as sinks, transformers, and sources of chemicals, often with seasonal patterns depending on the season. We term these passive wetlands, because there has been no purposeful addition of wastewater or active design of discharges into these wetlands. Many of these studies of the nutrient retention of passive wetlands are summarized in Table 8.1.

Freshwater Marshes. One of the first studies that identified freshwater wetlands for their role as nutrient sinks was on Tinicum Marsh near Philadelphia (Grant and Patrick, 1970). That study found decreases in phosphorus, inorganic nitrogen, and BOD as water flowed through this tidal freshwater wetland. Lee et al. (1975) summarized several years of research on the effects of freshwater marshes on water quality in Wisconsin and found that the marshes acted as nutrient sinks during the summer and sources in the spring. Mitsch (1977) found that 49% of the total nitrogen and 11% of the total phosphorus were removed by a floating water hyacinth marsh in north-central Florida, again with seasonal patterns of greater nitrogen uptake during summer months. Similar results were found by Klopatek (1978) in Wisconsin, Simpson et al. (1978) in New Jersey, and Peverly et al. (1982) in New York.

Northern Peatlands. Overall nutrient budgets for northern peatlands are relatively rare. Crisp (1966) illustrated an overall nutrient budget for a blanket bog-dominated watershed in the Pennines region of England. Outputs of nutrients exceeded inputs from precipitation, suggesting that the erosion of peat in the watershed often makes these systems nutrient sources in their natural conditions. Hemond (1980) found that annual nitrate input to a floating mat sphagnum bog in Massachusetts exceeded nitrate output by a factor of three, suggesting nitrate uptake by the plants or denitrification. Verry and Timmons (1982) found that a black spruce peatland in Minnesota retained 50% of the nitrogen and 61% of the phosphorus that flowed into it.

Forested Swamps. Kitchens et al. (1975) demonstrated a significant reduction in phosphorus as waters flowed through a swamp forest in south Carolina. Mitsch et al. (1979) developed a phosphorus budget for an alluvial cypress swamp in southern Illinois and found that 10 times more phosphorus was deposited with sediments during river flooding than was returned from the swamp to the river for the rest of the year. Studies by Kuenzler et al. (1980) and Yarbro (1983) found that a significant portion of phosphorus was retained by swamps in North Carolina. Kemp and Day (1984) and Peterjohn and Correll (1984) both illustrated the roles of wetlands as systems receiving agricultural runoff. The former study suggested that a Louisiana forested swamp acts as a transformer system, removing inorganic forms of nitrogen and serving as a source of phosphate and organic nitrogen and organic phosphorus. In the latter study in Maryland, a 50-m-wide riparian forest removed an estimated 89% of the nitrogen and 80% of the phosphorus that entered from upland runoff. Elder and Mattraw (1982) and Elder (1985) concluded that forested riparian ecosystems were nutrient transformers rather than sinks along the Apalachicola River in northern Florida, with a net uptake

Table 8.1 Examples of Studies Investigating Wetlands as Passive Sinks of Nutrients

Type, Location	Period	Nutrient Sink[a] N	P	Reference
(TIDAL) FRESHWATER MARSHES				
Tinicum Marsh, PA	Summer	Yes	Yes	Grant and Patrick (1970)
Hamilton Marsh, NJ	Yearly	S	S	Simpson et al. (1978)
(NONTIDAL) FRESHWATER MARSHES				
4 marshes, WI	Yearly	Yes	S	Lee et al. (1975)
Waterhyacinth marsh, FL	9 mo	Yes	S	Mitsch (1977)
Theresa Marsh, WI	Yearly	S	S	Klopatek (1978)
Brillion Marsh, WI	Yearly	Yes	Yes	Fetter et al. (1978)
Managed marsh, NY	Yearly	I	I	Peverly (1982)
Phragmites marsh, Denmark	Yearly	Yes	No	Jørgensen et al. (1988)
NORTHERN PEATLANDS				
Watershed, Pennines, UK	Yearly	No	No	Crisp (1966)
Forested Peatland, MI	Yearly	No	Yes	Richardson et al. (1978)
Thoreau's Bog, MA	Yearly	Yes	—	Hemond (1980)
Black Spruce Bog, MN	Yearly	Yes	Yes	Verry and Timmons (1982)
FORESTED SWAMP				
Riverine cypress swamp, SC	Winter; Spring	Yes	Yes	Kitchens et al. (1975)
Cypress-tupelo swamp, so. IL	Yearly	—	Yes	Mitsch et al. (1979)
Floodplain swamp, NC	Yearly	—	Yes	Kuenzler et al. (1980)
Floodplain swamp, NC	Yearly	—	Yes	Yarbro (1983)
Cypress strand, FL	Yearly	—	Yes	Nessel and Bayley (1984)
Swamp forest, LA	10 mo.	Yes	No	Kemp and Day (1984)
Riparian Forest, MD	Yearly	Yes	Yes	Peterjohn and Correll (1984)
Floodplain forest, FL	Yearly	Yes	Yes	Elder (1985)
Tupelo swamp, NC	Yearly	Yes	No	Brinson et al. (1984)
TIDAL SALT MARSH				
Delaware	Yearly	—	No	Reimold and Diaber (1970)
Georgia	Yearly	—	No	Gardner (1975)
Flax Pond, NY	Yearly	S	S	Woodwell and Whitney (1977), Woodwell et al. (1979)
Great Sippewissett Marsh, MA	Yearly	S	—	Valiela et al. (1978), Teal et al. (1979)
Sapelo Island, GA	Yearly	S	—	Whitney et al. (1981)
Lousiana	Yearly	Yes	Yes	Delaune et al. (1981), Delaune et al. (1983)

Source: Partly from Van der Valk et al., 1979 and Mitsch and Gosselink, 1986.

[a] S = seasonal sink; I = inconsistent sink.

of ammonium nitrogen and soluble reactive phosphorus and a net export of organic nitrogen.

Salt Marshes. Coastal salt marshes have the longest history of nutrient budget studies, and they have provoked the most controversy about the source–sink question for wetlands. Early studies by Reimhold and Daiber (1970) and Reimhold (1972) in Delaware and Gardner (1975) in Georgia described salt marshes as sources of phosphorus. Aurand and Daiber (1973) found a net import of of inorganic nitrogen into a Delaware marsh, whereas Stevenson et al. (1977) found a net discharge of both nitrogen and phosphorus from a Chesapeake Bay salt marsh. A comprehensive study of the Great Sippewissett Marsh in Massachusetts indicated an approximate balance between inputs and outputs of nitrogen, with tidal exchange dominating all other fluxes (Valiela et al., 1978; Teal et al., 1979) whereas a similar study on the annual exchanges of nitrogen between a salt marsh and coastal waters in Long Island, New York, found that the marsh was a source of nitrogen during the growing season and a sink of nitrogen in the winter and spring (Woodwell and Whitney, 1977; Woodwell et al., 1979).

Nixon (1980) developed a critical evaluation of the source-sink question for salt marshes with the following general conclusion:

> On the basis of very little evidence, marshes have been widely regarded as strong terms (sources or sinks) in coastal marine nutrient cycles. The data we have available so far do not support this view. In general, marshes seem to act as nitrogen transformers, importing dissolved oxidized inorganic forms of nitrogen and exporting dissolved and particulate reduced forms. While the net exchanges are too small to influence the annual nitrogen budget of most coastal systems, it is possible that there may be a transient local importance attached to the marsh–estuarine nitrogen flux in some areas. Marshes are sinks for total phosphorus, but there appears to be a remobilization of phosphate in the sediments and a small net export of phosphate from the marsh.

8.3.2 Ecotechnological Design of Wetlands

Over the past 15 years, the idea of purposeful application of wastewater to natural wetlands for water quality has been explored with many experiments and demonstrations. Closely allied with this concept is the design of artificial wetlands for purposes such as flood control, water quality enhancement, and even land stabilization. We consider all of these artificial, yet purposeful, manipulations of ecosystems as ecological engineering or ecotechnology. To some, the idea of using wetlands for receiving wastewater is as a *treatment* system; to others it is considered a *disposal* alternative. Regardless of what it is called, wastewater recycling in wetlands has been experimentally investigated in a number of studies, many of which are summarized by Nichols (1983) and Godfrey et al. (1985).

Northern peatlands were investigated in studies at Houghton Lake and other communities in Michigan by researchers from the University of Michigan (Richardson et al., 1978; Kadlec, 1979; Kadlec and Kadlec, 1979; Kadlec and Tilton, 1979; Tilton and Kadlec, 1979). A full-scale operation for disposal of 380,000 L/day (100,000 gal/day) of secondarily treated wastewater into a rich fen near Houghton Lake led to significant reductions of ammonia nitrogen, nitrate nitrogen, and total dissolved phosphorus as the water passed from the point of discharge.

A team of researchers from the University of Florida (Odum et al., 1977; Ewel and Odum, 1978, 1979, 1984) investigated the use of several cypress (*Taxodium*) domes in north-central Florida as systems of wastewater recycling and water conservation. After 5 years of experimentation, in which secondarily treated wastewater was added to the cypress domes at a rate of 2.5 cm/wk, results indicated that the wetland filtered nutrients, metals, microbes, and viruses from the water before it entered the groundwater. Productivity of the canopy pond cypress trees increased, although subcanopy growth did not increase as much (Ewel and Odum, 1978). The uptake of nutrients in these systems was also enhanced by a continuous cover of duckweed on the water surface, by the retention of nutrients in the cypress wood and litter, and by the adsorption of phosphorus onto clay and organic particles in the sediments.

8.4   WETLAND DESIGN PARAMETERS

The use of both artificial and natural wetlands for water quality enhancement requires the development of several design parameters:

1. *Wetland retention time.* The total amount of water or chemicals flowing into a wetland (Q), divided by the area of the study wetland (A), gives an estimate of the loading rate of a wetland:

$$L = \frac{Q}{A} \tag{8.1}$$

When we divide the loading rate by the average depth (d) of the wetland, we calculate the turnover rate (t^{-1}) as

$$t^{-1} = \frac{L}{d} \tag{8.2}$$

or alternately express the retention or residence time of the wetland (t) as the reciprocal of the turnover rate. The retention time (t), expressed then as

$$t = \frac{A \times d}{Q} \tag{8.3}$$

is one of the most important variable in the use of wetlands as wastewater treatment systems (Hammer and Kadlec, 1983). The longer the retention time of the wetland, the more time the water is in contact with the biological active sediments, and the greater the rates of physical processes such as sedimentation. But longer retention times require either a larger wetland area or a deeper wetland. When average depth becomes too great, the system may no longer function as a wetland and the role of vegetation and sediments is diminished.

2. *Chemical loading rates.* In addition to the hydrologic detention time, the amount of chemicals entering the wetland per unit time is a very important variable. There is an inverse relationship between chemical loading rate and percent removal efficiency. The retention of phosphorus, as determined by Richardson and Nichols (1985), can be expressed as

$$R_{\%P} = f\left(\frac{1}{L_P}\right) \tag{8.4}$$

where $R_{\%P}$ is the percent retention of phosphorus in a wetland and L_P is the phorphorus loading in g P/m^2-yr.

Likewise, nitrogen retention is described by Richardson and Nichols as

$$R_{\%N} = f\left(\frac{1}{L_N}\right) \tag{8.5}$$

where $R_{\%N}$ is the percent retention or loss of nitrogen in a wetland and L_N is the nitrogen loading in g N/m^2-yr.

For example, if a wetland system is being designed for removal of phosphorus, then highest efficiencies (60–90%) have been measured in the ranges of loading rates of 1–5 g P/m^2-yr, but efficiencies reduce to about 30% removal at loading rates of 15 g/m^2-yr (Richardson and Nichols, 1985). If total amount of chemical retained is more important than efficiency (remember, nature does not maximize for efficiency!) then the lower efficiency case above retains 4.5 g/m^2-yr, and the higher efficiency system retains from 0.6 to 4.5 g P/m^2-yr. The chemical loading rate, then, depends on whether one selects for efficiency or total retention of chemicals.

3. *Seasonal patterns.* Wetlands often display seasonal patterns of chemical uptake and release, often in synchronization with the needs to control certain chemicals. The design should be made for the limiting situations,

either when the receiving body of water is most vulnerable (perhaps at low flow), or when the runoff from uplands is the greatest (possibly in the spring).

4. *Aging.* It has been observed that some wetlands have received artificial loadings of chemicals for many years and are still functioning as nutrient sinks. Studies of this type have included freshwater marshes in Wisconsin (Spangler et al., 1977; Fetter et al., 1978) and in forested wetlands in Florida (Boyt et al., 1977; Nessel, 1978; Nessel and Bayley, 1984). Other investigators have suggested that continual chemical loading of wetlands will lead to a reduction of the wetland's efficiency in a process now called *aging.* Such a "saturation" in a wetland after a number of years of receiving chemicals can sometimes result in a reduced capacity for pollutant retention (Richardson, 1985; Kadlec, 1985). It would therefore be prudent in the design of a wetland treatment system either to design for a larger wetland than is really needed (perhaps twice the size) or to have a second wetland designed to take over after a few years and give the original wetland a chance to recover to a state of seminatural conditions. The aging of natural wetlands receiving low loadings of nutrients, such as from non-point sources, is not as well understood.

8.5   CASE STUDY: LAKE ERIE COASTAL WETLANDS

Wetlands have always been a part of the shoreline of the Laurentian Great Lakes, expanding and retreating with changing water levels, yet always maintaining themselves as ecotones between the uplands and the lakes. As shorelines were stabilized and the land was drained for agriculture and urban development, these wetlands were mostly destroyed or significantly altered and their buffering capacity was diminished or lost altogether. Table 8.2 gives several of the estimates of present and presettlement wetlands around the Great Lakes. Herdendorf (1987) estimates that more than 4000 km² of extensive coastal marshes and swamps in the western Lake Erie basin have been cleared, drained, and filled to the point where only 150 km² remain, most artificially diked from open access to Lake Erie (Figure 8.4). It could be surmised that, had the surrounding wetlands remained intact, the rate of cultural eutrophication of some of the lakes such as Lake Erie may have been much less severe.

Few if any studies have been carried out on Great Lakes coastal wetlands, for there have been very few published works on wetland functioning of these ecosystems. This is particularly apparent when compared with the abundant literature available on coastal salt marshes. Much of what is known about Great Lakes coastal wetlands is included in the proceedings of a Great Lakes Coastal Wetlands Colloquium (Prince and D'Itri, 1985). The editors of that work conclude that:

In spite of general scientific opinion that wetlands are important to Great Lakes

Table 8.2 Area of Coastal Wetlands Along Laurentian Great Lakes

Region	Wetland Area (km²)	Reference
GREAT LAKES WETLANDS (EMERGENT AND AQUATIC BEDS)		
Total	700	Kroll et al. (1988)
Lake Erie	187	
Lake St. Clair	165	
Lake Ontario	133	
Lake Huron	128	
Lake Michigan	50	
Lake Superior	37	
WESTERN LAKE ERIE		
Presettlement	>4,000	Herdendorf (1987)
Present	150	
LAKES ONTARIO, ERIE, AND ST. CLAIR	615	Prince and D'Itri (1985)
TOTAL GREAT LAKES		
Present	1,209	Herdendorf et al. (1981)
Lake Erie-Lake St. Clair	119	
COASTAL WETLANDS—ONTARIO		
Presettlement	500	McCullough (1985)
Present	330	
Lake St. Clair	25	
Lake Ontario	6	
COASTAL WETLANDS—OHIO		
1979	59	ODNR (1982)
COASTAL WETLANDS—MICHIGAN		
1979	428	Jaworski et al. (1979)

ecosystems, they represent one of the least well understood parts of those systems. Moreover, they are greatly diminished in extent and quality along most moderately settled shorelines. Still, in 1981, around the heavily settled lower Great Lakes (Ontario, Erie, and St. Clair), about 61,480 hectares of coastal wetland remained. . . .

8.5.1 Research Goals

The goals of our research are to use an ecosystem approach to western Lake Erie coastal wetlands (1) to determine if and how these systems are serving

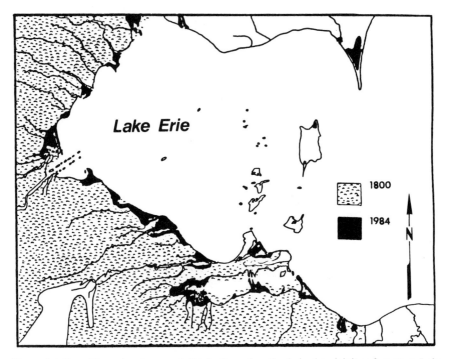

Figure 8.4 Presettlement and present distribution of wetlands in the vicinity of western Lake Erie (Herdendorf, 1987).

as a chemical and hydrologic buffers between the upland watershed and the downstream Lake Erie, and (2) to determine what design, for example, diked versus undiked, is the most effective for future wetland design along the lake. The overall project will answer a number of questions using field measurements and mathematical models of hydrologic, chemical, and ecological processes of the wetlands and adjoining watershed. This will ultimately determine the functioning of the remaining coastal wetlands along Lake Erie. Are these coastal wetlands chemical sinks, sources, or transformers for runoff and stream flow from the upland agricultural watershed? How valuable is their survival and protection to the enhancement of water quality of Lake Erie? What processes within the wetland are most important in the changes that occur? What are the seasonal patterns of nutrient dynamics in the wetland? How do the processes change with changing lake levels? What happens to the wetland when seiches occur, reversing the normal flow from the wetland to Lake Erie? If the wetland studied proves to be carrying out valuable functions, what are the design criteria for building similar wetlands in other watersheds along the Great Lakes?

8.5.2 Old Woman Creek Site

Old Woman Creek State Nature Preserve and National Estuarine Research Reserve located adjacent to Lake Erie in Erie County, Ohio is one of the

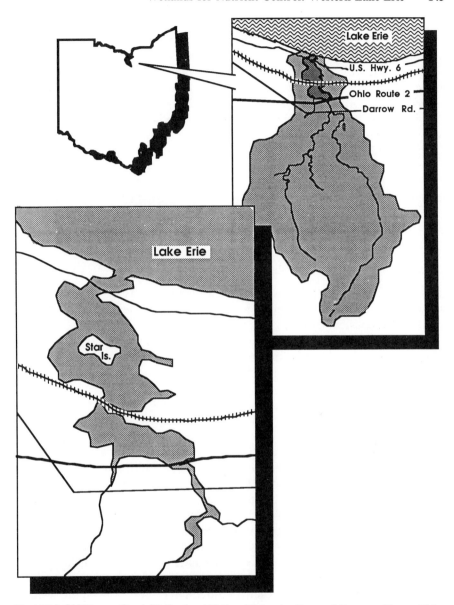

Figure 8.5 Old Woman Creek Wetland and National Estuarine Research Reserve, Huron, Ohio.

coastal freshwater wetlands that is being studied in more detail (Figure 8.5). The marsh itself is 30 ha in size and extends about 1 km south of the Lake Erie shoreline. It is approximately 0.34 km wide at its widest portion. Depths may reach up to 3.6 m in the inlet stream channel, but for the major portion of its area it is usually less than 0.5 m deep. The wetland has an outlet to Lake Erie that is often open but that can be closed for extended periods of time. Dramatic seiches on Lake Erie can reverse the flow, causing lake water to spill into the wetland. Aquatic habitats within the wetland include open water planktonic systems, embayment marshes with American lotus (*Nelumbo lutea*), white water lilies (*Nymphea tuberosa*), spatterdock (*Nuphar advena*), arrow arum (*Peltandra virginica*), and cattails (*Typha augustifolia*), and wooded wetlands in certain embayments. The major land use within the watershed (69 km^2) is agricultural. Sedimentation in the wetland was estimated to have been 0.76 mm/yr prior to agricultural development in the early 1800s and more than 10 times that (10 mm/yr) at present (Buchanan, 1982).

Owing to its status as a national estuarine research reserve, the marsh remains relatively undisturbed and is frequently used for nature education, recreation, and scientific study. Sanctuary facilities include a visitor's center and an aquatic ecology research laboratory on the site.

8.5.3   Ecosystem Nutrient Mass Balances

It is instructive to investigate a natural wetland such as Old Woman Creek to see how self-design has led to ecosystem development in tune with watershed and coastal processes over time. Some of the more important forcing functions that determine the wetland's ability to retain nutrients, particularly phosphorus, are the Lake Erie water level fluctuations, occasional seiches, and nutrient loading from the upstream watershed.

Lake Erie Water Fluctuations. Wetlands along Lake Erie in general, and Old Woman Creek Wetland in particular, are influenced by water level fluctuations of the Great Lakes. The water levels for Lake Erie over the past 125 years are shown in Figure 8.6. Over this period, there was a difference between low and high water level in Lake Erie of about 1.5 m. This difference is enough to affect significantly the structure and function of coastal wetlands, sending them "inland" during high water levels and allowing them to extend "lakeward" during low water levels. The Old Woman Creek wetland is in an equilibrium with this fluctuating water level, varying from a wetland dominated by emergent vegetation (during shallow water times) to one that is a plankton-floating leaved aquatic system (during high water level).

On a shorter time scale, the general direction of flow is from the wetland to Lake Erie, with the difference in elevation between the two bodies of water varying with storm events and short-term Lake Erie fluctuations. Figure 8.7 illustrates that the level of the wetland is generally at or near lake level. The difference between the two levels can be exacerbated when the mouth of the stream between the wetland and Lake Erie is closed, a rather frequent event.

Seiches. Large-period water level oscillations, owing to wind action, fre-

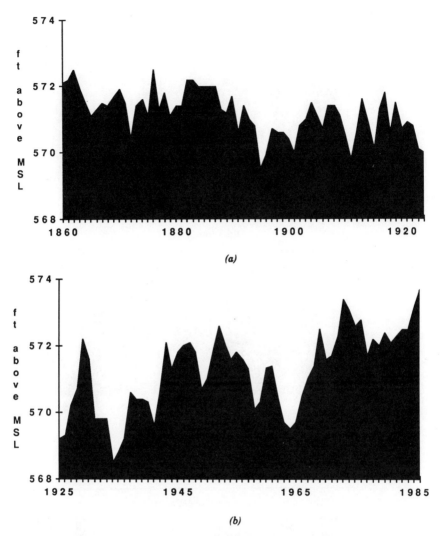

Figure 8.6 Water-level fluctuations of Lake Erie for (*a*) 1860–1924, and (*b*) 1925–1986. Data are given in feet above mean sea level.

quently occur on the Great Lakes. The coastal wetlands along the lakes are subject to water and chemical exchanges from seiches in much the same way that coastal salt marshes are subjected to tides, although these seiches are not as periodic as semidiurnal coastal tides. Sager et al. (1985), for example, measured 269 seiche events in one year on lower Green Bay on Lake Michigan with a mean amplitude of 19.3 cm and a mean period of 9.9 hr. Their study indicated that coastal marshes may be serving as sinks for total phosphorus and as transformers of nitrogen from dissolved oxidized forms

Lake Erie

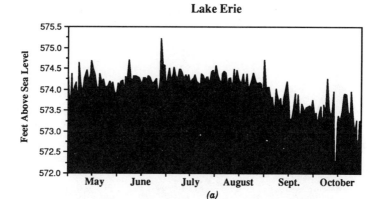

Old Woman Creek

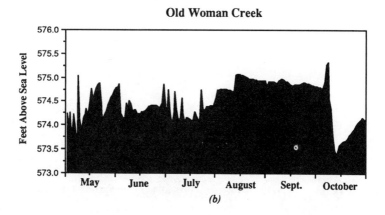

Estuarine Effects

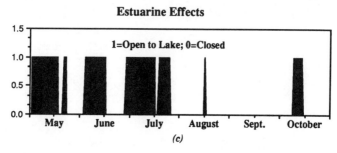

Figure 8.7 (*a*) Lake Erie water levels, (*b*) Old Woman Creek Wetland water levels, (*c*) opened or closed status of connection, and (*d*) difference for May–October, 1983.

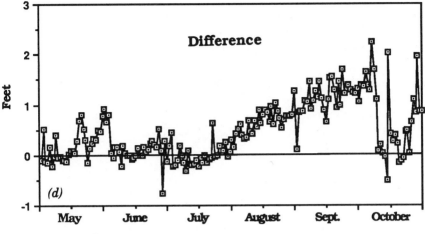

Figure 8.7 (continued)

to particulate and reduced forms. Seiches are a common occurrence along Lake Erie in the vicinity of Old Woman Creek, although the contribution of these events to the nutrient budget of the wetland is not known.

Loadings from the Upland Watershed. The 69-km^2 watershed that drains into Old Woman Creek wetland is primarily agricultural, with runoff containing relatively high levels of nutrients (Klarer, 1988). An estimated phosphorus loading of 0.5–1.0 kg/ha-yr for our watershed draining into Old Woman Creek as presented by Johnson et al. (1978), IJC (1980), and Novotny (1986) results in a calculation of 3500–7000 kg P/yr (12–23 g P/m^2-yr) discharged to the wetland. It should be noted that this loading rate is a conservative (low) estimate of the contribution of agricultural non-point sources to western Lake Erie. Many counties in Ohio in the western Lake Erie basin are estimated to have loading rates of 1.0–2.5 kg P/ha-yr. A study group of Great Lake's pollution called PLUARG (Pollution from Land Use Activities Reference Group) presents a range of 0.1–9.1 kg P/ha-yr (IJC, 1980).

Nutrient Retention by Wetland. Patterns of orthophosphate concentrations in steam flow into Old Woman Creek Wetland and at the bridge near the discharge to Lake Erie are shown in Figures 8.8 through 8.10. All figures clearly show that the concentrations of orthophosphate in the stream entering the wetland is higher, often considerably so, than the concentration of orthophosphate leaving the wetland. Figure 8.8 shows the significant difference in orthophosphate concentrations prior to, during, and after a series of storm events in May 1985. Klarer (1988) reports that concentrations of orthophosphate at the outflow of Old Woman Creek wetland were 20–49% of the concentrations at the inflow for all storm events measured in 1984–1985. When corrected for flow by using chloride as a conservative tracer,

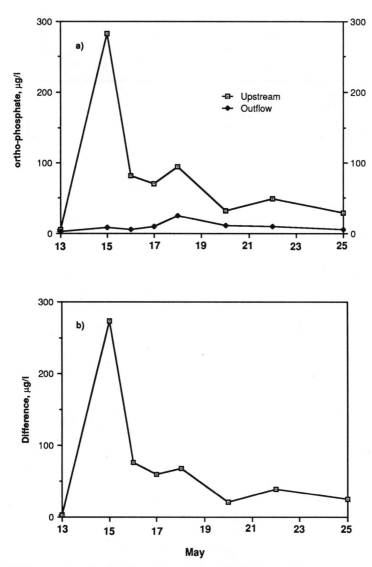

Figure 8.8 Pattern of ortho-phosphate concentrations in Old Woman Creek Wetland for storm events in May 13–25 1985: (top) at inflow (upstream) and outflow (mouth was open to Lake Erie); (bottom) decrease in orthophosphate concentration between inflow and outflow. Storm event of May 14–20 resulted in total of 7.9 cm of precipitation (unpublished data by Dave Klarer).

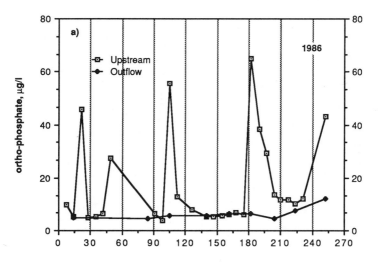

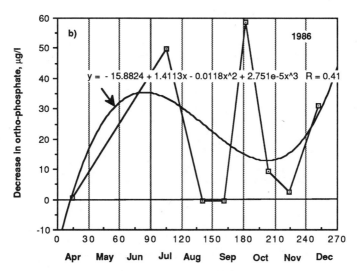

Figure 8.9 Pattern of orthophosphate concentrations in Old Woman Creek Wetland for 1986: (top) at inflow (upstream) and outflow; (bottom) decrease in orthophosphate concentration between inflow and outflow (unpublished data by Dave Klarer).

the outflow concentrations were 32–53% of the inflow. The difference between inflow and outflow concentrations over entire years (Figures 8.9 and 8.10) suggests a seasonal pattern with more uptake of orthophosphate by the wetland during summer months. The pattern is one that would be expected for a plankton and macrophyte dominated system. Heath (1986) had concluded that the uptake of phosphorus by Old Woman Creek Wetland was

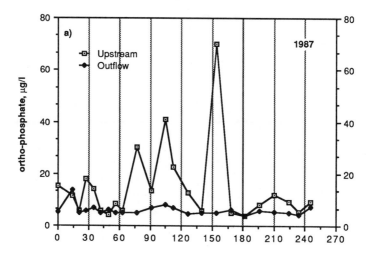

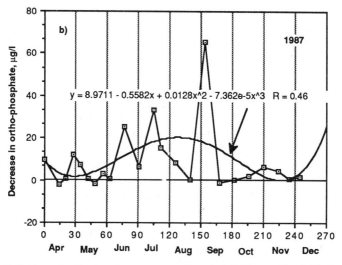

Figure 8.10 Pattern of orthophosphate concentrations in Old Woman Creek Wetland for 1987: (*a*) at inflow (upstream) and outflow; (*b*) decrease in orthophosphate concentration between inflow and outflow (unpublished data by Dave Klarer).

mostly biotic; these patterns support that view. Because flow data are not included with these measurements and because these are measures of only orthophosphate, not total phosphorus, a calculation of the net retention is not feasible. An estimate of wetland retention can be made from a well-established design criterion for wetland nutrient retention, the inverse relationship between nutrient loading between nutrient retention and nutrient

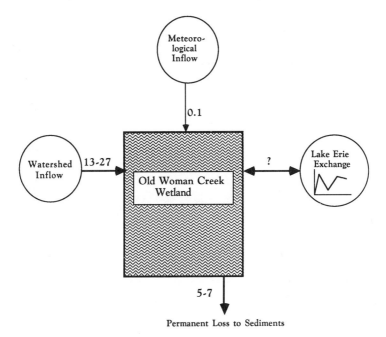

Figure 8.11 Ecosystem mass balance model of phosphorus in Old Woman Creek Wetland, showing estimated phosphorus flows in g P/m²-yr.

loading in a wetland (see Section 8.4). Higher loading rates yield lower percent removal whereas lower loading rates give a higher percent removal. Thus even if the range of loading has a great deal of uncertainty (as our estimate above does), the calculation of the total mass retained by the wetland converges on a narrower range. If a load from the watershed of 12–23 g P/m²-yr is assumed and the wetland nutrient loading model developed by Richardson and Nichols (1985) in Equation 8.4 is used, a first estimate can be made of the phosphorus retention of our wetland of 30–39%. This range translates to a nutrient retention of approximately 5–7 g P/m²-yr by the wetland or 1500–2100 kg P/yr total.

Ecosystem Models. Ecosystem models of wetlands will be used to guide field research efforts, to identify gaps in data, to investigate ecosystem behavior, and ultimately to aid in the management and possible design of Great Lakes coastal wetlands. Our approach has been to develop a hierarchy of models, ranging from a watershed approach through an ecosystem level approach to an approach that emphasizes ecosystem processes (Mitsch, 1988). Figure 8.11 illustrates an overall nutrient budget for Old Woman Creek at the ecosystem level with the numbers calculated above. The exchange of chemicals and water between Lake Erie and the wetland is difficult to determine. This is further complicated by water level fluctuations and seiches

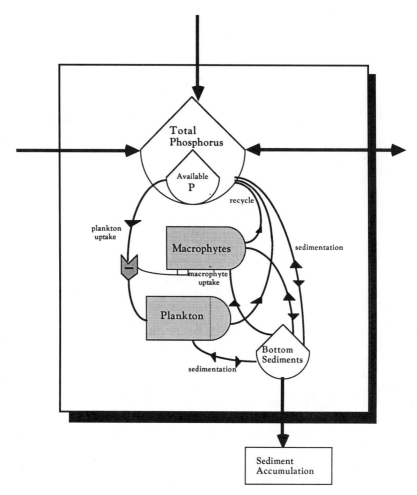

Figure 8.12 Ecosystem model showing system details of processes involved in wetland's role as phorphorus source, sink, or transformer.

on the lake. Figure 8.12 shows the processes in the wetland that contribute to its nutrient retention capability. Plant uptake, by both plankton and macrophytes, sedimentation, and resuspension are probably the most significant processes involved in the wetland retaining and releasing phosphorus.

8.5.4   Actual and Potential Role of Wetlands in Western Lake Erie

Our continued studies will demonstrate the role of coastal wetlands as buffer zones between the uplands and the Great Lakes. If the remaining wetlands along western Lake Erie function in a similar fashion to our case study wetland, and if our estimated phosphorus retention is correct, then it can be further estimated that the existing wetlands along western Lake Erie are retaining approximately 75–100 metric tons/yr. To put that possible retention into perspective, it should be noted that there was an estimated non-point phosphorus loading to western Lake Erie of about 2100 metric tons/yr for 1978–1980 (Yaksich et al., 1982). This suggests that the remaining wetlands are currently retaining about 3.5–5% of the non-point source loading to the lake. If we determine that these coastal wetlands are truly sinks for nutrients, it may be possible to suggest the construction of wetlands along the Great Lakes to take advantage of that function. For example, a program to develop 1000 km^2 of wetlands in the western Lake Erie shoreline and watershed (one-fourth of the extent of presettlement wetlands) could conceivably lead to a 24–33% reduction in non-point loading of phosphorus to the western basin. Modeling will aid in the management of the Great Lakes by demonstrating the temporal patterns of wetland productivity and nutrient exchange that occur with different hydrologic conditions, watershed uses, and lake levels changes. Economic valuation of coastal wetlands will be more feasible as a result of these kinds of studies, for we expect to demonstrate the long-term as well as short-term value of wetlands. Continued studies such as this one will also provide design criteria and cause for ecological engineers to protect other wetlands for their assimilative capacity in the landscape.

REFERENCES

ASWIPCA. 1984. America's clean water—The state's evaluation of progress, 1972–82. Interstate Water Pollution Control Administrators, Washington, DC.

Aurand, D. and F. C. Daiber. 1973. Nitrate and nitrite in the surface waters of two Delaware salt marshes. *Chesapeake Sci. 14:*105–111.

Boyt, F. L., S. E. Bayley, and J. Zoltek, Jr. 1977. Removal of nutrients from treated municipal wastewater by wetland vegetation. *J. Water Pollut. Control Fed. 49:*789–799.

Brinson, M. M., M. D. Bradshaw, and E. S. Kane. 1984. Nutrient assimilation capacity of an alluvial floodplain swamp. *J. Appl. Ecol. 21:*1041–1057.

Buchanan, D. B. 1982. Transport and deposition of sediment in Old Woman Creek Estuary, Erie County, Ohio. M.S. thesis. The Ohio State University, Columbus, OH, 198 pp.

Clark, E. H., J. A. Haverkamp, and W. Chapman. 1985. Eroding soils: The off-farm impacts. The Conservation Foundation, Washington, DC.

Cowardin, L. M., V. Carter, F. C. Golet, and E. T. LaRoe. 1979. Classification of Wetlands and Deepwater Habitats of the United States. U.S. Fish and Wildlife Service Publ. FWS/OBS-79/31, Washington, DC 103 pp.

Crisp, D. T. 1966. Input and output of minerals for an area of Pennine moorland: the importance of precipitation, drainage, peat erosion, and animals., *J. Appl. Ecol. 3:*327–348.

Delaune, R. D., C. M. Reddy, and W. H. Patrick, Jr. 1981. Accumulation of plant nutrients and heavy metals through sedimentation and accretion in a Louisiana salt marsh. *Estuaries 4:*328–334.

Delaune, R. D., C. J. Smith, and W. H. Patrick, Jr. 1983. Nitrogen losses from a Louisiana Gulf Coast salt marsh. *Est. Coast. Shelf Sci. 17:*133–142.

Elder, J. F. 1985. Nitrogen and phosphorus speciation and flux in a large Florida river-wetland system. *Water Resources Res. 21:*443–453.

Elder, J. F. and H. C. Mattraw. 1982. Riverine transport of nutrients and detritus to the Apalachicola Bay estuary, Florida. *Water Resources Res. 18:*849–856.

Ewel, K. C. and H. T. Odum. 1978. Cypress swamps for nutrient removal and waste-water recycling. In M. P. Wanielista and W. W. Eckenfelder, Jr., Eds., *Advances in Water and Wastewater Treatment—Biological Nutrient Removal.* Ann Arbor Science, Ann Arbor, MI, pp. 181–198.

Ewel, K. C. and H. T. Odum. 1979. Cypress domes: nature's tertiary treatment filter. In W. E. Sopper and S. N. Kerr, Eds., *Utilization of Municipal Sewage Effluent and Sludge on Forest and Disturbed Land.* The Pennsylvania State University Press, University Park, PA, pp. 103–114.

Ewel, K. C. and H. T. Odum. 1984. *Cypress Swamps.* University Presses of Florida, Gainesville, FL, 472 pp.

Fetter, C. W., Jr., W. E. Sloey, and F. L. Spangler. 1978. Use of a natural marsh for wastewater polishing. *J. Water Pollut. Control Fed. 50:*290–307.

Gardner, L. R. 1975. Runoff from an intertidal marsh during tidal exposure: regression curves and chemical characteristics. *Limnol. Oceanogr. 20:*81–89.

Godfrey, P. J., E. R. Kaynor, S. Pelczarski, and J. Benforado, Eds. 1985. *Ecological Considerations in Wetlands Treatment of Municipal Wastewaters.* Van Nostrand Reinhold, New York, 496 pp.

Grant, R. R. and R. Patrick. 1970. Tinicum Marsh as a water purifier. In J. Mc-Cormick, R. R. Grant, Jr., and R. Patrick, Eds., *Two Studies of Tinicum Marsh, Delaware and Philadelphia Counties, Pa.* The Conservation Foundation, Washington, DC, pp. 105–131.

Hammer, D. E. and R. H. Kadlec. 1983. Design Principles for Wetland Treatment Systems. U.S. Environmental Protection Agency Report EPA-600/2-83-026, Ada, OK, 244 pp.

Heath, R. T. 1986. Phosphorus Dynamics in the Old Woman Creek National Estuarine Sanctuary—A Preliminary Investigation. Final Report to Sanctuaries Programs Division, National Oceanic and Atmospheric Administration, Washington, DC, 105 pp.

Hemond, H. F. 1980. Biogeochemistry of Thoreau's Bog, Concord, Mass. *Ecol. Monogr. 50:*507–526.

Herdendorf, C. E. 1987. The Ecology of the Coastal Marshes of Western Lake Erie: A Community Profile. U.S. Fish and Wildlife Service Biol. Rep 85 (7.9), 171 pp.

Herdendorf, C. E., S. M. Hartley, and M. D. Barnes, eds. 1981. Fish and Wildlife Resources of the Great Lakes Coastal Wetlands within the United States. Vol. 1: Overview. U.S. Fish and Wildlife Service Report FWS/OBS 81/02-v1. Washington, DC, 469 pp.

International Joint Commission. 1980. Pollution in the Great Lakes Basin from Land

Use Activities. IJC Report to the Governments of the United States and Canada. International Joint Commission, Windsor, Ontario, 141 pp.

Jaworski, E., C. N. Raphael, P. J. Mansfield, and B. Williamson. 1979. Impact of Great Lakes Water Level Fluctuations on Coastal Wetlands. Final Research Report, Institute of Water Research, Michigan State University, East Lansing, Michigan. 351 pp.

Johnson, M. G. et al. 1978. Management Information Base and Overview Modelling. PLUARG Technical Report No. 002 to the International Joint Commission, Windsor, Ontario, 90 pp.

Jørgensen, S. E., C. C. Hoffmann, and W. J. Mitsch. 1988. Modelling nutrient retention by a reedswamp and wet meadow in Denmark. In W. J. Mitsch, M. Straskraba, and S. E. Jørgensen, Eds., *Wetland Modelling*. Elsevier, Amsterdam, pp. 133–151.

Kadlec, R. H. 1979. Wetlands for tertiary treatment. In P. E. Greeson, J. R. Clark, and J. E. Clark, Eds., *Wetland Functions and Values: The State of Our Understanding*. American Water Resources Association, Minneapolis, MN, pp. 490–504.

Kadlec, R. H. 1985. Aging phenomena in wastewater wetlands. In P. J. Godfrey, E. R. Kaynor, S. Pelczarski, and J. Benforado. Eds., *Ecological Considerations in Wetlands Treatment of Municipal Wastewaters*. Van Nostrand Reinhold, New York, pp. 338–347.

Kadlec, R. H. and J. A. Kadlec. 1979. Wetlands and water quality. In P. E. Greeson, J. R. Clark, and J. E. Clark, Eds., *Wetland Functions and Values: The State of Our Understanding*. American Water Resources Association, Minneapolis, MN, pp. 436–456.

Kadlec, R. H. and D. L. Tilton. 1979. The use of freshwater wetlands as a tertiary wastewater treatment alternative. *CRC Crit. Rev. Environ. Control 9*:185–212.

Kemp, G. P. and J. W. Day, Jr. 1984. Nutrient dynamics in a Louisiana swamp receiving agricultural runoff. In K. C. Ewel and H. T. Odum, Eds., *Cypress Swamps*. University Press of Florida, Gainesville, pp. 286–293.

Kitchens, W. M. Jr., J. M. Dean, L. H. Stevenson, and J. M. Cooper. 1975. The Santee Swamp as a nutrient sink. In F. G. Howell, B. Gentry, and M. H. Smith, Eds., *Mineral Cycling in Southeastern Ecosystems*. ERDA Symposium Series 740513, USGPO, Washington, DC, pp. 349–356.

Klarer, D. M. 1988. The role of a freshwater estuary in mitigating stormwater inflow. Old Woman Creek National Estuarine Sanctuary Tech. Report No. 5, Division of Natural Areas, Ohio Department of Natural Resources, Columbus, Ohio, 54 pp.

Klopatek, J. M. 1978. Nutrient dynamics of freshwater riverine marshes and the role of emergent macrophytes. In R. E. Good, D. F. Whigham, and R. L. Simpson, Eds., *Freshwater Wetlands: Ecological Processes and Management Potential*, Academic, New York, pp. 196–219.

Kroll, R., T. A. Bookhout, and K. E. Bednarik. 1988. Distribution and waterfowl use of the Great Lakes marshes. Abstract. Ohio Wildlife Meeting, Ohio Department of Natural Resources, Columbus, Ohio.

Kuenzler, E. J., P. J. Mulholland, L. A. Yarbro, and L. A. Smock. 1980. Distributions and budgets of carbon, phosphorus, iron, and magnanese in a floodplain swamp ecosystem. Water Resource Institute of North Carolina Report No. 157, Raleigh, NC.

Lee, G. F., E. Bentley, and R. Admundson. 1975. Effect of marshes on water quality. In A. D. Hasler, Ed., *Coupling of Land and Water Systems.* Springer, New York, pp. 105–127.

McCullough, G. B. 1985. Wetland threats and losses on Lake St. Clair. In H. H. Prince and F. M. D'Itri, Eds., *Coastal Wetlands.* Lewis Publishers, Chelsea, MI, pp. 201–208.

Mitsch, W. J. 1977. Waterhyacinth (*Eichhorinia crassipes*) nutrient uptake and metabolism in a north central Florida marsh. *Arch. Hydrobiol. 81:*188–210.

Mitsch, W. J., C. L. Dorge, and J. R. Weimhoff. 1979. Ecosystem dynamics and a phosphorus budget of an alluvial cypress swamp in southern Illinois. *Ecology 60:*1116–1124.

Mitsch, W. J., and J. G. Gosselink. 1986. *Wetlands.* Van Nostrand Reinhold, New York, 539 pp.

Mitsch, W. J. 1988. Ecological engineering and ecotechnology with wetlands: applications of systems approaches. In *Advances in Environmental Modelling: Proceedings 6th International Conference on the State of the Art of Ecological Modelling*, Venice, Italy, June 1987 Elsevier, Amsterdam.

Nessel, J. K. and S. E. Bayley. 1984. Distribution and dynamics of organic matter and phosphorus in a sewage enriched cypress swamp. In K. C. Ewel and H. T. Odum, Eds., *Cypress Swamps*, University Presses of Florida, Gainesville, pp. 262–278.

Nichols, D. S. 1983. Capacity of natural wetlands to remove nutrients from wastewater. *J. Water Pollution Control Fed. 55:*495–505.

Nixon, S. W. 1980. Between coastal marshes and coastal waters—a review of twenty years of speculation and research on the role of salt marshes in estuarine productivity and water chemistry. In P. Hamilton and K. B. McDonald, Eds., *Estuarine and Wetland Processes*, Plenum, New York, pp. 437–525.

Novotny, V. 1986. A review of hydrologic and water quality models used for simulation of agricultural pollution. In A. Giorgini and F. Zingales, Eds., *Agricultural Nonpoint Source Pollution: Model Selection and Application.* Elsevier, Amsterdam, pp. 9–35.

Odum, H. T., K. C. Ewel, W. J. Mitsch, and J. W. Ordway. 1977. Recycling treated sewage through cypress wetlands in Florida. In F. M. D'Itri, Ed., *Wastewater Renovation and Reuse.* Marcel Dekker, New York, pp. 35–67.

Ohio Department of Natural Resources. 1982. Ohio's Coastal Wetlands. ODNR, Columbus, Ohio, 7 pp.

Peterjohn, W. T. and D. L. Correll. 1984. Nutrient dynamics in an agricultural watershed: observations on the role of the riparian forest. *Ecology 65:*1466–1475.

Peverly, J. H. 1982. Stream transport of nutrients through a wetland. *J Environ. Quality 11:*38–43.

Prince, H. H. and F. M. D'Itri, Eds. 1985. *Coastal Wetlands.* Lewis Publishers, Chelsea, MI, 286 pp.

Reimhold, R. J. 1972. The movement of phosphorus through the marsh cord grass, *Spartina alterniflora* Loisel. *Limnol. Oceanogr. 17:*606–611.

Reimhold, R. J. and F. C. Diaber. 1970. Dissolved phosphorus concentrations in a natural salt marsh of Delaware. *Hydrobiologia 36:*361–371.

Richardson, C. J. 1985. Mechanisms controlling phosphorus retention capacity in freshwater wetlands. *Science 228:*1424–1427.

Richardson, C. J., D. L. Tilton, J. A. Kaldec, J. P. M. Chamie, and W. A. Wentz. 1978. Nutrient dynamics of northern wetland ecosystems. In R. E. Good, D. F. Whigham, and R. L. Simpson, Eds., *Freshwater Wetlands: Ecological Processes and Management Potential.* Academic, New York, pp. 217–241.

Richardson, C. J. and D. S. Nichols. 1985. Ecological analysis of wastewater management criteria in wetland ecosystems. In P. J. Godfrey, E. R. Kaynor, S. Pelczarski, and J. Benforado, Eds., *Ecological Considerations in Wetlands Treatment of Municipal Wastewaters.* Van Nostrand Reinhold, New York, pp. 351–391.

Sager, P. E., S. Richman, H. J. Harris, and G. Fewless, 1985. Preliminary observations on the seiche-induced flux of carbon, nitrogen and phosphorus in a Great Lakes coastal marsh. In H. H. Prince and F. M. D'Itri, Eds., *Coastal Wetlands.* Lewis Publishers, Chelsea, MI, pp. 59–68.

Simpson, R. L., D. F. Whigham, and R. Walker. 1978. Seasonal patterns of nutrient movement in a freshwater tidal marsh. In R. E. Good, Whigham, D. F., and R. L. Walker, Eds., *Freshwater Wetlands: Ecological Processes and Management Potential,* Academic, New York, pp. 243–257.

Smith, R. A., R. B. Alexander, and M. G. Wolman. 1987. Water-quality trends in the nation's rivers. *Science 235:*1607–1615.

Spangler, F. L., C. W. Fetter, Jr., and W. E. Sloey. 1977. Phosphorus accumulation-discharge cycles in marshes. *Water Resources Bull. 13:*1191–1201.

Stevenson, J. S., D. R. Heinle, D. A. Flemer, R. J. Small, R. A. Rowland, and J. F. Ustach. 1977. Nutrient exchanges between brackish water marshes and the estuary. In M. Wiley, Ed., *Estuarine Processes,* Vol. II, Academic Press, New York, pp. 219–240.

Tilton, D. L. and R. H. Kadlec. 1979. The utilization of a freshwater wetland for nutrient removal from secondarily treated wastewater effluent. *J. Environ. Qual. 8:*328–334.

Teal, J. M., I. Valiela, and D. Berla. 1979. Nitrogen fixation by rhizophere and free-living bacteria in salt marsh sediments. *Limnol. Oceanogr. 24:*126–132.

Valiela, I., J. M. Teal, S. Volkmann, D. Shafer, and E. J. Carpenter. 1978. Nutrient and particulate fluxes in a salt marsh ecosystem: tidal exchanges and inputs by precipitation and groundwater. *Limnol. Oceanogr. 23:*798–812.

Van der Valk, C. B. Davis, J. L. Baker, and C. E. Beer. 1979. Natural freshwater wetlands as nitrogen and phosphorus traps for land runoff. In P. E. Greeson, J. R. Clark, and J. E. Clark, Eds., *Wetland Functions and Values: The State of Our Understanding.* American Water Resources Association, Minneapolis, MN, pp. 457–467.

Verry, E. S. and D. R. Timmons. 1982. Waterborne nutrient flow through an upland-peatland watershed in Minnesota. *Ecology 63:*1456–1467.

Whitney, D. M., A. G. Chalmer, E. B. Haines, R. B. Hanson, L. R. Pomeroy, and B. Sherr. 1981. The cycles of nitrogen and phosphorus. In L. R. Pomeroy and R. G. Weigert, Eds., *The Ecology of a Salt Marsh*. Springer, New York, pp. 163–181.

Woodwell, G. M. and D. E. Whitney. 1977. Flax Pond ecosystem study: exchanges of phosphorus between a salt marsh and the coastal waters of Long Island Sound. *Mar. Biol. 41:*1–6.

Woodwell, G. M., C. A. S. Hall, D. E. Whitney, and R. A. Houghton. 1979. The Flax Pond ecosystem study: exchanges of inorganic nitrogen between an estuarine marsh and Long Island Sound. *Ecology 60:*695–702.

Yaksich, S. M., D. A. Melfi, D. B. Baker, and J. W. Kramer. 1982. Lake Erie Nutrient Loads, 1970–1980. Lake Erie Wastewater Management Study, U.S. Army Corps of Engineers, Buffalo, NY, 194 pp.

Yarbro, L. A. 1983. The influence of hydrologic variations on phosphorus cycling and retention in a swamp stream ecosystem. In T. D. Fontaine and S. M. Bartell, Eds., *Dynamics of Lotic Ecosystems*. Ann Arbor Press, Ann Arbor, MI, pp. 223–245.

9

RESTORATION OF RIVERINE WETLANDS: THE DES PLAINES RIVER WETLANDS DEMONSTRATION PROJECT

Donald L. Hey and Milady A. Cardamone

Wetlands Research, Inc., Chicago, Illinois

J. Henry Sather

Institute for Environmental Management, Western Illinois University, Macomb, Illinois

and

William J. Mitsch

School of Natural Resources, The Ohio State University, Columbus, Ohio

9.1 INTRODUCTION

Wetlands and their related scientific and engineering fields are gaining recognition. This is due to the growing realization that wetlands serve important functions within the riverine and lacustrine environments (Sather and Smith, 1984; Mitsch and Gosselink, 1986). Wetlands act as transitions between land and water, temporarily storing floodwaters, filtering and immobilizing sediments, absorbing excess nutrients, and providing nesting, breeding, and feeding areas for fish and wildlife. The loss of more than 50% of the freshwater wetlands in the United States (in some states the loss has been much greater; e.g., in Illinois 99% of the presettlement wetlands have been drained) has been severe. These losses have all but eliminated the natural ability of this country's streams and rivers to attenuate flood flows harmlessly, assimilate waste materials, and provide habitat for fish and wildlife.

Channelization and the associated draining of wetlands have increased and accelerated the amount of water moving downstream, exacerbating flooding. Although point source pollution controls (e.g., municipal wastewater treatment plants) have helped to relieve aquatic stress from stream pollution, non-point sources continue to overload surface waters with nutrients and toxic organic compounds. Urban and agricultural storm water runoff contributes more than 50% of the pollutants found in our streams and rivers. As a result of these conditions, flood damages continue to grow; many streams and rivers remain unfishable and unswimmable; and waterfowl populations continue to decline.

To avoid further wetland losses and to avoid the deleterious effects that such losses have on surface waters, federal, state, and local governments have developed, have enacted, and are enforcing preventive regulations. In addition, national and local conservation organizations are working to acquire and preserve these unique ecosystems. Although preservation and regulatory activities are important, they can at best only stem the tide of wetland losses. Such programs will not completely eliminate future losses because of the legal rights of landowners. Furthermore, these activities will not replace the millions of acres of wetlands that have been lost.

Preservation and regulatory programs must be combined with restoration and creation if we are to compensate for future losses and make up for those that have occurred in the past. Only through restoration will we be able to materially affect our water and wildlife resources. This is the case particularly in more developed areas where agriculture, industry, commerce, and housing have all but eliminated wetlands. The need for such special, unique landscapes is more pressing in these areas because they promise a solution to the environmental problems that blight the environment: flooding, poor water quality, and depauperate wildlife populations.

We must develop the design and construction techniques needed to rebuild our nation's wetland resources. We also need to develop management and maintenance programs to sustain these reconstructed landscapes. Although wetlands have been studied and used for a variety of purposes, no rigorous, comprehensive body of information is currently available from which to develop design criteria and management programs. The purposes of the case study described in this chapter, the Des Plaines River Wetlands Demonstration Project, are to start the development of this information base and to help develop the design and management criteria for wetland construction.

9.1.1 Past and Ongoing Research

Considerable wetland creation work has been carried out in coastal saltwater zones and tidal areas. Garbisch (1977) and Josslyn (1982) have made surveys of wetland restoration and marsh development projects in these areas. The focus of this work has been to reestablish vegetative communities for erosion control and habitat enhancement.

A second major area is the reclamation of lands disturbed by mining (Nawrot and Yaich, 1982; Kleinman and Erickson 1983, Ruesch, 1983; Brooks et al., 1985; Erwin and Best, 1985). Examples include lands that have been mined for coal in the eastern and midwestern states, lignite in the plains and western states, and phosphate in Florida. Each area presents unique problems and opportunities. In the coal mining regions, wetlands successfully treat acid mine drainage (see Chapter 12). However, Holbrook and Maynard (1985) argue that research is needed to develop more efficient designs. Wetland restoration in phosphate-mined areas has focused on revegetation to stabilize the ecosystem and to provide wildlife habitat.

The use of wetlands for wastewater treatment has gained significant attention in recent years (Hammer and Kadlec, 1983; see also Chapter 8). Initial work done in this area used existing wetlands. Still, the technology to construct wetlands for this purpose is in its early stages, with research oriented toward artificial systems such as shallow ponds with water hyacinths, and other aquaculture systems (Ewel et al., 1982). Bastian and Reed (1980) provided an engineering perspective on the use of wetlands for wastewater treatment and suggested that artificial or constructed wetlands present fewer difficulties than natural wetlands, particularly where inherent wetland values must be protected.

Created wetlands have been incorporated in more general efforts to rehabilitate streams. Examples include streams disturbed by mining, highway construction, and industrial development. Some of these methods can be applied to riverine wetland systems designed for habitat enhancement.

Some practical experience is being gained from work done in revegetating shorelines. Fowler and Hammer (1976) describe methods for seeding wet zones bordering reservoirs. Seed germination and application techniques along wet shorelines as described by O'Neill (1972) and Bell et al. (1974) provide an approach to vegetating stream channels.

What has not been done, but is needed, is the systematic and controlled evaluation of restoration/creation techniques. Optimal methods for building wetlands must be developed—not only for economic and functional reasons, but to develop fully the beauty of these landscapes.

9.2   THE DES PLAINES RIVER WETLANDS DEMONSTRATION PROJECT

9.2.1   Development

In 1982, the feasibility of using wetlands to solve water resource related problems was examined for northeastern Illinois. Funded by the Illinois Department of Energy and Natural Resources, the study resulted in a publication entitled "The Creation of Wetlands Habitats in Northeastern Illinois" (Hey et al., 1982), which concluded that wetlands could serve such a function and recommended the development of a demonstration project in Lake County, Illinois. Wetlands Research, Inc. was created in 1983, to organize and manage this demonstration and research project.

Wetlands Research, Inc. is a not-for-profit corporation which is governed by a seven-member board, three of whom are appointed by the president of the Lake County Forest Preserve District, a local government agency. The remaining four directors are appointed by the president of the Open Lands Project, a Chicago-based conservation organization. The purposes of the project, called the Des Plaines River Wetlands Demonstration Project, are to demonstrate how wetlands can benefit society both environmentally and economically and to establish design procedures, construction techniques, and management programs for restored wetlands. More specifically, the project goals are as follows:

- To demonstrate the benefits of wetlands restoration for flood control, water quality improvement, and fish and wildlife habitat enhancement;
- To formulate and evaluate restoration and management techniques;
- To create wildlife habitat and recreational areas; and
- To propose alternatives to existing environmental investment strategies and water management programs.

It is the intent of the project, as well, to show how all forms of economic development (e.g., urban and rural, industrial and commercial) can be the direct beneficiaries of restored wetlands.

The results of the project will lead to the following products:

- Design manual—describing the physical and biological design parameters needed to construct wetlands for identified functions.

Table 9.1 Research Technical Advisory Committee for Des Plaines River Wetlands
Demonstration Project

Dr. Gary W. Barrett, Miami University

Professor John Cairns, Jr., Virginia Polytechnic Institute and State University

David G. Davis, U.S. Environmental Protection Agency

Neil R. Fulton, Illinois Department of Transportation

Allan Hirsch, Dynamac Corporation

Marvin E. Hubbell, Illinois Department of Conservation

Robert Kadlec, University of Michigan

William J. Mitsch, The Ohio State University

Charles J. Newling, U.S. Army Corps of Engineers

Richard Novitski, U.S. Geological Survey

Paul G. Risser, University of New Mexico

J. Henry Sather (Committee Chair), Western Illinois University

Richard C. Smardon, ESF/SUNY

Arnold van der Valk, Iowa State University

Daniel E. Willard, Indiana University

James H. Zimmerman, University of Wisconsin-Madison

- Management and operations manual—a methods document for long-term monitoring and operation of created wetlands to maintain intended functions.
- Technology transfer—documents, scientific papers, and workshops targeting ecologists, engineers, citizens and public policy makers.
- Documentary film—describing the project from inception to completion, showing how wetlands can be built and how they function.
- Restored site—a living example of restoration technology will continue to function beyond the time limits of this project, leaving a restored ecosystem for recreational use, fishing, wildlife habitat, and further educational and research efforts.

With these objectives in hand and a grant from the U.S. Fish and Wildlife Service, Wetlands Research, Inc. began the collection of baseline information in 1985. More than 20 scientists and engineers were on the site qualifying and quantifying both biotic and abiotic conditions. Some of these scientists are members of the Research Technical Advisory Committee (Table 9.1). In October 1985, a workshop was held where these scientists and engineers created the design of what is, in fact, a living wetland restoration laboratory. Technical design drawings were then produced and a research project inaugurated.

9.2.2   Original Site Description

After a careful search for a suitable landscape within a river floodplain, a 182-ha site was selected for its degree of disturbance, location, and availability (Hey and Philippi, 1985a). Situated on both banks of the Des Plaines River, approximately 56 km north of Chicago, Illinois (Figure 9.1), the site encompassed land that has been drained for farming and grazing, mined for sand and gravel, and generally disturbed from its presettlement condition. Since 1960, the site has been owned by the Lake County Forest Preserve District and has been used for recreational purposes through the construction of a system of public access trails. Almost 1.7 km of the river stretches through the site. The once meandering, slow-moving rivulet that would have been found within the wide, wet floodplain covered with water-loving plants was leveed in part and confined within a channel formed by steep, tree-covered banks. Once the land was no longer used for agricultural purposes, drain tiles clogged and wet areas had again formed on the site, supporting monocultures of cattails.

Of the 182 ha, approximately 8 ha were made up of moving water at the annual mean river stage of 201 m (660 ft) above mean sea level (msl). In addition to the river and its backwater areas, the abandoned quarry lakes contributed 15 ha of open water. These three quarry lakes, ranging in depth from 7 to 12.5 m, were filled with groundwater with little or no connection to the river. Existing marsh wetlands covered approximately 17 ha (Hey, 1987). More than 58 ha of wooded area covered the site, including 14 ha of unmanaged oak–savanna. The remaining timber consisted of young (15–20 years) cottonwoods, elms, and hawthorns. Willows, cottonwoods, and other scrub trees lined the riverbanks. Old fields (52 ha) and 32 ha of ruderal area, including roads and trails, completed the site. Many of the old field rows were still visible and generally supported reed canary grass in the more wet areas and woody shrubs on higher ground.

Vegetation over the site had been greatly affected by heavy grazing, agriculture, fire deprivation, and lowered groundwater caused by farm tiling. Plant communities were dominated by those Eurasian species that characteristically move into an area after it has been severely disturbed by human activities and that, once established, prevent the return of the native species. Trees and shrubs moved into the abandoned fields and thicket-forming shrubs dominated the old grazed-out savannas.

The site is flat, with high ground being the oak–savanna areas and spoil piles and levees created by gravel mining and the construction of the railroad line lying along the east border. Almost 170 ha (92%) of the site falls within an elevation range of 660–669 feet above msl, exposing it to inundation when river levels rise.

The river level at mean flow is 201 m (660.0 ft) above msl. High flows have reached an elevation of 203 m (666.0 ft) above msl as measured at Wadsworth Road bridge on the northern boundary of the site. With a bottom

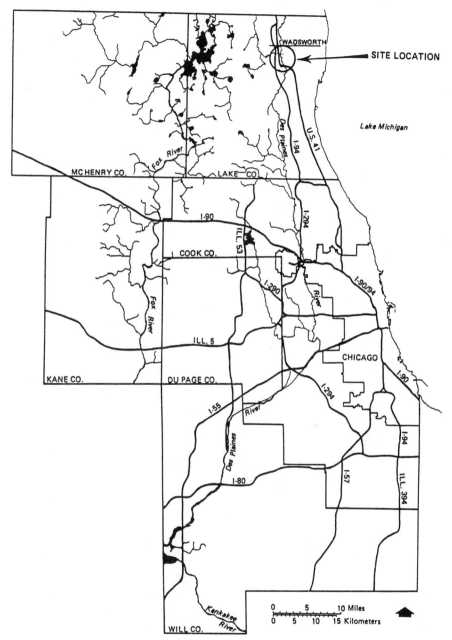

Figure 9.1 Site location of the Des Plaines River Wetlands Demonstration Project, Northeastern Illinois.

slope of 0.25 m/km, the river maintains a low-energy average velocity of less than 0.33 m/sec. The river begins in Wisconsin and drains approximately 374 km^2 comprised of 80% rural and 20% urban areas. Mill Creek, the only major tributary entering the river on the project site, drains an area of 177 km^2, which is mainly agricultural and horse-grazing country. A preliminary macro-scale water budget prepared from data collected in 1986 illustrates that little water is lost to groundwater.

9.2.3 Water Quality

The primary water quality problem of the Des Plaines River is associated with turbidity. With a mean concentration of 59 mg/L, more than 4500 tons of suspended solids enter the site per year via the Des Plaines River and Mill Creek. Seventy-five percent of these solids are inorganic and 95% are less than 63 μm in size. The resulting turbidity limits light penetration, thus inhibiting aquatic plant growth and sight feeding fish, such as pike (90% of the fish biomass is carp). Added to the overall unsightliness of the water is the wash of silt that coats the vegetation on the stream's banks.

Other observed water quality problems include violations of the Illinois state water quality standards for iron, copper, and fecal coliforms. Although not detected in amounts exceeding the U.S. Federal Food and Drug Administration's criteria, dieldrin, DDT, and PCBs were found in fish flesh samples. According to the results of benthic surveys, the stream is classified as semipolluted.

Owing to the stream's silty substrate and the limited stable surfaces offered by the riverbed and aquatic emergent plants, microbial populations in the stream are suppressed, perhaps by two to three orders of magnitude. This inhospitable habitat for microorganisms is further harmed by higher flows, relative to presettlement conditions, passing through the site from the upstream watershed due to wetland losses and channelization. The degraded stream conditions have endured close to a century and will continue long into the future unless corrective steps are taken.

9.2.4 The "Living" Laboratory

To test the corrective measures, a living wetland and floodplain laboratory was designed to meet the research needs (Hey and Philippi, 1985b). The ultimate plan for the site is shown in Figure 9.2. Macro-research efforts will observe changes over the whole site. Micro-research activities will concentrate experimentation on the experimental wetland areas to be built within the northern portion of the site (Figure 9.2). Descriptions of the design and construction activities are provided according to general geographical areas of the site as shown in Figure 9.2.

River Environs Areas.   Along the main stem of the river, three old river

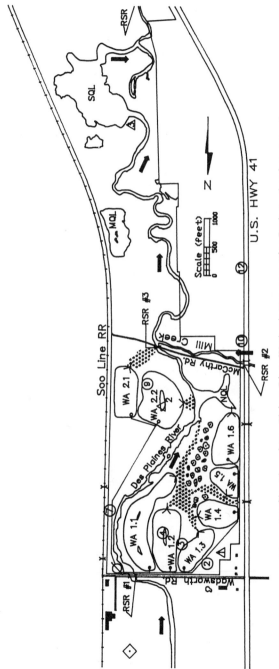

Figure 9.2 Site Plan of Des Plaines River Wetlands Demonstration Project, showing research and monitoring stations. QL, quarry lake; WA, wetland area; ◇, climate station; >‒, Parshall flume; ○, groundwater well; ∅, pumping station; △, precipitation gauge.

167

channels that had been leveed will be reopened to provide more flow-through area and add flood attenuation capacity. Within these river environs grades will be gentle and excavated ground will be revegetated with native Illinois wetland species. The entire length of the river through the site will have the banks cleared of woody and scrub vegetation and revegetated with native species, allowing the river to move freely across the floodplain. The site is confined by the railroad on the east and a U.S. highway on the west, so that lateral movement of the river can be contained over time. It is desirable to have the opportunity to observe the natural movement of the river channel and its banks.

Quarry Lakes.   The three abandoned quarry lakes will be interconnected with the river in two different ways. The southernmost lake was connected by actually routing the river directly into and out of the quarry through the construction of inlet and outlet channels in June 1986. In this way, this quarry acts as a sediment trap during the construction period and will continue to collect sediments for water quality improvement as a part of the permanent restoration of the site. The north and middle quarries will be opened to the river through the construction of a single channel, allowing these quarries to respond to river flows as backwater areas. Fish will be free to move into and out of the lakes and the river access to the lakes will provide additional flood storage capacity on the site.

The now steeply sloped edges of the quarries will be either filled in or excavated back to create aquatic shelves. Submergent and emergent vegetation will be planted to encourage habitat development for fish and shorebirds.

Experimental Wetland Areas.   Intended as microcosms of larger systems, eight experimental wetland areas have been designed with effective surface areas ranging from 1.6 to 4.7 ha. Water from the Des Plaines river will be pumped into each of these areas through an irrigation piping system. With the control of flow rates and water levels, a number of experimental variations can be explored as initial conditions for creating wetlands. Each cell will have maximum side slopes of 1:10 with level bottoms. Shape has been determined somewhat by site conditions and has been encouraged to be as far from round as the limits of construction methods will allow. Inlet flow control involves a variable-speed pumping system, a valve on the main line from the pump station, and a valve on each inlet feeder line. A stoplog dam to control water level will be placed at each experimental wetland area outlet. The design of these areas has allowed for detention times based on depths from zero to 1 m. All these experimental areas lie within the floodplain of the Des Plaines and, as such, U.S. Army Corps of Engineers' "404" permit conditions require that their enclosing berm heights not exceed the stage elevation of the mean annual flood.

Each of the first four wetlands that have been constructed in 1986 (WAs

1.3, 1.4, 1.5, and 1.6 in Figure 9.2) have had topsoil from the previously wet areas placed to depths of 10–15 cm. The vegetation in these areas had been dominated by cattails with some bulrush. As a seed bank, this topsoil will provide initial vegetation. Research strategy to be applied to each of these wetland areas with regard to water depths and vegetation is discussed in Section 9.3.1.

Uplands.   Much of the upland areas will be maintained and improved by keeping the larger trees and by clearing the understory to reestablish the savanna system. These forested areas with prairie grasses, downed tree limbs, and leaf litter buildup will provide foraging, nesting, and resting areas for small mammals, amphibians, and reptiles. One small woodland pond is to be constructed and managed as fishless for the protection of salamanders and woodland frog larvae, allowing adult transformation. The diversity of these wooded uplands is necessary for the encouragement of bird populations. Eight bird "guilds," or groups of bird types that share habitat needs, have been defined for consideration in the habitat design. These include migrant shorebirds, colonial herons, breeding waterfowl, migratory waterbirds, marsh-breeding birds, wet-meadow birds, woodland birds, and hedgerow/thicket birds. Although the majority of these bird guilds will find their most used habitat in the experimental wetland areas, the quarry lakes, and the marsh/prairie areas (passive wetlands) proposed for the site, the presence of wooded uplands interrelating with these other ecosystems is necessary for building successful habitats. Other smaller upland areas will be created as part of the quarry lake improvement work. Developing a dry prairie habitat made up of grasses, such as big bluestem, switch grass, Indian grass, and others on these upland areas will attract small mammals (moles and jumping and white-footed mice). An increased small mammal population, in turn, will attract predators (both bird and mammal) and enhance these communities. A wooded corridor along the eastern edge of the project site will be maintained as a large mammal habitat and wildlife refuge area.

Passive Wetland Areas.   Engineering and biological efforts that set initial conditions and periodic management activities intend to create wetlands that are rich in species and habitats. Designed to function as mudflats for shorebirds, the construction of the passive wetland areas shown in Fig. 9.2 will provide invaluable information on the efficacy of experimental methods used for clearing, grading, and installing native species over large expanses of wet/saturated soils.

Large expanses of land in the southern portion of the site will be cleared of existing shrubs and woody vegetation, spoil piles will be leveled, and native Illinois prairie plants will be installed. Designed with culverts and flap gates, these lands will receive water from the river during high stages, trapping it in the passive wetlands when the river level falls. Manual control of the flap gates can then release water at appropriate times depending on management intent.

Public Use Facilities.   With trails placed through and around both passive and experimental wetlands, people will be provided with the opportunity to enjoy the beauty of the restored wetlands and educational facilities, and simply have a place to ride bicycles, horses, or snowmobiles, walk, run, and play. An existing fishing pier will be renovated and additional fishing walls will be constructed on one of the quarry lakes. Rebuilding the existing canoe launch near the bridge on the north boundary of the site will encourage use of the river for recreation. A parking area now located near the existing canoe launch will be supplemented with a larger parking area on the northwest corner of the site. An interpretative center will eventually be constructed in this area.

Evaluation of public use of the site will be carried out through surveys, annual photographic documentation of site changes used in pictorial surveys, and periodic public meetings.

9.2.5   Construction Status

Construction began in April 1986. The first year resulted in completion of four experimental wetland areas, excavation, and grading of a drainage swale to outflows from the experimental wetlands back to the Des Plaines River, partial clearing of the understory in the northernmost oak–savanna on the west bank, routing of the river through the southern quarry lake, and clearing of most of the vegetation on the west side of the river and along the river's edge north of the confluence of Mill Creek and the river. In 1987, construction efforts were concentrated on installing monitoring instrumentation, planting prairie vegetation in the areas that had been cleared during the winter, and shaping the edge of the southern quarry lake.

Heavy earth-moving equipment has been used extensively. Approximately 380,000 m^3 (500,000 yd^3) of earth must be excavated and relocated to create the experimental wetland areas, the channel connections to each quarry lake, the restored river channels, and the reduced side slopes of the quarry and river banks. A donated low ground pressure bulldozer has done much of the tree and shrub clearing work and replacement of topsoil. The wet soil conditions over most of the site have required both creative planning and construction methods and maximizing work at times of low river levels. Earth bridges were built down the middle of proposed channels to provide access for both excavating equipment and earth hauling trucks. These bridges were then excavated along with the channel.

Revegetation techniques have included the use of a rangeland grass drill on ground that has been minimally prepared with no fertilization, hand casting of seed, and hand planting of grass seedlings.

9.3 RESEARCH AND MONITORING PROGRAM

9.3.1 Research Strategy

The research program covers a broad range of topics (Sather and Smith, 1985). The strategy is hierarchical with general questions about the interactions of stream, terrestrial, and wetland environments being asked and answered at the scale of the entire project site (Figure 9.2), termed *macro-system research*, and detailed questions specific to design and operation of wetlands being asked and answered at the scale of individual experimental wetlands, termed *micro-system research*. Examples of macro-research would be construction techniques, stream morphology, or river water quality. The design criteria for constructed wetlands will be developed from the micro-research work done on a smaller scale. This research will be carried out in highly regulated, experimental wetlands, designed specifically to afford a wide range in hydrologic conditions. In these experimental areas the aquatic environment will be manipulated and monitored in accordance with specific research protocol. Flow will be controlled by pumps, valves, and weirs and by the research regime. In the macro system, hydrologic conditions will depend upon the vagaries of precipitation and stream flow.

For the macro system, the research design incorporates inflow of water and solids to the site (RSR 1 and 2, Figure 9.2) as indicated in Figure 9.3. The movement of water and solids will be monitored through the modified river channels, passive wetlands, and the quarry lakes. The flow regime, of course, will be affected by the pumping station (Figure 9.2) used to lift water from the river to the experimental wetland areas for the micro-research program. Within the river environs, the soils, plant communities, microbes, fish, and wildlife will be monitored. Stream morphology will be carefully measured and compared to base conditions to detect changes. Hydrology, of course, is an important part of the overall scheme. Minimizing stage fluctuations, attenuating flood waves, and increasing detention time are important to the overall management objectives and are important hydrologic design conditions.

The monitored experimental variables are the fundamental factors that must be properly combined in order to replicate selected wetland functions. The research program will relate the experimental variables to the design criteria in order to establish design procedures and management programs. Successive experimental designs will be tested in order to evaluate a wide range of conditions to optimize the relationship between variables and criteria. The cost of achieving each set of experimental variables and related experimental criteria will be analyzed.

The experimental design for the micro system is similar to that of the macro research program. As shown in Figure 9.4, the only difference is in the time scale, spatial detail, and controlled hydrologic conditions afforded by the experimental wetlands areas.

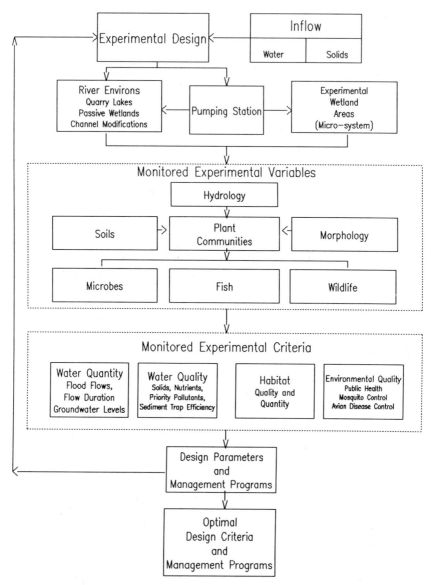

Figure 9.3 Macro-system research program for Des Plaines River Wetlands Demonstration Project.

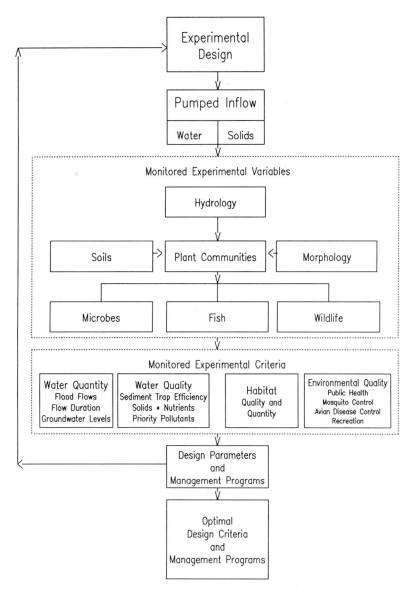

Figure 9.4 Micro-system research program for Des Plaines River Wetlands Demonstration Project.

The micro-system research will focus more specifically on the design of wetland areas concentrating on substrate, water depth, detention time, wetted surface, and microbial and plant communities.

A detailed list of the monitored experimental variables is given in Table 9.2. A companion list of the monitored experimental criteria is given in Table 9.3.

Some of the research questions that will be addressed at the macro-system level include the following:

1. What are the overall changes in water quality across and throughout the Des Plaines River site?
2. What changes affect the sediment load and suspended solids of the river?
3. To what extent and by what mechanism have flood flows been attenuated?
4. What are the changes in flow duration?
5. What effects have the site modifications had on the groundwater levels and quality?
6. What changes have occurred in the fish and wildlife populations?
7. What are the public health implications?
8. What are the impacts of the project on mosquito populations, and other factors related to human health?

At the micro-system research level, several sets of questions will be addressed.

Hydrology

1. What is the relationship between water detention time and the efficiency of nutrient and sediment removal?
2. What are the relationships among flow rates (total throughput), detention time, and the total amount of retained nutrients and sediments?
3. To what extent do localized, high velocities cause significant entrainment of nutrient and sediments?
4. What are the evapotranspiration rates relative to plant communities, depth, and flow rates?

Soils

1. What effects do soil clay, aluminum, and iron content have on retention of phosphorus?
2. What are the important factors in the loss of nitrogen through denitrification?

Table 9.2 Monitored Experimental Variables for Des Plaines River Wetlands Demonstration Project

Macro-System	Monitoring Frequency[a]	Micro-System	Monitoring Frequency[a]
Hydrology			
Stream flows	Continuous	Flow rates	Continuous
Morphology	Quarterly	Detention time	Calculated
Groundwater	Monthly	Velocity distribution	Quarterly
		Groundwater	Weekly
Soils			
Material outflows	Calculated	Texture	Quarterly
Accumulation/loss	Semi-annual	Redox potential	Daily
Sediment texture	Semi-annual	Vertical permeability	Annual
Sediment chemistry	Semi-annual	Organic content	Annual
		Iron and aluminum	Monthly
		Nutrients	Weekly
Morphology			
Cross-section characteristics	Semi-annual	Edge slope	Annual
Stream meanders	Semi-annual	Edge/area ratio	Annual
Quarry bathymetry	Annual	Channel development	Quarterly
Plant Communities			
Distribution	Seasonal	Biomass	Seasonal
Compositions	Seasonal	Nutrient analysis	Seasonal
Dominance	Seasonal	Composition and dominance	Seasonal
Changes	Seasonal	Productivity	Seasonal
		Stem density	Seasonal
		Root/rhizome biomass	Seasonal
		Decomposition rates	Seasonal
		Transpiration rates	Synoptic
Microbes and Invertebrates			
Species	Seasonal	Denitrification	Synoptic
Distribution	Seasonal	Plate counts (bacteria)	Seasonal
Abundance	Seasonal	Invertebrate counts	Seasonal
Dominance	Seasonal	Invertebrate diversity	Seasonal
		Mosquito larvae	Seasonal

Table 9.2 Monitored Experimental Variables for Des Plaines River Wetlands Demonstration Project (*continued*)

Macro-System	Monitoring Frequency[a]	Micro-System	Monitoring Frequency[a]
	Fish		
Species	Seasonal	Species	Seasonal
Distribution	Seasonal	Abundance	Seasonal
Abundance	Seasonal		
Dominance	Seasonal		
	Wildlife		
Species	Seasonal	Species	Seasonal
Distribution	Seasonal	Distribution	Seasonal
Abundance	Seasonal	Abundance	Seasonal
Dominance	Seasonal	Dominance	Seasonal

[a] "Seasonal" means monthly during the growing season.

3. How does the organic content of the sediment influence the nutrient retention?

Morphology

1. What is the relationship between wetted surface and water volume that optimizes sediment and nutrient retention?
2. How can short-circuiting channels be avoided?
3. What is the relationship between edge and water surface that optimizes nutrient removal?

Plant Communities

1. What is the role of vegetation in hydrologic budget?
2. What are the effects of hydrology, particularly flow rates, detention time, and water depth on plant community productivity?
3. How do different plant species affect local water velocities?
4. What is the importance of vegetation in the changes in water quality, particularly nutrients?
5. What plant species are most effective in reducing nutrients and sediments?

Microbes and Invertebrates

1. How do benthic invertebrates serve as indicator organisms of habitat value, particularly for waterfowl?

Table 9.3 Monitored Experimental Criteria for Des Plaines River Wetlands Demonstration Project

Macro-System	Monitoring System[a]	Micro-System	Monitoring Frequency[a]
Hydrology			
Longitudinal stage	Calculated	Flow duration	Calculated
Lateral stage	Calculated	Water budget	Calculated
Water Quality			
Sediment outflow	Quarterly	Suspended solids	Weekly
Suspended solids	Weekly	Dissolved solids	Weekly
Dissolved solids	Weekly	Nutrients	Weekly
Heavy metals	Weekly	Heavy metals	Weekly
Pesticides/Herbicides	Semi-annual	Pesticides/Herbicides	To be determined
Dissolved oxygen	Weekly	Dissolved oxygen	Daily
Chemical oxygen demand	Weekly	Chemical oxygen demand	Weekly
pH	Weekly	pH	Daily
Temperature	Continuous	Temperature	Continuous
Groundwater quality	Weekly	Groundwater quality	Weekly
		Constituent budgets	Calculated
		Coliforms	Monthly
Habitat			
Mammals	Annual	Mammals	Annual
Birds	Semi-annual	Birds	Semi-annual
Reptiles	Annual	Reptiles	Annual
Amphibians	Annual	Amphibians	Annual
Fish	Annual	Fish	Annual
Environmental Quality			
Disease vectors		Disease vectors	
Wildlife	Annual	Wildlife	Seasonal
Human	Annual	Human	Seasonal
Mosquitos	Seasonal	Mosquitos	Seasonal
Boating	Annual		
Trial activities	Quarterly		
Aesthetics	Semi-annual		

[a] "Seasonal" means monthly during the growing season.

2. What microbe communities and their concentrations optimize nutrient removal?

Fish and Wildlife

1. Which fish species survive under various hydrologic regimes?
2. What effects do the fish populations have on water quality?
3. How do mammals such as beaver and muskrat change the hydrologic variables?
4. What plant communities, densities, and specific species are most appropriate as habitat for various species of wildlife?
5. How is the value of created wetlands as habitat enhanced by vegetation type and hydrologic conditions?

9.3.2   Monitoring Facilities

Integral to research and ongoing management of created wetlands are the identification and placement of appropriate monitoring facilities. The demands of the project, however, have required procedures beyond those of traditional environmental engineering. Traditional measuring techniques can be used for water quality and the hydrologic measurements of flow, stage, velocity, and others. However, by the introduction of complex biological components, monitoring the wetlands for this demonstration project has required the development of creative and experimental techniques that will guide both the proposed research, and future design and management of wetlands.

Some of the questions that must be answered are: how will we know when a wetland has developed and is functioning as intended? How soon in the life of a created wetland can we predict the efficacy of that wetland? What must be measured to assure continued optimal functioning of created wetlands? What must be measured to indicate the need for management action? When and how often must these measurements be taken? The demonstration project has designed monitoring facilities and programs to begin to answer these questions; these are described below according to fields of knowledge. Generally, all data that will be continuously recorded will be transmitted by either radio transmitting devices or hard wired communications equipment directly to an on-site computer. Until this system is installed, data will be recorded on data analoggers, punched tapes, or circular inked charts.

Hydrology. Surface water monitoring stations have been installed at four locations throughout the site to record river stage on a continuous basis (Figure 9.2). Regular measurement of discharge is taken at each station to develop stage–discharge relationships. Stage monitoring will continue through the life of the project.

Table 9.4 Identified Toxins in the Project Watershed[a]

HERBICIDES	INSECTICIDES
Atrazine (P, L)	Chlordane (S)
Metribuzon (P, L, S)	DDT (S)
Dicamba (P, L)	Silvex (S)
Pendimethalin (P, L)	Dieldrin (S)
Trifluralin (P, L)	Lindane (S)
Glyphosate (P, L)	Diazinon (S)
Alachlor (S, L)	Fonofos (L)
Metalaclor (S, L)	Phorate (L)
Cyanazine (S, L)	Terbufos (L)
Butylate (L)	
METALS	PRIORITY POLLUTANTS
Lead (S)	Cyanides (S)
Zinc (S)	Phthalates (S)
Copper (S)	Asbestos (L)
Chromium (S)	Pentachlorophenol (S)
Arsenic (S)	Endosulfans (S)
Cadmium (S)	Phenol (S)
Nickel (S)	PCBs (S)
	Petroleum Byproducts (S)

[a] Reason for including constituent: P = persistence, L = loadings, S = sampling.

For each experimental wetland area, hydrologic monitoring will be similar to that of the overall site. Stage–storage relationships will be developed so that continuous monitoring of water depths within the wetland area can provide storage information. Continuous monitoring will be done with pressure-sensitive water depth meters. Each experimental wetland area will be equipped with inlet flowmeters and Parshall flumes with continuous flow measurement on all outlet structures.

Water Quality.   Both surface water quality and groundwater quality have been measured in the pre-start-up phase of the project. Samples are taken at each surface water monitoring site as described above. Groundwater samples are taken at selected monitoring wells throughout the site. Table 9.3 lists the surface water quality parameters that are sampled on a weekly basis. These measurements will continue through the length of research. A specific study to identify more exotic pollutants, such as those found in pesticides and herbicides, was initiated in the pre-start-up stage by scanning agricultural practices underway within the basin to help narrow the types of chemicals to be tested for. A preliminary list is included in Table 9.4. Studies will identify detailed analyses that will be carried out over one full year cycle.

This information will provide initial condition data. Periodic repeats of this study will be done both in the river and in the outflows from each experimental wetland area.

Geology. The most significant geological monitoring that was initiated in the pre-start-up phase and that will continue throughout the project is the fluvial geomorphological work. Twenty-eight cross-sections have been permanently established along the Des Plaines and Mill Creek from above the project site to more than 3 km downstream of the site. Each section will be measured for bank–full width, bank–full depth, and channel texture to determine lateral movement of the channels and changes in these parameters as the result of project construction and operation. In addition, erosion grids and pins will monitor the sediment deposition and scouring rates and patterns. Particle size distribution analyses are carried out on sediment samples to determine textural qualities of sediments.

Groundwater movement and level changes are monitored through several nests of monitoring wells transversing the site (Figure 9.2). Seasonal measurements are taken at all of these macro-research wells. Several additional groundwater monitoring well nests (36 wells, 4 wells in each nest) are installed in and near the experimental wetland areas. These wells will be monitored monthly for water level and water quality from pre-start-up through the life of the project.

Climatology. One complete weather station has been installed (Figure 9.2) as a base station on the upland area to collect climatic conditions of the general environment of the project site. A second station will be located in or near the experimental wetland areas to capture the localized micro-climatic conditions of the lowlands. The weather patterns in northeastern Illinois are inconsistent within short distances, requiring several precipitation gauges on the site. Other than precipitation, the climatic variables to be monitored include wind speed, wind direction, global radiation, direct radiation, diffuse radiation, air temperature, relative humidity, soil temperature, and soil moisture.

Vegetation. Research plots are proposed in four areas of the site to determine how best to establish diverse and stable native Illinois wetland plant communities. These plots will explore the effects of gradient, soil moisture, periods of inundation, and management on community establishment. In addition to these research plots, existing vegetational communities will be sampled by establishing transects through various community types, including oak–savanna, riparian marsh, fen, and forested wetland. Control communities will also be sampled to compare the difference between areas that are managed by regular burning and those that are left untouched. The sampling protocol includes the identification of plants and seedlings along defined transects on an annual basis.

Soils.   Soil characterization, including chemical and physical parameters in areas of vegetation research, will show analysis of the development of soils over time as vegetation is established and the impact of changing soil conditions on vegetation development. Soils are monitored through periodic sampling for chemical and physical parameters and soil type classification, and through in-place soil moisture and temperature probes.

Fish.   Baseline quantification of fish species and biomass was completed within the river channel and quarry lakes. Periodic sampling will be done to compare changes in species type and relative biomass. Other fish research has included the study of the effects of carp on turbidity in the river. In the summer seasons of 1986 and 1987 this fish research has shown correlation between carp activity and river turbidity. This study will be periodically repeated once the carp control program has been implemented.

Macroinvertebrates.   Baseline monitoring has included the quantification of aquatic macroinvertebrate communities. The methods for quantifying are experimental and are expected to be refined over the life of the project. Sediment sampling and measurements were employed in several sites in and near the project site. These areas included the Des Plaines River, Mill Creek, the Middle Quarry Lake (as yet undisturbed by construction), a nearby existing cattail marsh, and a second wetland area on the site. Species type and relative abundance were determined. Generally, on the project site, macroinvertebrate populations are low in abundance. Changes in macroinvertebrate populations could become a significant indicator of higher species development. Annual sampling is anticipated at the macro-research scale. The use of macroinvertebrate sampling in the experimental wetland areas may prove to be a valuable indicator of the health and effectiveness of the wetlands. Therefore, sampling will occur seasonally in these wetland areas.

Mosquitoes.   Because mosquitoes are commonly associated with wetlands mosquito populations will be carefully monitored. By identifying potential breeding areas and the type and frequency of mosquitoes appearing throughout each growing season, a potential health problem could be prevented. If populations were to become a problem, controls would be used. During the first two seasons of construction and research, no controls have been necessary.

Microorganisms.   Already identified as the key factors in water quality treatment, the monitoring of microorganisms is essential in the developing experimental wetland areas. Baseline studies have identified the type and quantity of microorganisms expected to be found in the area. Monthly sampling of natural sediment and hydrophyte substrates and the introduction of artificial substrates for determining colonization rates will be carried out in the experimental wetland areas. Periodic sampling in various aquatic environments throughout the project site will also be completed to show changes over time.

Migratory Waterfowl.   Accounting of migratory waterfowl species began in the first year of baseline data collection (1985). Annual surveys will systematically show changes in populations associated with changes in habitat.

Mammals.   Periodic sampling to identify mammal species type and quantity will be carried out for comparison with the baseline work already completed. Although mammal populations are not anticipated to be significant indicators on the efficacy of wetland development, changes in their presence show impacts of wetland development.

Public Use.   As a valued recreational area for local people, the use of the site for this purpose and others will be monitored through various mechanisms. Baseline work included on-site surveys of users and photographic documentation for use in preference surveys. Annual photographic documentation at permanently identified locations throughout the site will continue. A documentary film developed from annual (or more frequent) filming at key on-site locations will provide a historical and educational account of wetland development and functioning.

9.4   FUTURE PLANS AND CONCLUSIONS

The experimental wetlands (micro system) and the entire Des Plaines Wetland site (macro system) will be monitored and experimented with to determine the optimal ecological engineering design for riverine wetlands. Simulation models will be used to guide the research and "experiment" with the site and its wetlands. Ultimately we hope to demonstrate that riverine wetlands, with an optimal design to their hydrologic, chemical, and biotic conditions, will prove to be effective in the restoration of the floodplains and rivers of the American midwest.

REFERENCES

Bastian, R. B. and S. C. Reed. 1979. Aquaculture systems for wastewater treatment: Seminar proceedings and engineering assessment. U.S. EPA 430/9-80-006. Washington, DC.

Bell, A. L., E. D. Holcombe, and V. H. Hicks. 1974. Vegetating stream channels— a multipurpose approach. *Soil Conservation 40(5):*16–18.

Brooks, R. P., D. E. Samuel, and J. B. Hill, Eds. 1985. *Wetlands and Water Management on Mined Lands.* The Pennsylvania State University, University Park, PA.

Cairns, J., Jr. 1980. *The Recovery Process in Damaged Ecosystems*. Ann Arbor Science, Ann Arbor, MI.

Erwin, K. L. and G. R. Best. 1985. Marsh community development in a central Florida phosphate surface-mined reclaimed wetland. *Wetlands 5:*155–200.

Ewel, K. C., M. A. Harwell, J. R. Kelly, H. D. Grover, and B. L. Bedford. 1982. *Evaluation of the Use of Natural Ecosystems for Wastewater Treatment*. Ecosystems Research Center, Cornell University, Ithaca, NY.

Fowler, D. K. and D. A. Hammer. 1976. Techniques for establishing vegetation on reservoir inundation zones. *J. Soil Water Conservation 31:*116–118.

Garbisch, Jr., E. W. 1977. Dredged Material Research Program. Environmental Concern, Inc. St. Michaels, MD.

Hammer, D. E., and R. H. Kadlec. 1983. Design principles for wetland treatment systems. U.S. Environmental Protection Agency Report, EPA-600/2-83/026,Ada,OK, 244 pp.

Hey, D. L. 1987. *Constructing Wetlands for Stream-Water Quality Improvements*. American Water Resources Association, pp. 123–136.

Hey, D. L. and N. S. Philippi. 1985a. Baseline Survey, Vol. 2. Wetlands Research, Inc. Chicago, IL.

Hey, D. L. and N. S. Philippi. 1985b. Design, Construction Specifications and Site Management, Vol. 3. Wetlands Research, Inc. Chicago, IL.

Hey, D. L., J. M. Stockdale, D. Kropp, and G. Wilhelm. 1982. Creation of wetland habitats in northeastern Illinois. Illinois Department of Energy and Natural Resources, DOC. No. 82/09, Springfield, IL, 117 pp.

Holbrook, J. A. and B. R. Maynard. 1985. Wetlands and Water Management on Mined Lands. Needed Wetland Research: A Federal Perspective. U.S. Bureau of Mines. Pittsburgh, PA, 195 pp.

Josselyn, M. 1982. Wetland Restoration and Enhancement in California. Tiburon Center for Environmental Studies. Tiburon, CA.

Kleinman, R., and P. E. Erickson. 1982. Full-scale field trials of a bactericidal treatment to control acid mine drainage. In 1982 Symposium on Surface Mining, Hydrology, Sedimentology and Reclamation, Dec. 5–10, Lexington, KY, pp. 241–245.

Mitsch, W. J. and J. G. Gosselink. 1986. *Wetlands*. Van Nostrand Reinhold, New York, 539 pp.

Nawrot, J. R., and S. C. Yaich. 1982. Wetland development potential of coal mine tailing basins. *Wetlands 2:*179–180.

O'Neill, D. J. 1972. Alkali bulrush seed germination and culture. *J. Wildlife Management 36:*649–652.

Ruesch, K. J. 1983. A survey of wetland reclamation projects in the Florida phosphate industry. Florida Institute of Phosphate Research, Publ. No. 03-019-011, Bartow, FL, 39 pp.

Sather, J. H. and R. D. Smith. 1984. An overview of major wetland functions and values. Western Energy and Land Use Team. U.S. Fish and Wildlife Service, FWS/OBS-84/18, Washington, DC, 68 pp.

Sather, J. H. and R. D. Smith. 1985. Research Plan, Vol. 4. Wetlands Research, Inc. Chicago, IL.

10

ECOLOGICAL ENGINEERING FOR TREATMENT AND UTILIZATION OF WASTEWATER

Ma Shijun

Center of Eco-environmental Studies, Academia Sinica, Beijing, China

and

Yan Jingsong

Nanjing Institute of Geography and Limnology, Academia Sinica, Nanjing, China

10.1 INTRODUCTION

The volume of wastewater discharged increases day by day with growing populations, rising living standards, and the development of industrial and agricultural production. For example, in China, the daily discharge volume of wastewater from cities and towns is about 99.6 million m^3 at present and is forecast to attain 270 million m^3 in the year 2000. The contents of wastewater are highly diverse. In general, wastewater contains six kinds of pollutants, namely, suspended solids, organic matter, nutrients, water-soluble salts, heavy metals, and bacteria and viruses. Most of these are biodegradable, such as organic matter and nutrients, and a few of them are nondegradable, such as heavy metals and some manmade organic compounds. The main bulk of these pollutants are originally utilizable resources. To discard such wastewater with many utilizable resources in it not only pollutes the environment and destroys the steady state of an ecosystem when the inflow of pollutants exceeds the environmental carrying capacity or homeostatic limit in that ecosystem but also abandons many useful materials.

Water is a basic element for human life, but at present, the condition of many fresh bodies of water has become disappointing to humankind. For example, because pollution does great damage to water quality, the water becomes unavailable to serve for the necessities of life, industry, and agriculture; many species of organisms are reduced in quality and quantity; some of the resources of aquatic life are harmed or even extinguished; and water loses its scenic or recreational value as it becomes turbid, perhaps with an offensive odor.

10.2 METHODS AND PATHWAYS USED FOR TREATMENT AND CONTROL OF WASTEWATER

In general, there are four kinds of methods and pathways for treatment and control of wastewater.

10.2.1 Natural Self-Purification of Water Bodies

Untreated wastewater may be directly discharged near sources of pollution into natural flowing bodies of water such as streams, rivers, ponds, lakes, and seas. This pathway for sewage treatment mainly depends on the natural self-purification of water bodies. Although it is very convenient and is usually practiced in most regions, especially in developing countries, the actual result of implementation of this method is mostly negative, with respect to both environmental social efficiency and economic benefits.

Most of the streams, rivers, ponds, and lakes in the vicinity of cities or towns are regarded as unlimited sewers. This kind of pathway for treating wastewater is unaccommodating and should be improved.

10.2.2 Artificial Chemomechanical Regeneration System

In this system wastewater is held in a limited space and then a series of physical and biochemical treatment processes are used for purification. Such treatment processes usually consist of three steps:

Primary Treatment: Use of precipitation of suspended solids to reduce the turbidity, BOD, COD, and so on of the wastewater.

Secondary Treatment: Biochemical processes for oxidative or reductive degradation of biodegradable organic pollutants. These include anaerobic and facultative ponds and aerated lagoons.

Tertiary or Advanced Treatment: Biological or chemical processes to reduce or remove more phosphates, nitrites, organic matter, and so on.

Many natural bodies of water that receive wastewater may be protected from pollution by artificial chemomechanical regeneration systems. The operation of these systems results in higher efficiency of purification for wastewater with marked environmental efficiency as well. On the other hand, to establish and keep such systems in operation, large amounts of investment and energy are needed, so their economic value is much lower compared with other methods. Such systems are hard to set up, especially in underdeveloped regions, but they have been applied extensively in many developed countries.

10.2.3   Control of Pollution Sources

Local environmental protection policies consisting of standards for the discharge of wastewater and for concentrations of various pollutants are formulated according to the environmental carrying capacity of bodies of water. In order to conform to the standards of discharge of wastewater, in addition to employing a train of sewage treatment systems or techniques, it is important to improve the management, equipment, and techniques in the factories producing the wastewater in order to decrease its volume and concentration of pollutants. This pathway will result in not only increased production and actual economic benefit but also reduced pollution.

10.2.4   Ecological Engineering for the Conversion of Wastewater into Resources

A series of techniques based on fundamental ecological principles established in a local region to adjust the structure and function in polluted bodies of water to maintain the ecobalance is called *an ecological engineering system*. This system can be thought of as increasing the amount and quality of water purification by the multilayer and gradational utilization of originally abandoned resources in the wastewater. Such systems have been designed and are operating in both developing and developed countries for the conversion of wastewater into utilizable resources.

Among the different approaches to sewage treatment, ecological engineering is the one with the highest economic benefits, because its capital investment and consumption of materials for building and operation are less than for all other methods. Furthermore, a great deal of material wealth and many other economic benefits can be obtained by regeneration and retrieval from wastewater of many abandoned resources. Through the processes of concentration, transformation, and decomposition of pollutants in polluted bodies of water, water is purified into utilizable clean water for the necessities of life, industry, and agriculture and the difference between supply and demand of freshwater resources is mitigated. Thus economic, ecological, environmental, and social values all can be attained through the practice of ecological engineering systems.

10.3   BASIC PRINCIPLES OF ECOLOGICAL ENGINEERING FOR THE CONVERSION OF WASTEWATER INTO UTILIZABLE RESOURCES

Ecological engineering for the conversion of wastewater into utilizable resources is only one among many kinds of ecological engineering. Its establishment simulates the homeostatic mechanisms of ecosystems to adjust the structure and function of an ecosystem or co-ecosystem to an ecobalance and steady state. But because there are often some slight differences in the nature and main objectives of different types of ecological engineering, special emphasis is placed here on ecological principles that every kind of ecological engineering depends on. These important basic ecological principles are described in the following sections.

10.3.1   Law of Tolerance

The presence or abundance and functioning of an organism may be controlled or limited by a variety of essential environmental factors, whether biotic, chemical, physical, or a combination of these. For each species there is a range in an environmental factor within which the species functions at or near an optimum. There are extremes, both maximum and minimum, near which the functions of a species are inhibited then curtailed. The upper and lower limits are termed "range of tolerance." Species vary in their limits of tolerance to the same factor. The following terms are used to indicate the relative extent to which organisms can tolerate variations in environmental factors. The prefix *steno-* means that the species, population, or individual has a narrow range of tolerance and the prefix *eury-* indicates that it has a wide range. Thus "stenohaline" and "euryhaline" are used in respect to salinity, "stenooxybiotic" and "euryoxybiotic" in respect to dissolved oxygen, "stenothermal" and "eurythermal" in respect to temperature, and so on.

10.3.2   Coordination of Structures with Functions in the Ecosystem

As a result of long-term evolution and development of ecosystems, any homeostatic ecosystem in nature has internal structural organization with definite adjustment ability, causality multiplication, and closed loops. The components of an ecosystem or a sub-ecosystem of a co-ecosystem have established mutual interdependence, interconnection, and interinhibition. They are conditioned by a definite order of time and space, a given layer of structures, and a quantitative ratio of materials and energy transfers in the normal metabolic processes of the ecosystem. For example, substances are transferred and transformed between the environment and aquatic plants in a lake or a pond only through a series of physical, chemical, and biological processes and proper layers in structures. These processes include removal of the substance from the water at the interface between the organisms and water, then entrance into the organism across the cell membrane, assimilation and utilization within an organism by the reaction of enzyme systems, and finally excretion of the waste of dissimilation from the interface between water and organism. The amount of various elements absorbed by plants in phytosynthesis is always in a fixed ratio; for example, in phytoplankton the ratio of oxygen to carbon to nitrogen to phosphorus is about $140:41:7.2:1$.

For another example, based on multilayer trophic structures of the food chain, the transference, transformation, decomposition, concentration, and regeneration of substances occur according to a definite order in time and space within a certain structure, such as primary producer, primary consumer, secondary consumer, tertiary consumer, and decomposer. Moreover, these structures must have a relatively steady ratio in quantity so that the passage of energy or substances through trophic levels within a stable food chain conforms to the principle of energy dissipation, thus forming ecological pyramids of biomass and energy.

10.3.3 Harmonic Mechanism of Mutual Restriction and Compensation

Ecosystems are considered as mutual-loop control systems with adjustment. Their harmonic mechanism exists in the restriction, transference, and compensation of substances and energy in multilayer structures within an ecosystem. Many components within an ecosystem or a sub-ecosystem within a co-ecosystem in a region are mutually dependent. Just as with organisms or populations, ecosystems have the ability of self-maintenance and self-adjustment to resist outside stress or interference, especially from certain human activities, and have the ability to maintain the ups and downs or the steady state of the ecosystem. This is termed *homeostatic mechanism* or *elasticity*. But there is a limit to the elasticity, namely, a homeostatic plateau or elastic limit that indicates the maximum stress that can be reached in a structural material without causing permanent destruction of the original dynamic equilibrium in the ecosystem. In order to maintain the ups and downs or steady state and environmental self-purification, it is necessary to have certain important kinds of components in a suitable quantity ratio within an ecosystem or a co-ecosystem.

10.3.4 Harmonization of Mutual Adaptation Between Structures and Functions in an Ecosystem

In an ecosystem, the structures are the framework and channels for completing the functions, whereas the functions are the basis for maintaining the existence of structures and for transferring and transforming substances, energy, and information for structure development. The harmony between structures and functions is a prerequisite for achieving the steady state of an ecosystem.

Some human activities, such as discharging sewage and waste in excess of the elastic limit of a body of water, damage the harmony between environment and organism or humans, destroy or interfere with the original optimal ecobalance for this ecosystem, and cause environmental pollution.

When structures and functions harmonize well within an ecosystem, organisms not only establish many mutually conditioned relationships among them (as, for example, in food chains), but also have become specialized in their living habits, such as two species (1) occupying different ecological niches in the ecosystems and (2) making gradational utilization, according to their own particular needs, of the substances offered by the natural world. In the metabolic processes of a homeostatic ecosystem, waste and remnants from primary production by a species (or agent) A is used as the raw and processed materials of secondary production by another species (or agent) B in quantity and quality; the waste and remnants from the secondary production are used as the raw and processed material of tertiary production by the species (or agent) C, and so on. Production wastes are utilized according to a set of such patterns until all are fully utilized or until the waste and remnants from the final production cannot be utilized. Analyzed from economic and environment protection views, this *multilayer and gradational utilization* of substances, remnants, and waste in a homeostatic ecosystem is not only efficient in utilizing natural resources but also results in less or no waste; namely, it causes less pollution or is free of pollution.

It is desirable in ecological engineering for the conversion of wastewater into utilizable resources to apply the above principles of structures and functions, with their full use of waste and resulting high economic benefits for (1) adjusting the disharmonic structures and functions in a polluted body of water, and (2) establishing a semiartificial ecosystem to combine sewage treatment and full utilization of the wastewater. These principles should also be applied to agroindustrial production systems if these are to make full use of limited natural resources to speed up the development of human economy while protecting the ecological environment (Ma, 1983).

10.4   CATEGORIES OF ECOLOGICAL ENGINEERING FOR THE CONVERSION OF WASTEWATER INTO UTILIZABLE RESOURCES

A complex technological-process system of production modeled on the principles of ecosystems would be an example of ecological engineering (Ma, 1985, 1986). It consists of several physical or chemical technological processes interconnected with one another to constitute the whole system. According to the nature and the main objectives of engineering, ecological engineering for the conversion of wastewater into utilizable resources should fall roughly into one of the following four categories.

10.4.1   Pollution-free Technology, the Self-purifying and Substance-regeneration System of Environment in the Industrial City

Waste heat discharged from electric power plants, chemicals such as carbon dioxide, sulfur, and nitrogen released from the combustion of fossil fuel in various factories, and certain heavy metals in the waste liquor of process industries are common sources of environmental pollution. To recover and purify such wastes is a social responsibility that must be taken seriously in urban and industrial areas. Use has been made of waste heat from electric power plants and other factories as heat sources for houses in winter in many urban areas. If greenhouses requiring different temperature gradients are built near factories that discharge waste heat, their waste heat can be used, based on the heat-diffusing coefficient, to cultivate certain aquatic plants and fishes for overwintering, accelerating breeding, and increasing production. For example, in Suzhou and Yangzhou, China, the Nile tilapia (*Tilapia nilotica* L.), which cannot overwinter in nature in eastern China, is cultured in greenhouses built near electric power plants. The waste heat is used for overwintering, advancing maturity and breeding, and increasing production of this fish. In Shenyang, Northeast China, the waste heat from some factories is used to cultivate water hyacinths (*Eichhornia crassipes*), which can be incorporated into food for domestic animals, whose excrement is then applied to fields and gardens as manure. This utilization of waste heat will help to reduce the carbon dioxide that is released from the combustion of fossil fuels in the factory.

In Jianjian, China, wastewater that is discharged from a pharmaceutical factory after producing antibiotics contains a concentration of organic matter of more than 20,000 ppm. That wastewater is used as the raw material to produce yeast. This secondary production not only is productive and economical but also decreases environmental pollution.

The engineering systems formed by such pollution-free (or at least low-pollution) technologies are designated *pollution-free engineering* (Figure 10.1). According to some authorities, aquatic plants can absorb and concentrate trace heavy metals. Research and studies are being carried out by the authors in Wuxi, China, where the water hyacinth is being regarded as a living medium to recover silver from wastewater discharged from a film studio.

An important aspect of ecological engineering is that comprehensive pollution-free production in the factory can be attained by combining many kinds of specialized workshops or factories into a unified technological process or production line. This is based on simulating the principles of endless cycling of materials and incessant regeneration of organisms in the ecosystem.

10.4.2   Treatment and Utilization of Sewage by Soil Systems

The farmland, grassland, and forest ecosystems, all based on soil, not only can degrade and purify many kinds of pollutants in sewage, but also can

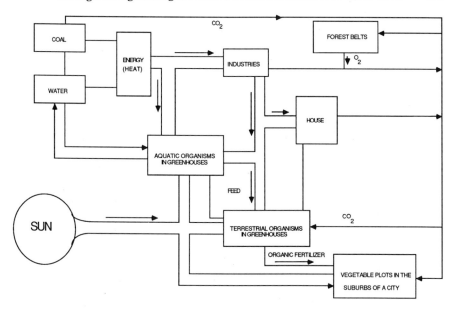

Figure 10.1 A model of the pollution–free technological-process self-purifying system in industrial cities (Ma, 1985).

produce many useful products by utilizing these pollutants. There are numerous species of microorganisms that can decompose many kinds of organic materials in soil. Inorganic salts, especially nutrient salts decomposed from organic materials and contained in sewage, then are absorbed by plants that grow in soil. This results in the production of many useful farm products, forage grasses, or forest products. The soil itself also seems to be a natural sieve, effectively filtering out most of the pollutants in sewage, namely, suspended solids, degradable organic pollutants, nutrient salts, and some pathogenic bacteria and viruses. The rate of removal may exceed 90% for BOD, total nitrogen, suspended solids, and colon bacilli, and nearly 50–80% for phosphorus.

In China, the area used for sewage irrigation is more than 1 million ha. The use of sewage for irrigation mitigates the scarcity of water resources, especially in dry or semiarid regions, and supplies needed water for the development of industry and agriculture. It also increases farm production and reduces fertilizer expenses by utilizing nitrogen, phosphorus, potassium, and some trace elements from the sewage, which are required by plants. The farm ecosystems in a sewage irrigational region play an important role in decreasing pollution in the river system in that region because they have the capacity of purification described above.

Some heavy metals and organic materials such as pesticides, which are degraded with difficulty by the soil system, become concentrated in the farm products and enter the food chain when their content in sewage is greater than the standard for surface water. In this case, pretreatment for sewage should be undertaken to avoid harm to humans and still keep the benefits of utilizing the wastewater. It is inadmissible for such sewage to be used to irrigate farms directly. Such pretreatments may be oxidation in waste stabilization (aerobic treatment) ponds or anaerobic ponds and sediment ponds. It is also acceptable for sediment in discharge channels to be used instead of that in sediment ponds to bring self-purification of the channel into full play. Sewage that contains a large quantity of nondegradable pollutants should be discharged directly to timber forests or reed marshes for irrigation to avoid polluting food.

A case study on ecological engineering of wastewater land treatment system may be selected from west Shenyang, where 0.432 million metric tons of effluent per day are applied to a meadow. The brown earth has a cation exchange capacity (CEC) of 17–23 meq/100 g in the topsoil and a slow rate of infiltration of 30–40 mm/day for the subsoil.

Based on intensive studies in the laboratory and field in 1982–1985 (Gao et al., 1985; 1986), a model for applying ecological engineering to wastewater land treatment systems has been suggested. The main features of this application are as follows:

1. Pollution source control is made a prerequisite so as to guarantee the applicability of ecological engineering. This is accomplished by means of the integrated parameter $TOC/BOD_5 = 0.8$ to restrict the nondegradable organic pollutant industrial discharge to the sewage system.
2. The remains of nondegradable organic pollutants are degraded by enhancing anaerobic and multistage facultative lagoons.
3. Modification of soil–plant systems with various ecological structures, including willow trees, energy sorghum high in sugar content, and rice, is scheduled, instead of monoculture, to bring municipal water resources into harmony with nature.
4. Based on studies of the assimilative capacity of soil–plant systems, the environmental protection strategy in terms of the total amount of control for individual pollutants is identified.
5. Using lateral percolation, purified effluent from the soil–plant system is used to develop fishery ponds for higher economic benefits.
6. Sludge accumulated on the bottom of fishery ponds can be recovered by the soil–plant system to maintain a favorable nutrient cycling.

Domestic and municipal wastewater in west Shenyang contains 14–28 mg N/L and 0.13–1.53 mg P/L. The purification efficiency of land treatment systems for nitrogen and phosphorus in wastewater and the uptake rates of these nutrients by crops show that crop yield increased 20–80% after irrigation, and the purification efficiency of land treatment system was 94.12% for nitrogen and 98.12% for phosphorus. The final effluent from the soil–

plant system delivered to the receiving body of water is expected to match the latest standard of rank 2–3 of surface water quality standards approved by the People's Republic of China.

Another case study on ecological engineering of wastewater land treatment systems is the sewage land treatment system for the Haolin River Mining Area of semi-arid grass region (Feng et al., 1984; Sun et al., 1986). The optimum plan put foward is to set up a multiple land treatment system of sewage reservoirs–forest–lawn. The guiding ideology establishing this system is that, on the ecological engineering principle, the sewage can become innocuous and a reused resource through the multi-functional metabolic processes of a multiple land treatment system.

10.4.3 Regional Multifunctional Self-purification of Sewage

There are many regulating processes coexisting with complex structures in natural ecosystems, such as concentration and diffusion, synthesis, and addition. Under normal conditions, natural ecosystems do not easily collapse because of excessive accumulation of some particular substances or substance, for even if a part of the original structure of the ecosystem is changed by excessive accumulation of substances, replacement by species adapting themselves to the new conditions would normally occur. This phenomenon is designated "self-purification."* The modeling of a technological system with complex functions should be regarded as an important approach to dealing with and preventing industrial pollution. Figure 10.2 illustrates the hypotheses of the application of ecosystem principles to environmental protection for the self-purification of sewage. Ecological engineering includes interlocking food chains, the flow of materials and energy in three directions, and the input and output of different kinds of materials and energy in various ways.

For instance, much attention has been devoted to the stocking of fish cultivated in biological sewage stabilization pond systems, and many studies have been proposed in recent years (e.g., Schroeder, 1975; Neil, 1976; Reid, 1976; Allen and Carpenter, 1977; West and Henderson, 1978). These studies mainly show that nutrient salts such as nitrogen, phosphorus, and potassium are retrieved from the sewage by fish, through the intermediate activities of bacteria, algae, and other plankton. Since the 1950s, in some regions of China, many obvious improvements in efficiency of purification of sewage and of recovery of utilizable bioresources have been made through fish culture with sewage from cities (Institute of Hydrobiology, Academia Sinica, 1965).

Since the first purification lake was built at the junction of the Ruhr and Lini Rivers in 1919, many purification lakes have been constructed in China,

---

* In this book this process is also referred to as "self-design" or "self-organization" (see Chapters 1 and 6).

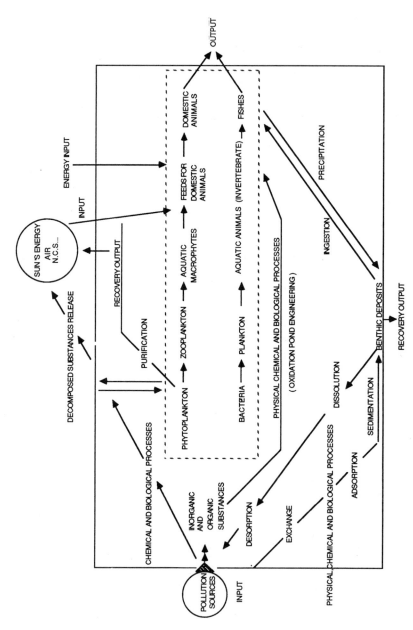

Figure 10.2 Scheme for the self-purification process of sewage (Ma, 1985).

Germany, Britain, and elsewhere. Such a lake is a small reservoir built for the treatment of wastewater to ensure good water quality of a river. In the process sedimentation and biodegradation fully develop the self-purification processes and improve the water quality. A purification lake is also one of the schemes of ecological engineering for regional, multifunctional self-purification of sewage. Through its operation, a local section of river pollution can be controlled. For example, in Qiqihar, northeastern China, pollution of the Nen River has been controlled by using storage and treatment of wastewater in a purification lake (Clayfield and Holloway, 1976; Woods et al., 1984; Imhoff, 1984).

In Yaer Lake, situated in Echeng County, Hubei Province, China, purification efficiency has been demonstrated on wastewater polluted by organic pesticides including parathion, malathion, dimethioate, and BHC from organophosphate pesticide manufacturing processes. Aquaculture utilization has also been demonstrated by applying sewage stabilization pond systems including anaerobic ponds, facultative ponds, aerobic ponds, and fish culture ponds. In order to mobilize fully the self-purification of a body of water for sewage treatment, the main principle is to utilize the coordination among four kinds of chief factors in the aquatic ecosystem, namely, producers (algae, aquatic vegetation), consumers (fish and other aquatic animals), decomposers (bacteria), and abiotic fctors (solar energy, water with other chemical components). The bacteria–algae symbiotic system plays an important role in this sewage stabilization pond system. Bacteria decompose and biodegrade organic materials, including organophosphates and chlorides, and purify the wastewater in a preliminary fashion. In this procedure, bacteria produce CO_2, NH_4^+, PO_4^{3-}, and so on, which are absorbed by algae and other aquatic vegetation and assimilated in their bodies under sunlight. The oxygen released by the photosynthesis of algae is supplied for the needs of oxidizing organic materials by bacteria. Then through transfer and transformation among various trophic levels, a great quantity of algae and other plankton is utilized in the procedure of purification by rearing fries. The energy fixed by algae appears finally in the form of fish harvesting, and the water is further purified. This is described in more detail in the case study. (See Section 10.5.)

10.4.4   Conversion of Wastewater into Utilizable Resources

This type of ecological engineering includes a set of techniques for treatment and utilization of wastewater based on a semiartificial ecosystem or combined production system, through the cycling and regeneration of substances including waste and pollutants. This applies the principles of endless cycling of materials and incessant regeneration of organisms in the ecosystem to establish a beneficial cycle. Examples include the multifunctional agroindustrial combined production system (Figure 10.3) and the multilayer and

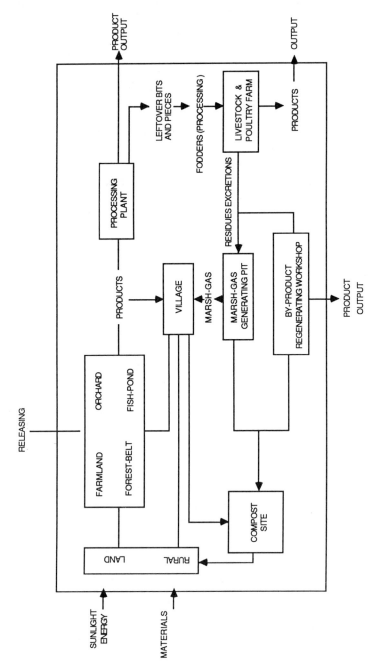

Figure 10.3 Model of the general structure of an agroindustrial combined production (Ma, 1985).

gradational utilization of wastewater with a water hyacinth–fish pond described in detail below.

10.5   CASE STUDY: THE CONVERSION OF WASTEWATER INTO UTILIZABLE RESOURCES USING WATER HYACINTHS

Wastewater treatment using water hyacinths (*Eichhornia crassipes*) has been applied in China, the Philippines, Burma, the United States, India, and Thailand. The studies on water hyacinth have been always on the level of autecology and population, including growth, production, reproduction, absorption of nutrients and metals, and decomposition of some organic pollutants (Gopal, 1986). Since 1982, using the above principles of ecological engineering, we have attempted to use these studies as a basis for focusing on the ecosystem level in our studies (Yan, 1986, 1987). Ecological engineering for the conversion of wastewater into utilizable resources using water hyacinth as the major agent—a semiartificial ecosystem—was established in 1984 and has been operating so far.

10.5.1   Site Description

The experimental site is about 2386 m long and 40–100 m wide in Fumentang, a section of stream in an eastern suburb of Suzhou (Figure 10.4). It receives a mixture of wastewater and surface runoff from this city. The annual discharge, which contains about 28–55 ppm of COD, 1.76–6.70 ppm of nitrogen, and 0.16–0.18 ppm of phosphorus, is about 10 million m^3/yr through the sector. The residence time of the discharged water from the city in the experimental region is about 8–32 hours. During May to December, 1984, approximately 2.7 ha of water hyacinths was planted, distributed mainly between measured sectors I–III in this section of stream and occupying about 30% of the total area of this semiartificial ecosystem. Four monitoring stations for sampling were set up between the inlet and outlet, namely, sections I–IV, with the distances from the inlet being 0, 627, 1100, and 2386 m, respectively. During the experimental period from April to December, the concentrations of chief pollutants and the species, density, and standing crop of phytoplankton, zooplankton, benthos, bacteria, and periphyton on the roots of water hyacinth were determined or measured each month at four sections. The current velocity and discharge were measured twice a day.

10.5.2   Main Principles and Mechanisms for This Type of Ecological Engineering

The major principle applied in this type of ecological engineering is that adjusting the structure and function of a semiartificial ecosystem results in

Table 10.1 The Number (10^6 Individuals/m^2) of Microinvertebrates Attached to the Roots of Water Hyacinth (I), Living in the Water Column Under the Water Hyacinth (II), and Living in the Water Column Without Water Hyacinth (III)

Species	September 1984			October 1984			November 1984		
	I	II	III	I	II	III	I	II	III
Vorticella microstoma	170	14	4.0	90	1.7	2.8	28	14	4.0
Vorticella convallaria	21	+	+	5.5	0.19	+	6.3	2.0	180
Epistylis urceolata	120	2.4	+	30	+	+	15	–	–
Epistylis articulata	21	–	–	5.5	–	–	1.3	–	–
Stentor multiformis	12	2.4	2.4	0.4	0.04	0.92	15	+	1.2
Stentor amethystinus	1.6	+	+	+	–	–	0.33	–	–
Hemiophrys fusidens	0.4	–	–	0.85	–	–	0.38	–	–
Litonotus fasciola	1.1	–	–	18	–	–	1.2	–	–
Thylakidium truncatum	1.8	–	–	2.5	–	–	–	–	–
Oxytricha fallax	0.25	0.4	–	+	–	–	–	–	–
Euplotes muscicola	2.0	–	–	0.25	–	–	0.65	–	–
Prorodon teres	0.8	–	–	3.5	–	–	–	–	–
Cyclidirem glaucoma	+	–	2.4	+	+	–	+	–	–
Aspidisca costata	+	–	–	+	–	–	+	–	–
Frontonia leucas	2.4	–	–	+	–	–	+	–	–
Actinophyrs sol	+	7.2	0.6	+	1.9	0.6	–	6.4	4.0
Actinosphaerium eichhorni	+	72	0.5	+	2.6	2.2	+	+	+
Halteria grandinella	–	64	14	–	2.2	2.8	–	4.8	5.2
Cyclidium glaucoma	–	24	8.4	+	0.4	2.0	–	+	+
Colpoda cucullus	+	–	–	2.8	–	–	2.2	–	–
Paramecium multimicronucleatum	+	–	–	5.8	0.04	0.3	4.5	1.2	1.7
Paramecium aurelia	+	–	–	–	–	–	–	+	–
Spirostomum minus	1.2	–	–	1.1	–	–	5.5	–	–
Nebela vitraea	0.5	+	–	0.8	+	–	+	+	–
Diffugia globulosa	0.5	–	–	–	–	–	–	–	–
Diffugia avellana	0.8	–	–	2.3	–	–	–	–	–

Pelomyxa Palustris	3.5	+	+	+	+	–	+	+	+
Peranenema trichophorum	1.0	–	–	+	–	–	–	–	–
Bodo edax	+	14	8.8	+	1.6	3.2	+	+	+
Brachionus falcatus	2.0	–	–	0.33	–	–	–	–	–
Brachionus capsuliflorus	+	+	–	+	2.6	0.056	–	–	–
Brachionus calyciflorus	1.0	+	+	1.1	+	0.18	–	–	–
Rotaria tardigrada	0.5	–	–	0.5	+	–	8.0	–	–
Mytilina ventralis	+	–	–	0.13	+	–	–	+	–
Colurella adriatica	5.5	7.2	2.4	20	+	–	3.0	–	–
Filinia longiseta	–	+	+	+	2.9	0.48	–	–	–
Monostyla unguitata	21	6.4	–	3.5	–	–	–	–	–
Rotaria neptunia	1.0	9.6	0.48	1.2	–	–	1.0	+	+
Conochilus unicornis	10	0.48	2.4	2	+	+	0.25	–	–
Philodina	3.0	–	–	–	–	–	–	–	–
erythrophthalm									
Keratella quadrata	–	2.4	2.4	–	0.08	0.25	–	–	–
Aeolosoma hemprichii	0.2	–	–	0.1	–	–	–	–	–
Oligochaeta sp.	–	–	–	0.19	–	–	0.016	–	–
Nematoda sp.	0.018	–	–	0.08	–	–	+	–	–
Chudors sphaericus	0.0073	–	–	0.003	–	–	–	–	–
Camptocercus rectirostris	0.0025	+	–	+	–	–	–	–	–
Thermocyclops taihokuensis	–	–	+	+	+	0.048	–	+	+
Total standing crop, g/m²	41.4 ± 0.8	16.0 ± 0.3	4.92 ± 0.23	20.1 ± 0.55	1.61 ± 0.1	1.58 ± 0.08	3.1 ± 0.23	2.84 ± 0.10	1.75 ± 0.10
Total number, 10⁷/m²	60.00	13.68	4.06	43.25	2.30	3.00	17.75	0.89	0.73

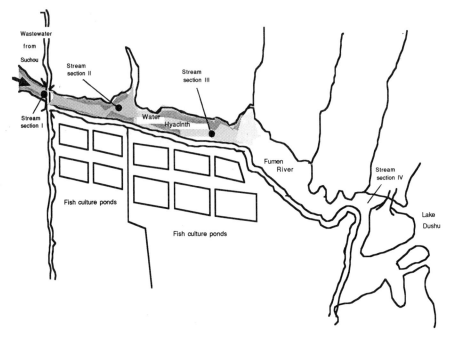

Figure 10.4 Site for water hyacinth wastewater treatment system in an eastern suburb of Suzhou, China.

a high quantity of pollutants being transformed into utilizable resources and transferred from wastewater to another place. Planting water hyacinths is a key ecotechnique in this semiartificial ecosystem.

Because a great quantity of water hyacinths was planted, their well-developed and abundant root system suspended in the subsurface layer of water provided enlarged matrices for periphyton. This resulted in greatly increasing the number of species or strains, density, and standing crop of bacteria and microinvertebrates. For example, our survey in September, October, and November 1984 showed that the number of strains, density, and standing crop of bacteria in the area with water hyacinths were 6.2, 20–30, and 30 times more than those in the area without water hyacinths, respectively. The number of microinvertebrate species, density, and biomass in the area with the floating vegetation were 2.3, 11.5, and 17.5 times those in the area without water hyacinths, respectively (Table 10.1). The number of bacteria reached 72 strains and number of invertebrates 50 species. Most of the additional species are the same species or genera as those in an activated sludge system of a sewage plant. Organic pollutants, including detritus, phenol, and petroleum, can be directly fed on by them, increasing the efficiency of purification of organic pollution. Moreover, the roots of the water hyacinth and most of the periphyton attached to the roots can secrete certain materials that promote the coagulation of suspended organic matter on the roots of this floating vegetation, resulting in decreased concentrations of organic pollutants.

10.5.3 Laboratory Experiments

The results of three laboratory experiments, which illustrate the mechanisms mentioned above, are described as follows. In the first experimental group, in which water hyacinths are planted in the sewage, concentrations of COD, total phosphorus, total nitrogen, and inorganic nitrogen and phosphorus more or less reduce with time (Figure 10.5a–c). In the second experimental group, in which some artificial matrices with some microorganisms transplanted from the roots of water hyacinth are put into the sewage, concentrations of COD, organic nitrogen and organic phosphorus reduce with time, but those of inorganic nitrogen and inorganic phosphorus increase with time. In a control group, the variations in concentrations of COD, total nitrogen, total phosphorus, and inorganic nitrogen and inorganic phosphorus are smaller than those in the first and second experimental groups. The results show that in the first experimental group, the concentrations of inorganic nitrogen and inorganic phosphorus increase at first owing to the decomposition of organic material by microorganisms attached to the roots of water hyacinth and living in the water, but then they are reduced owing to absorption by this vegetation. In the second experimental group, the concentrations of inorganic nitrogen and inorganic phosphorus increase owing to the decomposition of organic materials by the microorganisms attached to the artificial matrices and living in the water, but the increased inorganic nitrogen and phosphorus remain in the water owing to the absence of absorption by any organism. In the control group, the concentrations of COD and organic nitrogen and phosphorus vary less, owing to a deficiency of decomposers on the matrices; inorganic nitrogen and phosphorus also vary less owing to a lack of primary producers to absorb the inorganic nutrients.

10.5.4 Ecosystem Mass Balance

In the semiartificial ecosystem, the inorganic nitrogen and phosphorus brought in by the sewage and decomposed from organic pollutants by microorganisms are absorbed by the water hyacinth as a source of nutrients. According to our experiments regarding survival in the field, a growing period from May to November in Suzhou region, China, results in an average yield of 90–100 metric tons/ha (9000–10,000 g/m^2-yr). Such a production absorbs 1580 kg nitrogen, 358 kg phosphorus, and 198 kg sulfur per hectare from the water in which they grow (Zhang et al., 1986). This decreases the concentrations of nitrogen, phosphorus, and sulfur in the water and increases the efficiency of purification. Then water hyacinths with microorganisms and organic pollutants attached or coagulated on root surfaces are harvested

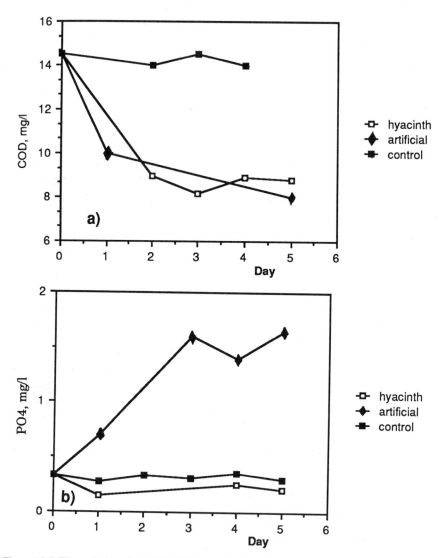

Figure 10.5 The variation of (*a*) COD, (*b*) inorganic phosphate, and (*c*) total inorganic nitrogen in laboratory experiments with water hyacinths, artificial matrices, and control.

from this semiartificial ecosystem to serve as feed in fish culture ponds, duck farms, pig farms, and oxen farms.

Our survey collected data on the experimental site from April to December, a period when water hyacinths grow luxuriantly in this semiartificial ecosystem (Figures 10.6, 10.7). From May to November the concentrations of COD, TN, NH_4^+, TP, PO_4^{-3}, organic nitrogen and phosphorus in the outlet are obviously less than that in the inlet. When water hyacinths were absent in the semiartificial ecosystem, from December to April of the second

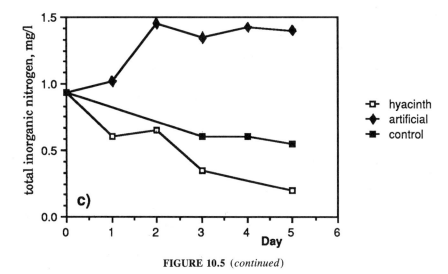

FIGURE 10.5 (*continued*)

year, the differences in concentration of COD, nitrogen, and phosphorus between the inlet and outlet were very small or even negative. It was calculated that a total of 507 tons of COD, 67.4 tons of nitrogen, and about 4 tons of phosphorus were removed, as well as 2500 tons of fresh weight of water hyacinth to use as feed and green fodder for fish, ducks, oxen, swine, and snails. The results show that the efficiency in this ecological engineering application is striking.

10.5.5   Economic Benefits

Based on the multilevel trophic structure and multilayer and gradational utilization of nitrogen, phosphorus, and organic materials, through the decomposition, concentration, regeneration, transformation, and transference of substances, especially pollutants, this semiartificial aquatic ecosystem can be combined with a fish culture pond, duck farm, pig farm, or oxen farm to establish a multifunctional combined production system. Thus the disadvantages of pollution are converted to advantages by transforming sewage into green fodder. This semiartificial ecosystem in combination with other production systems may be envisoned as a natural sewage treatment plant without electric power consumption. Not only are many environmental efficiencies obtained, but also some economic benefit and social efficiency are provided. For instance, in 1984 when this experiment had just begun, the output value of water hyacinth used as green fodder was 70,000 yuan (RMB) (approx. $18,000) whereas the cost for investment and management and operation is less than 10,000 yuan (RMB) (aprox. $2500). Because of the wastewater treatment by this ecological engineering method, the water of Dushu Lake, which receives mainly the mixture of wastewater and runoff

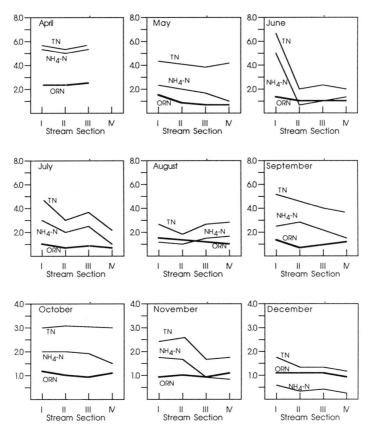

Figure 10.6 Changes in concentrations (mg N/L) of total nitrogen (TN), organic nitrogen (ORN), and ammonia-nitrogen (NH₄–N) in four sections of semiartificial water hyacinth ecosystem at Suzhou, China from April to December, 1984.

water from Suzhou city through our experimental river section, has been up to standards of water quality for cultivated fish, and the quality of fish cultured in this lake has improved. Meanwhile, the water treatment promoted the development and increasing production of fish cultures in ponds and lakes, as well as poultry and livestock on land.

10.5.6   Fish Culturing

The results of fish culture in an experimental pond of 0.3 ha with the use of water hyacinths as the only green fodder demonstrate its high food value (Zhang et al., 1986). The gross and net production of fish in this pond are 27.7 and 22.1 tons per hectare. The food requirement is about 11.0 kg of water hyacinth with 1.3 kg of fine feeds for every 1 kg of herbivorous fishes [e.g., grass carp (*Ctenopharyngodon idella*) and Wuchang fish (*Mega-*

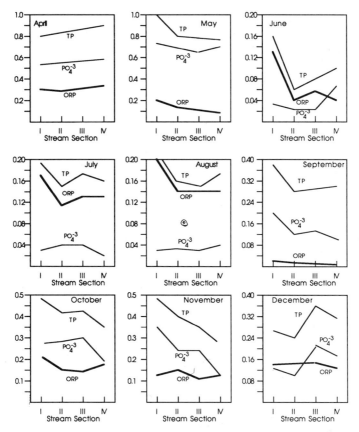

Figure 10.7 Changes in concentrations (mg P/L) of total phosphorus (TP), organic phosphorus (ORP), and inorganic phosphorus (PO_4^{3-}) in four sections of semiartificial water hyacinth ecosystem at Suzhou, China from April to December, 1984.

lobrama amblycephala], and after consumption of water hyacinth by the herbivorous fish for growing 1 kg of body weight the residue and excrement can promote an increase of 1 kg of production of other fishes such as silver carp (*Hypophthalmichthys molitrix*), big head carp (*Aristichthys nobilis*), common carp (*Cyprinus carpio*), tilapia (*Sarotherodon nilotica* ♂ × *Sarotherodon mossambica* ♀) and crucian carp (*Carassius auratus*). The annual net rate of increasing weight of fishes cultured in this experimental pond was 4.2 times.

10.5.7 Heavy Metal Uptake and Residuals

Many heavy metals are concentrated and accumulated easily in water hyacinths from very low concentrations in water. Heavy metal enrichment in this plant varies with its aquatic habitat. In the case of higher concentrations

Table 10.2 Heavy Metal Residual (ppm) and Enrichment Coefficient (times) of Dry Weight of Water Hyacinth Grown in Experimental Ecological Engineering Site in Suzhou, China, October, 1984

Metal	Roots	Leaves	Stalks	Whole Body
Cu				
ppm	81.3–172.0	15.1–2.84	17.2–34.4	29.8–70.2
times	10,072–24,002	1,598–3,966	1,762–3,804	3,441–9,804
Pb				
ppm	38.4–71.9	8.5–23.1	5.3–115.0	13.0–37.7
times	30,734–124,545	6,296–40,909	2,994–54,545	7,345–150,455
Zn				
ppm	1,856.0–2,294	246.0–338.0	308.0–464.0	421.0–975.0
times		6,345–14,259	8,951–14,259	10,767–40,046
Cd				
ppm	1.05–2.89	0.81–1.73	0.05–1.63	0.22–2.41
times	826–2,429	623–1,454	38–1,370	1,277–1,819
Hg				
ppm	0.3–0.64	0.11–0.30	0.03–0.16	0.17–0.25
times	15,600–22,875	5,000–10,345	1,154–6,667	6,358–12,500
Cr				
ppm	47.8–115.0	5.76–10.4	5.2–8.24	11.3–37.7
times	23,431–62,162	2,224–6,320	2,544–4,178	5,539–20,378
Ni				
ppm	33.8–63.9	3.3–6.2	5.4–6.3	8.2–22.2
times	4,599–11,350	449–917	136–1,036	1,116–3,757
Co				
ppm	15.0–28.8	2.0–8.0	1.1–4.8	3.6–11.0
times	1,659–3,678	329–875	122–526	398–1,631
As				
ppm	15.2–35.0	1.3–7.4	2.2–4.8	4.1–29.3
times	7,103–20,000	607–4,253	978–2,424	1,916–13,022

of certain metals in water, the water hyacinth has a higher residual amount than in waters with lower concentrations of these metals. Water hyacinths with high residual amounts of metals are not fit for use as fodder. However, the rivers in Suzhou and suburbs of other cities are mainly polluted by organic pollution and only slightly polluted by heavy metals. According to our surveys and measurements, the heavy metal concentrations are only several to tens of parts per billion. The heavy metal residual amount and enrichment of water hyacinth grown in our experimental semiartificial ecosystem are shown in Table 10.2. Its residual amount and enrichment in the roots are

Table 10.3 Average Residual Metals Amount (ppm) in Fresh Weight of Fishes and Ducks Fed with Water Hyacinth as Whole Body (1) or Without Roots (2) as Sole Green Fodder

Metal	Grass Carp		Wuchang Fish		Ducks (1)		
	(1)	(2)	(1)	(2)	Fresh	Liver	Egg
Cu	5.21	4.44	3.98	4.85	16.4	96.6	6.4
Pb	1.51	1.51	1.69	1.72	0.78	0.97	0.29
Zn	35.2	40.8	45.9	39.4	17.0	389.0	12.5
Cd	0.037	0.027	0.064	0.049	0.117	1.24	0.003
Hg	0.062	0.052	0.068	0.061	0.051	0.063	—
Cr	0.349	0.332	0.537	0.494	0.269	0.076	0.048
Ni	0.693	0.705	0.775	0.961	0.51	0.45	0.08
Co	0.228	0.211	0.323	0.323	0.47	0.32	0.03
As	0.18	0.18	0.142	0.222	—	—	—

more than those in the leaves and stalks (Yan, 1986, 1987; Day and Zhang, 1988).

Our experiments included measurement of residual metals in cultured fishes and ducks fed with water hyacinths from our semiartificial ecosystem, with and without roots, as the sole green fodder. The results of the measurement of residual metal amounts (ppm) in fresh weight of whole bodies of fishes, including grass carp and Wuchang fish, and the flesh, liver, and eggs of duck (*Anas platyrhynehos*) are listed in Table 10.3. The residual amount of most kinds of heavy metals in these fishes and ducks fed with water hyacinth without roots is always 2–10% less than that found in the same species and same body part when fed with whole water hyacinth. However, all of the residual amounts of heavy metals do not exceed the limitation of the criteria for food (Yan, 1986, 1987; Day and Zhang, 1988). Thus the water hyacinth is demonstrated to be a safe and valuable green fodder when it lives and grows in an aquatic environment only slightly polluted by heavy metals (with concentrations below 1000 ppb).

10.5.8   Key Ecotechniques

According to the results of our studies on the interrelations among main elements of this semiartificial ecosystem, six ecotechniques for increasing the purification efficiency are suggested.

1. *Ensure sufficient planting area of water hyacinths for the absorption of nutrients and for supplying matrices for sessile microorganisms.* The area may vary and depends on (1) the expected degree of purification of wastewater discharged into the semiartificial ecosystem, (2) the discharge volume and pollutants of the inflow, and (3) growth rates and production of water hyacinths including the rate of absorption of nutrients in the local conditions.

A model for estimating the planting area of water hyacinths in this ecological engineering application has been established as follows:

$$A_w = \frac{(C_3 - C_1)Q - (C_3 V + C_4 P_n)A}{C_2 V_1 + C_4 P_n - (BGE + BRA(MM_a + ZZ_a + C_5)} \qquad (10.1)$$

where A is the total area of water surface in the semiartificial ecosystem; A_w is the area required for planting water hyacinth to attain at least the expected level of purification; B is the expected average biomass of water hyacinth; C_1 is the concentration of a given pollutant in the discharged water at the inlet; C_2 is the concentration of a given pollutant in the water discharged at the output before establishment of this ecological engineering; C_3 is the expected concentration of a given pollutant in the output; C_4 is the concentration of organic substances produced by phytoplankton; C_5 is the coagulation and adsorption rate of organic pollutants by the water hyacinth; E is the absorption rate of nitrogen, phosphorus, or other pollutants by water hyacinths; G is the growth rate of water hyacinth; M is the average biomass of sessile microorganisms on the roots of water hyacinth; M_a is the degradation rate of organic substances by those microorganisms; P_n is the mean net production (mg/l) of phytoplankton; Q is the discharge per unit of time at the input; R is the proportion of roots to whole body of water hyacinth; V is the evapotranspiration in the local region; Z is average standing crop of periphyton and microinvertebrates attached to the roots of the water hyacinths; and Z_a is the degradation rate of organic pollutants by those sessile periphyton.

2. *Adjust the density of the water hyacinth population to maintain its rapid population growth rate and net production.* More nitrogen, phosphorus, sulfur, and other pollutants are absorbed and more organic pollutants are degraded with a more rapid growth rate and higher net production of water hyacinths. It is one of the important conditions for the use of water hyacinths in this ecological engineering application that the plant have a rapid growth rate and high production. However, these may vary with the density of the plant. For example, when the density approaches to the carrying capacity, its growth rate becomes lower. In order to maintain and promote a rapid growth rate and high production of the water hyacinth, its population density should be adjusted at the time of the change in its density during its growth period. This is in accordance with the equation

$$\frac{dW}{dt} = r - \frac{2rW}{K} \qquad (10.2)$$

which is derived from

$$\frac{dW}{dt} = \frac{rW(K - W)}{K} \qquad (10.3)$$

when the first-order derivative value equals zero. W is water hyacinth bio-mass; r is the intrinsic growth rate, and K is the carrying capacity. dW/dt has a maximal value near $\frac{1}{4}rK$ where $W = \frac{1}{2}K$. In other words, when the population density of the water hyacinth is nearly half of the carrying capacity, its net production is highest. The results of our experiments and field surveys show that the carrying capacity of the water hyacinth population is about 24 kg/m², with its highest net production occurring in a density of 10–15 kg/m² during July to September in our semiartificial ecosystem. It requires 7–10 days for water hyacinths to increase in density from 10 to more than 15 kg/m². During this period, the net production of this plant is about 0.5–0.8 kg/m² day.

3. *Harvest the water hyacinth on time.* It is necessary to adjust the plant density and transfer some nutrients absorbed by the water hyacinth from this semiartificial ecosystem to another place to avoid secondary pollution by saprophytism and decomposition. This requires that the plant harvest be carried out on time. Generally, a 7–10-day interval between harvests is proper for maintaining its density approaching $\frac{1}{2}K$, nearly half of the value of its carrying capacity, in summer. In other seasons a slightly longer interval is needed. The area of planting water hyacinth is divided into seven to ten subregions, and they are harvested in turn. One-third to one-half of the standing crop is harvested and removed from this semiartificial ecosystem every 7–10 days, one after another in rotation.

4. *Utilize the water hyacinth fully.* Full utilization of this floating vege-tation, the prerequisite for removing them, is another key for ecological engineering. One of the ways that water hyacinth is used is as feed for fishes, ducks, oxen, pigs, and snails. As mentioned above, this results in economic benefit and environmental efficiency.

5. *Add artificial matrices into this semiartificial ecosystem.* One of the purifying mechanisms for organic pollutants in this ecological engineering is the enlarged matrices for sessile microorganisms that can degrade the organic pollutants provided by the root system of the water hyacinth. The species of strains, density, and production of microorganisms increase and result in higher purification efficiency for organic pollutants. However, the more fertile the water is, the smaller the volume and surface of the roots. Generally, the roots of this floating vegetation are only about one-fourth or one-fifth of the whole body in relatively fertile water, such as most sewage. It is imperative to add some artificial matrices in the semiartificial ecosystem if the water is very fertile, in order to increase the number of sessile mi-croorganisms and to raise the purification efficiency for organic pollutant removal.

6. *Inoculate bacteria to accelerate biodegradation of special organic pol-lutants such as phenol and petroleum oil and its products.* The strains of bacteria that are able to grow on phenol as sole carbon sources were iden-tified as *Bacillus meraterium*, isolated from the roots of the water hyacinth.

Through our experiments with bacteria cultured in various media with different contents of phenol, results show that these strains of bacteria have a great capability for utilizing and biodegrading phenol. If water polluted by phenol is put into a semiartificial ecosystem, it is better to inoculate these strains of bacteria on the roots of water hyacinths or artificial matrices. In the same way, if the water is polluted by another special organic pollutant, such as petroleum oil or its products, organophosphates, or BHC, then another strain of bacteria that can rapidly biodegrade that special kind of pollutant should be inoculated.

10.5.9   Conclusion and Future Research Projects

The results of our first step of studies on the conversion of wastewater into utilizable resources by the use of water hyacinth as the major agent were crowned with success in both theory and practice. The economic benefit and ecological (environmental) and social efficiency are all obvious through practice. These arise from the harmonization of structures with functions in this semiartificial ecosystem after rearrangement by humans according to the main ecological principles. The interrelations among major elements in this semiartificial ecosystem are summarized in Figure 10.8.

This ecological engineering application has some advantages over common sewage treatment plants. They are chiefly (1) the obvious efficiency of purification of wastewater, (2) operation without electric power consumption, (3) lower investment for construction, (4) less subsidy of energy and materials, (5) higher benefit of output, through conversion of sewage into green fodders and fertilizer, namely, the conversion of waste into benefit. However, there are also some disadvantages. These mainly relate to the body of water and seasonality: (1) a certain area of water is needed for planting water hyacinth; (2) there is a high efficiency of purification only in warm seasons when the water hyacinth is growing, and there is no effect on purification in cold seasons where water hyacinth can not overwinter in nature in some regions; (3) if the water hyacinth accumulates high levels of residual heavy metals in a heavily polluted water body, it is not fit for use as feed for fish, livestock, and poultry. In order to overcome these disadvantages future research is being designed. For example, we need studies on artificial selection of other aquatic plants that can grow in colder weather and have a high decontamination efficiency and production of resources. Studies are also needed on the use of water hyacinths that have accumulated large amounts of heavy metals.

10.6   CASE STUDY: TREATMENT OF WASTEWATER FROM AN ORGANOPHOSPHATE PESTICIDE AND BHC MANUFACTURING PROCESS

Another type of ecological engineering was established in Yaer Lake, Echeng County, Hubei Province of China, for the treatment of wastewater

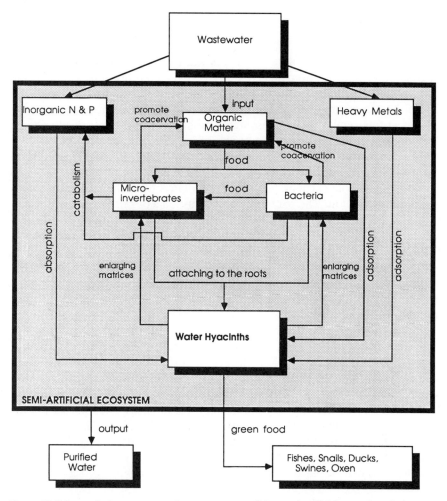

Figure 10.8 Interrelations among major components of the semiartificial water hyacinth wastewater treatment system.

polluted by some organic pesticides including parathion, malathion, dimethate, *para*-nitrophenol, diethylthiophosphate, and BHC from the pesticides manufacturing process. This semiartificial ecosystem consists of a series of five ponds. The first and second ponds have an area of 26.6 and 23.0 ha, respectively, and are mainly anaerobic ponds. The third pond of 36.6 ha is chiefly a facultative pond. The fourth, of 99.9 ha, is an aerobic pond. The fifth, with 213 ha, is a fry culture pond. The depth of this pond is about 3 m. The total volume of water is almost 5.6×10^6 m^3. The input discharge of wastewater from some chemical factories is about 54,000–70,000 tons/day. The total COD is 2.4–8.4 tons of oxygen per day.

10.6.1   Mechanism of Biodegradation of Organophosphate Pesticides and BHC

Among these organic pollutants in this semiartificial ecosystem, the biodegradability and physiological toxicity of parathion are much higher and more predictable than for the others, so that the metabolism, transport, and fate of parathion are used as the main representatives of those organic pollutants in this semiartificial ecosystem and have been most studied (Zheng, 1982; Zheng et al. 1982a, b; Tan et al. 1982a, b, c; Sun et al., 1982).

The biodegradation of organophosphate and BHC is closely related to bacteria. Two strains of bacteria that are able to grow on parathion and *para*-nitrophenol as sole carbon sources were isolated from the oxidation ponds in Yaer Lake region. The bacteria were identified as *Pseudomonas* sp. CTP-01 and CTP-02, respectively. Parathion was rapidly degraded by *Pseudomonas* sp. CTP-01 to produce diethylthiophosphate and *para*-nitrophenol, the latter being further metabolized by *P.* sp. CTP-02. The last becomes stable inorganic phosphorus and then it is absorbed by algae and other aquatic plants to be used as one of their nutrients. The oxygen needed for aerobic bacteria can be mainly provided by photosynthesis of algae and other aquatic plants. The biodegradation rate of parathion and *p*-nitrophenol by *Pseudomonas* sp. CTP-01 and CTP-02 is 6–11 g/L-day. It is much higher than that of alkaline hydrolysis at pH 10 at indoor temperatures. The enzymatic hydrolysis of parathion was also investigated (Zheng, 1982b). Cell-free enzyme preparations of *Pseudomonas* sp. CTP-01 hydrolyze parathion at a maximum rate of 1×10^4 nmoles/mg Prot.-min. The optimal temperature was 45–50°C. The activity was completely lost at 80°C. The optimal pH for the activity of cellfree enzyme preparation was 7.0–7.5. The copper ion (Cu^{2+}) causes activation of the enzyme. In the presence of 10^{-3} M Cu^{2+}, activity increased about 20 times (Zheng et al., 1982b). A study on the inducible synthesis of parathion hydrolase in *Pseudomonas* sp. CTP-01 (Tan et al., 1982a) showed that parathion hydrolase may rapidly hydrolyze the P–O bond of the parathion molecule. Experiments show that the parathion-induced formation of the enzyme in *Pseudomonas* sp. CTP-01 is responsible for parathion degradation. Methyl parathion and *para*-nitrophenol also serve as inducers of the enzyme. Response of stationary phase cells to substrate induction is much faster than the response of cells at the exponential phase of growth. For *Pseudomonas* sp. CTP-02, the optimal temperature of *para*-nitrophenol degradation is 35°C and optimal pH is 7.5. The rate of *para*-nitrophenol degradation can be expressed as

$$\frac{dc}{dt} = k_1 t - k_2 \tag{10.4}$$

where k_1 and k_2 are degradation constants and t is instantaneous time. When the cultures of *Pseudomonas* sp. CTP-02 were supplied with *para*-nitro-

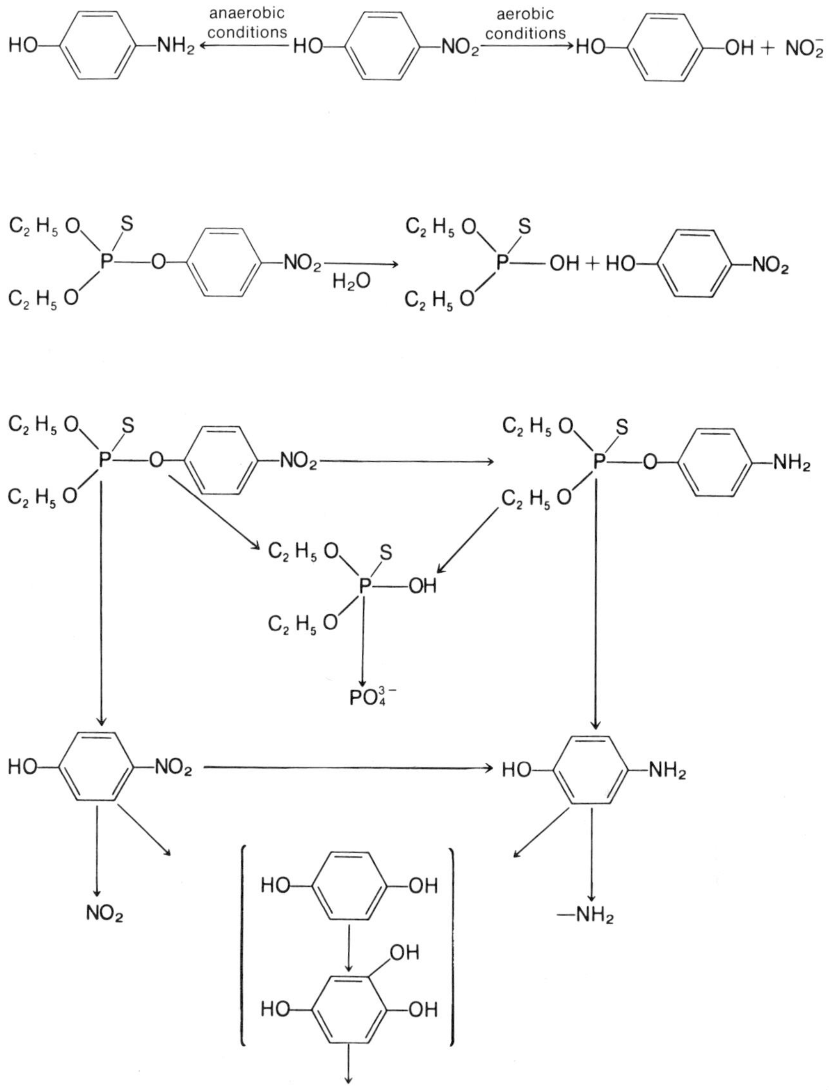

Figure 10.9 Parathion used as a carbon source of bacteria by splitting decomposition of ortho-aromatic hydrocarbon.

phenol, stoichiometric quantities of nitrite were released. The results show that the aromatic nitro group is detached before the ring fission. The fate of parathion in the algae–bacteria system may be synthesized as in Figure 10.9.

The effect of parathion and its degradation products on photosynthesis of *Scenedesmus obliquus*, which is a dominant species of algae in this semiartificial ecosystem, was investigated by Tan et al. (1982c). The toxicity of *para*-nitrophenol is much stronger than that of the sodium salts of nitrophenol, diethylthiophosphate, and parathion; their concentrations that inhibit photosynthesis of *Scenedesmus obliquus* by 50% are 16, 47, and 100 mg/L, respectively. This means that the algae probably have high limits of tolerance to these toxicants and can bear sudden changes in the wastewater treatment procedure. Studies of a simulated algae–bacteria system of *Pseudomonas* sp. CTP-02 and *Scenedesmus obliguus* using *para*-nitrophenol as substrate indicate that the oxygen needed for aerobic bacteria can be provided by the photosynthesis of algae.

REFERENCES

Allen, G. H. and R. L. Carpenter. 1977. The cultivation of fish with emphasis on salmonids in municipal wastewater lagoons.

Clayfield, G. W. and D. M. Holloway. 1976. Experimental purification lake on the River Thames. *Water Pollution Control 75*(3):341.

Day, Quan and Zhang, Yushu. 1988. Absorption of heavy metals by the second accumulation of fishes after fed on water hyacinth. *Jour. of Fisheries of China.* *12*(2):135:144.

Feng Zong-wei and Sun Tiehong. 1984. Planning selected of sewage treatments in mining area of Huolin River Region. Pamphlet, Shenyang, China. 46 pp.

Gao, Zhengmin, Qi, E. S., Zhang, F. Z., Li, P. J., Zhao, Z. S., Dai, T. S., Hang, S. H., Qu, X. Y. 1985. The purifying function of soil-plant system and its role in practice for ecological engineering. Pamphlet, Shenyang, China. 16 pp.

Gao, Zhengmin, Qi, E. S., Zhang, F. Z., Li, P. J., Zhao, Z. S., Dai, T. S., Han, S. H., Qu, X. Y. 1986. Studies on ecological engineering land treatment system for wastewater from West Shenyang Area. in Gao, Zhengmin, ed. *Studies on Pollution Ecology of Soil–Plant Systems*; China Science and Technology Press, Beijing, China, pp. 124–143.

Gopal, B. 1986. *Waterhyacinth*. Elsevier, Amsterdam.

Imhoff, K. R. 1984. The design and operation of the purification lake in the Ruhr Valley. *Water Pollution Control 83*(2):243.

Institute of Hydrobiology, Academia Sinica. 1965. The research report of cultivation of fry in domestic sewage. Symposium of Sixth Congress of the Committee of West Region Fishery Research. Sci. House, Beijing.

Ma Shijun. 1983. *Ecological Engineering*. Beijing Agricultural Sciences No. 4, pp. 1–2, Agr. Publ. House. Beijing, China.

Ma Shijun. 1985. Ecological engineering: application of ecosystems principles. *Environ. Conservation. 12*(4):331–335.

Ma Shijun and Pong Tienjie. 1986. The progression of rural industry and utilization of its wastes. Proceedings Symp. Econ. Devel. Rural Indus. Environ. Protection, Environmental Sciences Publ. House. Beijing, China. 18 pp.

Neil, J. H. 1976. The harvest of biological production as means of improving effluents from sewage lagoons. Research Rpt. 38. Canadian Agreement Great Lake Water Quality (Can.).

Reid, G. V. 1976. Algae removal by fish production. Water Resources Symposium No. 9, Ponds as a wastewater treatment alternatives, Center for Research in Water Resources, University of Texas, Austin, 417 pp.

Schroeder, G. L. 1975. Productivity of sewage fertilized fish pond. *Water Res. (G.B.)9:*269.

Sun, M. J., Zhang Yongyuan, Tan Yuyun, and Zhang Jinjun. 1982. Transport and fate of BHC in aquatic environment. 3. BHC accumulation in *Scenedesmus obliquus* and *Monia*. *Acta Hydrobiol. Sinica.* 7(4):527–531.

Sun, Tiehang, Jiang, F. Q., Chang, S. J., Dai, T. S., Zhang, F. Z., Liu, J. S., Yang, C. F., Hang, S. H., and Liu, H. L. 1986. A feasible analysis on sewage land treatment system for the Huolin River Mining Area of semi-arid grass region. In Gao, Zhengmin ed, *Studies on Pollution Ecology of Soil–Plant Systems*, China Science and Technology Press, Beijing, China. pp. 92–105.

Tan, Y. Y., Zhang Yongyuan, Sun Meijuan and Zhang Jinjun. 1982a. The dynamics of *p*-nitrophenol degradation by *Pseudomonas* sp. CTP-02. *Acta Hydrobiol. Sinica* 7(4):507–512.

Tan, Y. Y., Zhang Yongyuan and Sun Meijuan. 1982b. Effect of parathion and its degradation products on photosynthesis of *Scenedesmus obliquus* and p-nitrophenol degradation in algae–bacteria systems. *Acta Hydrobiol. Sinica.* 7(4):513–519.

Tan, Y. Y., Zhang Yongyuan, Sun Meijuan and Zhang Jinjun. 1982c. The inducible synthesis character of parathion hydrolase in *Pseudomonas* sp. CTP-01. *Acta Hydrobiol. Sinica* 7(4):521–526.

West, E. H. and B. Henderson. 1978. Feed fish effluent and reel in saving. *Water Wastes Eng.* 15(6):38.

Woods, D. R., M. B. Green and R. C. Parish, 1984. Lea-Marton purification lake: operational and river quality aspects. *Water Pollution Control 83*(2):226.

Yan, Jingsong. 1986. The main principles and types of ecological engineering for the conversion of waste water into resources. *Rural Eco-environment 8:*40–44.

Yan, Jingsong. 1987. Studies on the ecological engineering for the conversion of wastewater into utilizable resources. *Bull. of Nanjing Inst. of Geography, Acad. Sinica. 1:*12–14.

Zhang, Yushu, Day, Q. Y., Li, X. Q., and Zhang, X. G. 1986. Studies on the water hyacinth used as fodder for cultured fishes. *Jiangshu Ecology.* 2:129–136.

Zheng, Y. Y. 1982. BHC accumulation and elimination in fish. *Acta Hydrobiol. Sinica* 7(4):533–538.

Zheng, Y. Y., D. Zhuang, M. J. Sun, Y. Y. Tan, Q. Z. Zhang, and J. Q. Li. 1982. Static and dynamic simulation tests of oxidation pond for treatment of waste water from organophosphate pesticide manufacturing process. *Acta Hydrobiol. Sinica* 7(4):488–498.

Zheng, Y. Y., D. Zhuang, M. J. Sun, Y. Y. Tan, Q. Z. Zhang, and J. Q. Li. 1982a. Mechanism of biodegradation of organophosphate pesticides in aquatic ecosystem: 1. Enzymatic hydrolysis of parathion. *Acta Hydrobiol. Sinica* 7(4):499–506.

11

FUNCTION OF A LAGOON IN NUTRIENT REMOVAL IN LAKE BIWA, JAPAN

Akira Kurata

Lake Biwa Research Institute, Otsu, Shiga Prefecture, Japan

and

Masaru Satouchi

Shiga Prefectural Junior College, Hikone, Shiga Prefecture, Japan

11.1 INTRODUCTION

Lake Biwa is the largest freshwater lake in Japan and is located in the center of the main island. Recently eutrophication has proceeded extensively in the lake. Since the first occurrence in 1977, red tides by the flagellate *Uroglena americana* have occurred annually from April to June in almost the entire lake and have caused serious trouble for people receiving their water supply from this lake. In order to reduce problems caused by eutrophication, various ordinances have been enacted and many improvements have been made by responsible authorities. However, eutrophication still progresses slowly but steadily by the increase of nutrient loading from a number of sources, especially non-point sources such as domestic wastewater and cultivated field runoff. Nowadays, therefore, removal of nutrients from the water inflowing to the lake is one of the most important approaches to solving the environmental problems of the lake.

There are many large and small natural lagoons around the lake. These lagoons play a significant role in removal of nutrients loading to the lake, as reported in an earlier paper (Kurata, 1983). The self-purificative function of natural lagoons and wetlands adjacent to lakes and reservoirs must be considered and utilized further for wastewater treatment and discharge from cultivated fields.

In this chapter, an attempt is made to clarify quantitatively the function of nutrient removal in the ecosystem of natural lagoon adjacent to Lake Biwa.

11.2   SITE DESCRIPTION

Some 30 large and small natural lagoons are distributed around Lake Biwa. Most of them are connected to the lake through narrow channels and have the same water level as the lake. The study area is in Nishinoko Lagoon, which is one of the largest lagoons, though shallow (maximum 2.0 m depth). Surrounded by paddy fields, the surface area is approximately 2.9 km^2 and the shoreline reaches 18.45 km. In this lagoon, *Phragmites communis* grows well with very high densities, covering about 167 ha. The total catchment area of the lagoon is approximately 83.9 km^2, and 77.3% of it is flatland. Although about a dozen rivers flow into this lagoon, the catchment area of Hebisuna River, which is the main inflowing river, occupies 72.8% of the total catchment area. Chomeiji River is the main, almost the only, outflowing river from the lagoon.

Water sampling stations were set up as a rule at the mouths of 15 inflowing rivers or channels to the lagoon. Locations of the stations are shown in Figure 11.1. Because of weak currents or low flow, the water of the lagoon sometimes reversed its flow at Stations A2, B3, and C2. Generally, water current was very weak and the water stagnated at the reed zone of the lagoon. A description of each sampling station is shown in Table 11.1. Among the inflowing water at Stations D1, D2, D3, and D4, the water volume at D2 in the river Hebisuna dominates the inflowing water and is equivalent to 80% of the total inflowing water volume of these four rivers.

Collection of water samples and the measurement of water current at each station were carried out every week for an entire year, from April 12, 1983 to April 11, 1984.

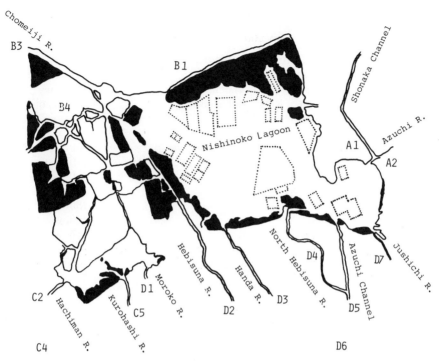

Figure 11.1 Location of sampling stations in Nishinoko Lagoon adjacent to Lake Biwa, Shiga Prefecture, Japan. Solid areas represent *Phragmites communis* zones. Dotted lines represent freshwater pearl cultivation farms.

11.3 RESULTS AND DISCUSSION

11.3.1 Water and Nutrient Budgets

The water of this lagoon is supplied by a dozen or more inflowing rivers, precipitation, and groundwater. During the investigation period, the inflowing water volume and precipitation in the catchment area of each river were measured. Data on the groundwater volume in the lagoon were obtained from the municipal bureau of Ohmihachiman city. Approximately 60% of the total inflowing water volume is supplied by the Hebisuna River. Discharge of the River and precipitation in its catchment area are shown in Figure 11.2. The inflowing water volume of each river was closely correlated to precipitation in its catchment area. If there was rain in the catchment area of the lagoon, the water level of the lagoon responded within a very short time. The total inflowing water volume throughout the year was estimated at 140×10^6 m^3.

Water samples were collected at each station and assayed for many limnological parameters in order to define the nutrient load from the catchment

Table 11.1 Description of Sampling Stations on Nishinoko Lagoon, Lake Biwa, Japan

Abbreviation	Location and Designation	Water Current
A1	Runoff from reclaimed field	Inflow
A2	Azuchi River	Inflow, outflow
B1	Pumping up to reclaimed field	Outflow
B3	Chomeiji River. Hakuo Bridge	Inflow, outflow
B4	Maruyama Bridge. Flow into Chomeiji River	Outflow
C2	Hachiman River. Honen Bridge	Inflow, outflow
C4	Channel for irrigation from April to August	Inflow
C5	Kurohashi River	Inflow
D1	Moroko River	Inflow
D2	Hebisuna River	Inflow
D3	Handa River	Inflow
D4	Channel for irrigation	Inflow
D5	Channel for irrigation from April to August	Outflow
D6	Northern stream of Hebisuna River	Inflow
D7	Jushichi River	Inflow

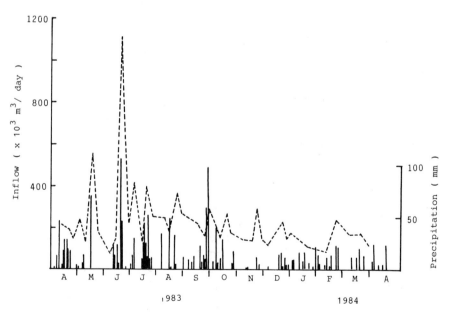

Figure 11.2 Discharge of Hebisuna River (– – – –) and precipitation (——) in its catchment area during study.

Table 11.2 Water Quality of Rivers Flowing into and out of Nishinoko Lagoon (Annual Average in mg/L)[a]

Station	COD	Total Nitrogen	Total Phosphorus
Inflow			
A1	4.97	1.99	0.132
A2	4.45	2.35	0.143
B3	4.29	1.41	0.037
C2	4.70	2.02	0.204
C4	4.45	2.80	0.162
C5	3.49	2.36	0.116
D2	2.79	1.78	0.090
D6	3.57	1.75	0.098
D7	4.73	2.11	0.149
Average	3.37	1.95	0.104
Outflow			
A2	5.66	2.04	0.124
B1	4.03	1.20	0.055
B3	3.17	1.31	0.054
C2	4.56	1.95	0.136
D5	4.98	1.49	0.081
Average	3.70	1.41	0.060

[a] The values are weighted by water volume (annual loading substances/annual water volume). Station numbers refer to Figure 11.1.

area to the lagoon. Annual average values of COD, total nitrogen, and phosphorus of the water of each inflowing and outflowing river are shown in Table 11.2. In this table, the values are represented as the value weighted by water volume. The inflowing water volume is dominated by the Hebisuna River as described above. The water quality of the Hebisuna River at Station D2 had the lowest levels of most parameters compared with those of the other stations. Usually the total nitrogen and phosphorus concentrations of outflowing rivers, especially at Station B3, which is located in the main and almost only outlet of this lagoon, were low compared with those of inflowing rivers. The annual change in total nitrogen and phosphorus concentrations is shown more clearly in Figure 11.3, comparing the main inflowing and outflowing waters of the lagoon. Except for a few cases, the outflowing water had lower levels than the inflowing water in both total nitrogen and total phosphorus concentrations throughout the year.

To clarify the function of nutrient removal from the inflowing water by the lagoon for the water body of Lake Biwa, diurnal and annual change of

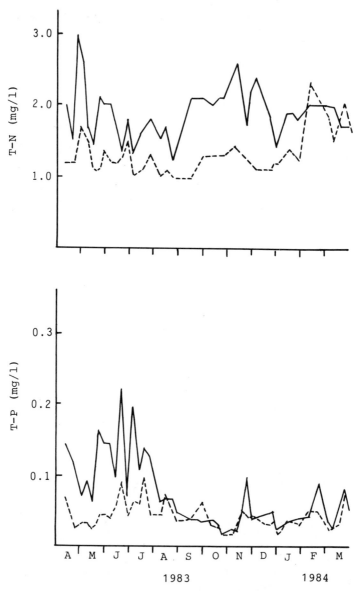

Figure 11.3 Annual changes of (top) total nitrogen and (bottom) total phosphorus concentrations in the Hebisuna River (——) and the Chomeiji River (– – – –) during the study period. The Hebisuna River is the major inflow and the Chomeiji River is the major outflow of Nishinoko Lagoon.

Table 11.3 Diurnal Change of Nutrient Balance at Each Station on Nishinoko Lagoon

Station	Volume ($\times 10^3$ m³) Inflow	Outflow	COD (kg) Inflow	Outflow	Total N (kg) Inflow	Outflow	Total P (kg) Inflow	Outflow
A1	40.2		199.9		80.1		5.32	
A2	20.6	15.6	91.6	89.2	48.3	32.1	2.94	1.96
B1		2.4		9.1		2.9		0.13
B3	10.5	262.2	45.0	792.1	14.8	348.6	0.39	13.54
C2	5.1	19.5	23.7	89.1	10.2	38.0	1.33	2.66
C4	5.7		25.5		16.0		0.93	
C5	72.4		249.0		176.4		8.23	
D2	196.8		503.6		363.5		15.31	
D5		84.0		417.8		125.3		6.80
D6	30.4		102.2		54.9		2.44	
D7	2.0		9.7		4.3		0.30	
Total	383.7	383.7	1,250.2	1,397.3	768.5	546.9	37.19	25.09

nutrient balance was calculated according to the measured water volume of inflowing and outflowing rivers. The results obtained are shown in Tables 11.3 and 11.4. In comparison with the inflowing and outflowing waters in this lagoon, the total COD value was 11% higher in outflow than in inflow. It is thought that the increased COD value is attributable to production within the lagoon because of such a high energy supply in the shallow water body compared with the lake and because of the nutrient-rich conditions in the lagoon surrounded almost completely by paddy fields. However, both the total nitrogen and phosphorus contents were low in outflowing water compared with inflowing water. This means that nutrients loaded to the lagoon were reduced by active uptake by different kinds of organisms, such as phytoplankton, benthic macrophytes, and reeds, in the course of passing

Table 11.4 Annual Nutrient Balance (April 1983–March 1984) at Each Station on Nishinoko Lagoon

Station	Volume ($\times 10^3$ m³) Inflow	Outflow	COD (kg) Inflow	Outflow	Total Nitrogen (kg) Inflow	Outflow	Total Phosphorus(kg) Inflow	Outflow
A1	14,729		73,177		29,328		1,946.0	
A2	7,533	5,765	33,515	32,636	17,685	11,757	1,076.1	713.7
B1		893		3,341		1,075		48.7
B3	3,836	95,971	16,459	289,900	5,415	127,586	142.7	4,956.2
C2	1,851	7,142	8,692	32,599	3,732	13,901	488.2	974.0
C4	2,097		9,334		5,872		340.0	
C5	26,516		91,145		64,565		3,013.3	
D2	72,062		184,318		133,026		5,602.1	
D5		30,731		152,915		45,858		2,490.5
D6	11,130		37,406		20,104		891.9	
D7	748		3,540		1,583		111.3	
Total	140,502	140,502	457,586	511,391	281,310	200,177	13,611.6	9,183.1

Table 11.5 Nutrient Loading from the Catchment Area of Rivers (kg/day)

	Catchment Area of Hebisuna R.		Catchment Area of Kurohashi R.		Catchment Area of Other Rivers		Total	
	N	P	N	P	N	P	N	P
Natural	9.9	0.35	1.1	0.04	1.0	0.04	12.0	0.43
Agricultural	117.1	16.26	15.5	1.68	29.5	2.83	162.1	20.77
Industrial	75.0	5.00	22.6	1.51	1.8	0.13	99.4	6.64
Domestic	100.3	24.08	49.6	11.75	26.5	5.18	176.4	41.01
Estimated loading	302.3	45.69	88.8	14.98	58.8	8.18	449.9	68.85
Inflow loading	419.5	17.79	193.0	9.19	133.1	8.28	768.6	37.19

through the lagoon. The reductions of nitrogen and phosphorus were 221.6 and 12.1 kg/day, respectively. Accordingly, self-purification rates of nitrogen and phosphorus in this lagoon were 28.8 and 32.5%, respectively. As shown in Table 11.4, $140,502 \times 10^3$ m^3 of water flowed into the lagoon and the same amount of water flowed out from the lagoon in a year, from April 12, 1983 to April 11, 1984. The total nitrogen and phosphorus loaded to the lagoon in a year were 281 and 13.6 metric tons, respectively. However, 81.1 metric tons of the former and 4.4 metric tons of the latter remained in the lagoon ecosystem. Thus the lagoon plays an important role of nutrient removal from the wastewater entering the lake.

11.3.2   Nutrient Sources

The kinds of nutrients loaded from the catchment area to the lagoon generally depend on socioeconomic conditions in the watershed, such as land use, agricultural and industrial activities, water utilization and management, and vegetation and population density. Nitrogen and phosphorus loadings of different sources from the catchment area of inflowing river are shown in Table 11.5. Each value from different sources is estimated according to the unit factor method of the Japan Society of Civil Engineers. As shown in the table, the dominant nutrient loading sources from the catchment area are agricultural runoff and domestic wastewater. The loading values estimated according to the unit factor method and the loading values actually measured are compared in this table. The estimated value of the total nitrogen load is low compared with the measured loading value. On the other hand, the estimated value of the total phosphorus load is high compared with the measured value. In the case of the unit factor method for nutrient load, the values are estimated by population density, wastewater treatment method, use of

Table 11.6 Populations of Cities and Their Catchment Areas and the Percentage of City Populations in the Catchment Areas

	Total Population (A)	Population of Catchment Area (B)	$\frac{B}{A} \times 100 \ (\%)$
Ohmihachiman city	62,953	26,209	41.6
Yohkaichi city	38,915	20,309	52.2
Azuchi town	10,780	10,115	93.8
Eigenji town	6,978	1,405	20.1
Gokasho town	9,752	1,251	12.8
Notogawa town	19,750	339	1.7

fertilizer, number of domestic animals, and so on. These socioeconomic factors vary daily, monthly, and seasonally. Overestimation or underestimation of loaded nutrients occurs simply because of this variability.

The population density, land use, and number of domestic animals in the catchment area are shown in Tables 11.6 and 11.7. In the upper stream of inflowing rivers to the lagoon, there are two cities and four towns. Some of the areas of these cities and towns extend to the catchment area of this lagoon. Population of the urban area included in the catchment of the lagoon is shown in Table 11.6. Total population of the catchment area is 59,628 persons, with about half in Ohmihachiman City, one of seven main cities in Shiga Prefecture. Nutrient loading of domestic wastewater to the lagoon is derived primarily from these urban areas. The total catchment area of the lagoon is 41.9 km² and about 86% of it is paddy fields, considered to be a comparatively heavy non-point source of nutrient loading.

11.3.3   Harvesting to Remove Nutrients

As shown in Figure 11.1, *Phragmites communis* grows well in very high densities, and it colonizes about 60 ha in the water area and 107 ha in the

Table 11.7 Land Use and Number of Domestic Animals in the Catchment Area of Two Main Rivers

River	Land Use (km²)					Domestic Animals		
	Total	Paddy Field	Orchard	Farm	Forest	Cattle	Swine	Poultry
Hebisuna	29.45	24.94	0.11	0.73	3.67	920	1322	81,268
Kurohashi	4.21	3.99	0.02	0.13	0.07	25	160	4,608
Others	8.22	7.03	0.04	0.47	0.68	63	146	2,449
Total	41.88	35.96	0.17	1.33	4.42	1008	1628	88,325

Table 11.8 Annual Removal of Nitrogen and Phosphorus from the Lagoon by the Harvest of *Phragmites communis*

Total area of *P. communis*	60 ha
Average number of *P. communis*/m^2	52
Nitrogen content in *P. communis*/m^2	36.6 g
Phosphorus content in *P. communis*/m^2	4.3 g
Total harvest of *P. communis* (dry wt)	910 t
Total nitrogen removal	16.4 t
Total phosphorus removal	2.3 t

land area in this lagoon region. *P. communis* is harvested every year during the winter for different kinds of daily life products, for example, sunshade screen, domestic screen, and fancy goods, and the standing crop of it reaches approximately 910 metric tons dry weight. Harvesting of so much *P. communis* grown in the lagoon area signifies the removal of the same amount of nutrients every year. Removal of nutrients by harvest of *P. communis* in the lagoon was investigated minutely and calculated; the results are shown in Table 11.8. As shown in the table, annual removal of nitrogen and phosphorus from this ecosystem is 16.4 and 2.3 metric tons, respectively.

In addition to this process, the shellfish *Hyriopsis schlegeli* is cultivated to produce freshwater pearls in the lagoon. A high biomass of phytoplankton is consumed for the growth of *H. schlegeli*, namely, for the production of freshwater pearls. It is estimated that approximately 700,000 *H. schlegeli* are harvested annually from the lagoon. This process is also effective in removal of nutrients from the lagoon. Moreover, the macrophyte *Elodea canadensis* is harvested during the early summer by many fishermen to improve environmental conditions for navigation and cultivation of freshwater pearls in the lagoon. These kinds of effective processes for the removal of nutrients from this ecosystem are summarized in Table 11.9. In all, 31.2

Table 11.9 Removal of Nitrogen and Phosphorus from the Lagoon by the Harvest of *P. communis*, *Hyriopsis schlegeli*, and *Elodea canadensis* and by Fish Catch

	Total Nitrogen (metric tons)	Total Phosphorus (metric tons)	Value of Annual Harvest (10^6 yen)
P. communis	16.4	2.3	180
H. schlegeli	4.0	0.27	1500
E. canadensis[a]	10.5	1.5	—
Fish catch[b]	0.3	0.02	8
Total	31.2	4.09	1688

[a] 3000 metric tons.
[b] 11.2 metric tons.

metric tons of nitrogen and 4.09 metric tons of phosphorus are removed throughout the year from the lagoon by different kinds of human activities, such as harvesting of reeds and macrophytes, freshwater pearl cultivation, and fish catch, although 81.1 metric tons of nitrogen and 4.4 metric tons of phosphorus from the loaded nutrients remain in the lagoon. Thus harvesting of nitrogen and phosphorus by the above methods contributed 38.5 and 93.0%, respectively, to the entire nutrient retention of the lagoon. Nitrogen, however, is undoubtedly reduced actively through denitrification in the reed zone by epiphytic microorganisms, as reported in an earlier paper (Kurata and Kira, 1986).

11.4   ECOLOGICAL ENGINEERING ROLE OF THE LAGOON

Nishinoko Lagoon plays an important role as a whole ecosystem in removing nutrients from the inflowing wastewater into Lake Biwa, as if it were a natural ecological wastewater treatment system. A simplified model of the system is shown schematically in Figure 11.4. This is a first estimate of an ecological engineering model applied to the environs of Lake Biwa from the point of view of nutrient removal by utilizing natural ecosystems. This model must be studied further in other types of lagoons and applied to more lake environments suffering from eutrophication.

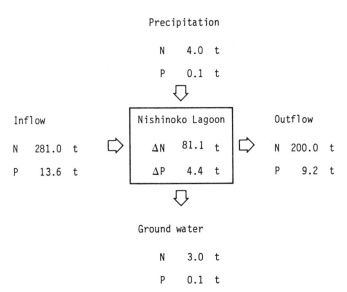

Figure 11.4 Nitrogen and phosphorus budgets for Nishinoko Lagoon adjacent to Lake Biwa, Japan.

REFERENCES

Kurata, A. 1983. Nutrient removal by epiphytic microorganisms of *Phragmites communis*. In R. G. Wetzel, Ed., *Periphyton of Freshwater Ecosystems*. Dr. W. Junk Publishers, The Hague, pp. 305–310.

Kurata, A. and T. Kira. 1986. Function of lagoon in nutrient removal in Lake Biwa. Proceedings of the Fifth Japan–Brazil Symposium of Science and Technology, The Japan Shipbuilding Industry Foundation, Tokyo, pp. 180–185.

12

DESIGN AND USE OF WETLANDS FOR RENOVATION OF DRAINAGE FROM COAL MINES

M. Siobhan Fennessy and William J. Mitsch

School of Natural Resources, The Ohio State University, Columbus, Ohio

12.1 INTRODUCTION

One of the more menacing types of water pollution in coal mining regions of the world is acid mine drainage. It reduces water quality for recreational, municipal, and industrial use and limits the economic value of affected waters (Yeasted and Shane, 1976; Down and Stocks, 1977). Acid mine drainage is formed when pyrite (FeS_2), a compound commonly associated with coal deposits, becomes exposed to air and water during the mining process. This causes the sulfide contained in pyrite to become soluble, initially producing large amounts of sulfuric acid. Ultimately, these reactions produce 1 mole of ferric hydroxide ($FeOH_3$) and 3 moles of H^+ (often present in the form of sulfuric acid, H_2SO_4) for each mole of pyrite oxidized. Both $Fe(OH)_3$ and H_2SO_4 are extremely toxic to aquatic life and severely minimize the domestic use of affected surface waters. More than 8000 km of the streams and rivers in the United States fail to meet water quality standards as a result of acid mine drainage. The major portion of this water originates in abandoned mined lands (Girts and Kleinmann, 1986).

The use of both natural and created wetlands as tools in the treatment of mine drainage from coal mines is increasing in the United States. The construction and use of wetlands in the reclamation of mined areas and the drainage associated with them is increasing in popularity as an ecological engineering alternative to conventional, chemically based reclamation methods. Wetlands have the advantages of being self-perpetuating, low-maintenance, and cost-efficient treatment systems. They serve as interface systems to lessen the impact of mine drainage from mined lands on the adjacent downstream areas (Mitsch et al., 1985). Brooks (1984) sees the presence of wetlands on mined lands as eliminating the need for perpetual chemical treatment. In addition to processing mine effluent, created wetlands may also provide a habitat for fish and wildlife.

12.1.1 Chemistry of Acid Mine Drainage

Water leaking from coal mines is characteristically acidic and high in concentrations of sulfate, calcium, magnesium, aluminum, iron, manganese, and zinc. Levels of iron, sulfate, and manganese are typically the most excessive (Curtis, 1977). This contaminated water, referred to as acid mine drainage (AMD), is formed when the pyrite ($FeS_{2(s)}$) associated with coal seams is uncovered and subjected to the oxidizing forces of water and air. Water flow out of mined areas carries iron as it becomes soluble in the pyritic oxidation process. In addition, several species of chemosynthetic bacteria including *Thiobacillus ferrooxidans, T. thiooxidans*, and *Ferrobacillus ferroxidans* are known to catalyze the oxidation of pyrite. These species are obligate chemoautotrophic bacteria that use the energy obtained from the oxidation of Fe^{2+} for metabolic functions (Lundgren, 1975; Letterman and Mitsch, 1978; Jaynes et al., 1984; Tarleton et al., 1984).

Pyrite remains stable under reducing conditions and its oxidation begins immediately upon contact with oxygen (Jørgensen, 1983). Singer and Stumm (1970) proposed two phases for pyrite oxidation, an initiator reaction:

$$FeS_2 + 0.5O_2 + 2H^+ \rightarrow Fe^{2+} + 2S^0 + 2H_2O \qquad (12.1)$$

and a propagation cycle:

$$4Fe^{2+} + O_2 + 4H^+ \rightarrow 4Fe^{3+} + 2H_2O \qquad (12.2)$$

$$FeS_2 + Fe^{3+} \rightarrow 2Fe^{2+} + 2S^0 \qquad (12.3)$$

They concluded that the oxidation of ferrous iron (Equation 12.2) is the rate-determining step in the generation of mine drainage through chemical oxidation. Although pyrite oxidation is exacerbated by the presence of ferric iron (Equation 12.3), its formation is also accelerated by the chemosynthetic, iron-oxidizing bacteria *Thiobacillus thiooxidans*, *T. ferrooxidans*, and *Ferrobacillus ferroxidans*. The amount of AMD contributed by the bacteria is thought to be proportional to their metabolism (Lundgren, 1975). Singer and Stumm (1970) demonstrated the importance of microbial activity in a comparison of mine water in which half was untreated and half was sterilized. The rate of Fe^{2+} oxidation was approximately 10^6 times higher in the unsterilized sample. In basic environments where pH is buffered by sedimentary bedrock, these acidophilic bacteria are thought to exist in acidic microsites (Olson et al. 1979). Although under these conditions pH may remain circumneutral, microbial activity still contributes to high levels of iron and sulfate.

The impact of *T. ferrooxidans* can be minimized in several ways. The addition of sodium lauryl sulfate (SLS) surfactant depresses bacterial activity, effectively reducing acid formation (Erikson et al., 1985). The activity of *T. ferrooxidans* is also negligible below the water table so acid production is eliminated in this zone. Inundation by only a few millimeters of water is enough to accomplish this (Nawrot, 1985).

Following pyrite dissolution, Fe(III) commonly forms a precipitate, as seen in the following equation:

$$Fe^{3+} + 3H_2O \rightarrow Fe(OH)_{3(s)} + 3H^+ \qquad (12.4)$$

The solid formed is an orange colored, gelatinous precipitate, commonly referred to as ferric hydroxide or "yellow boy" (Jaynes et al., 1984; Tarelton et al., 1984; Ackman and Kleinmann, 1985). Its chemical makeup is complicated; its chemical configuration is not well defined. Gotoh and Patrick (1974) cite hematite (Fe_2O_3) as the compound that most often occurs in waterlogged soils where $Fe(OH)_3 \cdot nH_2O$ is involved in iron redox equilibria. Nordstrom (1980) found that yellow boy may be jarosite [K-, Na-, or

$Fe(SO_4)_2(OH)_6$]. Brady (1982) found that the precipitate is ferrihydrite, a poorly ordered iron oxide with a layer structure and a composition of $5Fe_2O_3 \cdot 9H_2O$. This compound has a highly reactive surface and a strong affinity for sulfate. In fact, sulfate may act as a catalyst for the formation of ferrihydrite in mine waters. The precipitation of iron hydroxides is strongly pH dependent; in the small pH drop from 3.5 to 3.2 the solubility of iron increases by a factor of 10 (Jørgensen and Johnsen, 1981).

Manganese is very soluble in acidic waters and low redox environments such as those found in anaerobic sediments. Manganese is a persistent problem in mine runoff where the rate of its oxidation is strongly pH dependent, occurring very slowly at pH values less than 8.0 (Waltzlaf, 1985):

$$Mn^{2+} + \tfrac{1}{2}O_2 + 2H^+ \rightarrow Mn^{4+} + H_2O \tag{12.5}$$

and

$$Mn^{4+} + H_2O \rightarrow MnO^{2-} + 2H^+ \tag{12.6}$$

Precipitation of manganese from the water column using excess alkalinity (in the form of lime) is the most commonly used and least expensive method of removing manganese from water (Waltzlaf, 1985). Once iron and manganese have precipitated, they are not immobile in the sediments; both can be remobilized (Gilbin, 1985).

12.1.2   Effects of Acid Mine Drainage on Aquatic Life

Turbidity, chemical precipitates such as ferric hydroxide, and high concentrations of iron, manganese, sulfate, and other chemicals reduce both the species diversity and population sizes of benthic invertebrates and fish present in streams receiving acid mine drainage (Koryak et al., 1972; Dills and Rogers, 1974; Curtis, 1977; Letterman and Mitsch, 1978; Bosserman and Hill, 1985). Dills and Rogers (1974) found the lowest macroinvertebrate numbers at sampling stations nearest acid seeps. Immature chironomids, tubicifids, sialids, and some species in the order Tricoptera are tolerant of acid mine drainage, whereas larval forms in the orders of Plecoptera, Odonata, and Ephemeroptera are sensitive to it (Koryak et al., 1979). In areas where mine water becomes neutralized, the resultant deposition of ferric hydroxide keeps biomass values low (Koryak et al., 1972). Letterman and Mitsch (1978) found the greatest depression in numbers of benthic invertebrates in areas where net iron deposition was greater than 2–3 g/m^2-day. They found not only a decline in the number of individuals within the area of ferric iron deposition, but also a decrease in the mean biomass of from 14.0 to 1.0–1.5 g wet weight/m^2. A recovery of only 18% of the total biomass was observed as the stream improved in quality.

The impact of mine drainage on the fish community is species specific but significant, with the greatest impact on benthic fish species. These are

Table 12.1 Examples of Previous Studies Investigating the Use of Wetlands to Control Coal Mine Drainage

Location	Wetland Vegetation Type	Area of Wetland	Iron Concentration Inflow/outflow (mg/L)	Iron Concentration % Reduction	Reference
Ohio	Portable *Sphagnum* bog	7.1 m²		50–70	Kleinmann et al., 1983
Kentucky	*Eleocharis* spp., *Typha* spp., and *Dulichium* spp.	3.25 km²	4.5/3.3	27	Mitsch et al., 1985
Lab study	*Sphagnum recurvum*	5.5 m²		60–90	Gerber et al., 1985
West Virginia	*Sphagnum* spp., *Polytrichum* spp.	23 ha	26–73/<2	92–97	Weider and Lang, 1982
Lab study	Portable *Sphagnum* bog	7.1 m²	(1) 50/0–9 (2) 100/7–45	82–100 55–93	Burris et al., 1984
Maryland	*Sphagnum* spp., *Polytrichum* spp.	270 m²	40/4.4	89	Wieder et al., 1985
Pennsylvania, and West Virginia	*Sphagnum* spp. or *Sphagnum* and *Typha* spp. combined		(1) 24/0.5 (2) 8.7/1.2 (3) 24/0.6	98 86 97.5	Kleinmann, 1985

typically sedentary, poor-swimming fish who rely on benthic invertebrates for food (Letterman and Mitsch, 1978). Bosserman and Hill (1985) noted an absence of fish in areas receiving heavy loads of AMD. Species diversity and fish abundance were reduced in AMD streams for up to 20 years after mine abandonment (Hill, 1983). Mine drainage may affect fish in particular and aquatic biota in general in several ways (Everhart et al., 1975; from Hill, 1983): decreased dissolved oxygen levels in the water, increased acidity, reduced food availability, destruction of spawning grounds, accelerated aging of water bodies, clogging of migration channels, and physical injury to gills from suspended solids.

12.2 ECOLOGICAL ENGINEERING DESIGN FOR WETLANDS AS MINE DRAINAGE TREATMENT SYSTEMS

Conventional treatment of mine water consists of raising the pH of the effluent to 8 or 9 using limestone, then aerating to precipitate iron and manganese. Biological treatment by wetlands has been found to reduce the levels of virtually all contaminants found in mine water (Seidel, 1976; Brooks, 1984; Gerber et al., 1985; Kleinmann, 1985; Mitsch et al., 1985; Nawrot, 1985). Examples of studies that have investigated the role of wetlands for mine drainage control are given in Table 12.1. These studies have ranged from investigations of laboratory mesocosms to natural wetlands to constructed wetlands specifically designed to control mine drainage. Several design considerations are important in using wetlands to treat acid mine drainage and are discussed below.

12.2.1   Biochemical Processes

Wetland sediments, which are generally anaerobic below a thin oxidized surface layer and contain organic carbon compounds for microbial growth (e.g., lactate, pyruvate) (Gambrell and Patrick, 1978), provide a favorable environment for iron, manganese, and sulfate reduction. Both microbial and chemical reducing processes occur in the anoxic zone of sediments, transforming iron and sulfates to hydrogen and iron sulfides, including pyrite (Jørgensen, 1983, Kalin, 1987; Hedin et al., 1988). Sulfate (SO_4^{2-}) reduction to sulfide (S^{2-}) occurs as assimilatory sulfate reduction in the genera *Desulfovibrio* and *Delsulfotomaculum*. These genera employ sulfate as the terminal electron acceptor in anaerobic respiration, forming hydrogen sulfide and iron sulfides (Mitsch and Gosselink, 1986).

Sulfate reduction can be limited in freshwater systems (unlike marine systems) by low sulfate concentrations (Cappenberg, 1974). As Hedin et al. (1988) note, wetlands built for AMD treatment are not subject to this limitation. Reducing sediments should function to maximize sulfate reduction, eventually forming insoluble pyrite. The accumulation of reduced sulfur and pyrite in sediments will not affect aboveground wetland processes as the buildup of ferric hydroxides on sediment surfaces might. The conversion of Fe^{2+} to a stable solid may occur in the following sequence:

$$Fe^{2+} \rightarrow 5Fe_2O_3 \cdot 9H_2O \text{ (ferrihydrite)} \rightarrow FeOOH \text{ (goethite)} \rightarrow FeS_2.$$

(12.7)

Anoxia in wetland sediments provides conditions for the conversion of soluble metals to insoluble forms. Following this, humic materials of high molecular weight that are commonly found in the sediments act to immobilize the metals (Maltby, 1987). Translocation and metabolic processes are thought to concentrate various chemicals in plant tissues and after senescence they are also trapped in organic litter. The removal of iron and manganese from solution, although somewhat contingent upon hydrologic factors, depends more on the chemical conditions in the wetland (Gilbin, 1985). Chelation, precipitation, and microbial redox reactions are important in the removal of metals from water (Wolverton et al., 1976). The reduction of water velocity in a wetland, in part due to the presence of aquatic vegetation, causes a filtering and settling of suspended solids (Boto and Patrick, 1979; Brooks, 1984).

12.2.2   Loading Rate and Retention Time

The loading rate of mine water into a natural or constructed wetland determines its maximum treatment efficiency (Girts and Kleinmann, 1986). The volume and the concentration of pollutants in the influent, along with existing effluent standards and the rate of treatment, determine the optimal retention time of water in the wetland. Retention time is a factor of the flow path length and water volume to substrate area ratio (Mitsch and Gosselink, 1986). To achieve maximum contaminant reduction, precipitation, runoff, infiltration, and evapotranspiration must be considered. Severe evapotranspiration may drastically increase detention times, leading to highly anaerobic conditions, vegetation death, and eventual exposure of anaerobic sediment zones. High hydraulic loading increases flow so detention times are reduced. Major storm events and spring thaw may also cause flushing from the mine, resulting in high concentrations of sulfate and salts, scouring of the substrate, and channelization (Wieder and Lang, 1984; Wile et al., 1985; Girts and Kleinmann, 1986). Seven-day retention times and a hydrologic loading rate of 200 m^3/ha-day are thought to be optimal conditions for wetlands receiving wastewater (Wile et al., 1985).

12.2.3 Slope

Wetlands designed for iron and manganese removal require low-velocity sheet flow on a slope of less than 5% to maximize contact with vegetation and substrate (Brooks, 1984). The U.S. Environmental Protection Agency (EPA, 1985), however, recommends slopes of only 1% to optimize efficiency. Reduced efficiencies of nutrient and metal uptake coincide with reduced detention times and channelization within the wetland (Tilton and Kadlec, 1979; Brooks, 1984).

12.2.4 Substrate

The substrate used in the construction of a wetland depends on the type of vegetation to be planted. Hay or peat is commonly used as a substrate for moss wetlands. Cattail marshes are often set in beds of composted hay and manure on top of a layer of limestone (Girts and Kleinmann, 1986). Sediments with a high organic matter content have been shown to absorb more metals than those with a high mineral content (Gilbin, 1985). Mitsch et al. (1985) recommend replacement of topsoils in the wetland basin to preserve seed banks (a source of new plant life) although reclamation using topsoil is an expensive prospect. Limestone is often used to render coal refuse more hospitable for vegetation, but the alternative use of sewage sludge as a substrate will increase the organic matter content by 2.0–2.5 times more than limestone does, and at much less expense (Joost et al., 1987). The improved condition of the soil with the addition of sewage sludge is linked to superior growth of vegetation both in terms of biomass and the percent canopy cover (Topper and Sabey, 1986). The aquatic plants present also improve the soil structure by preventing erosion, acid runoff, and leaching of nitrates and metals into groundwater (Joost et al., 1987).

12.2.5   Vegetation

Establishment of emergent, perennial vegetation is generally accomplished by placing rhizomes directly into the subsurface zone. This zone should be saturated to prevent its acidification (Nawrot and Yaich, 1982; Warburton et al., 1985). *Typha* spp. (cattail) do very well under these conditions and many other graminoid species will volunteer (Brooks, 1984). Local stock typically has a higher viability than those obtained commercially owing to better ecotypic adaptations to that area. In one case, survival of a locally collected *Scirpus* sp. was 95% higher than survival of a commercial stock (Warburton et al., 1985). This is critical because removal of iron and manganese is directly proportional to the density of the vegetation in a wetland (Brooks, 1984).

Typha spp. wetlands are often used for acid mine drainage treatment because *Typha* is cosmopolitan in distribution, is acid tolerant, and thrives under diverse—even potentially phytotoxic—environmental conditions (Nawrot and Yaich, 1982). Adaptations such as high evapotranspiration rates, extensive rhizome development, and rapid biomass production provide the potential for the bioaccumulation of contaminants (Dykyjova and Kvet, 1978; Snyder and Aharrah, 1985; Wile et al., 1985). The ability of aquatic macrophytes to exclude soluble iron by precipitating iron oxides in the oxidized rhizosphere of their roots is thought to be a critical adaptive feature for their survival in environments such as acid mine drainage (Wheeler et al., 1985).

Typha wetlands have been demonstrated to be effective in removing iron and manganese from mine waters (Snyder and Aharrah, 1985). At a *Typha* wetland in Pennsylvania, iron concentrations dropped from 20–25 to 1 mg/L and the concentration of manganese decreased from 30–40 to 2 mg/L (Kleinmann, 1985). In this case, however, manganese removal was attributed to microbial activity. Evidence is conflicting as to the cause of metal removal from mine drainage. Nawrot and Yaich (1982) report higher levels of bioaccumulated metals in plants growing on slurry impoundments than those growing on unmined areas. Gilbin (1982) contends that uptake of metals by vegetation does not represent a major outflow from the system. He found only a small portion of the annual metal input to a marsh present in plant biomass. Metal utilization and transformation by microorganisms and deposition by adsorption on suspended solids are more likely the cause of metal removal. *Typha* spp. roots may provide a substrate for these microorganisms.

Wetlands dominated by *Sphagnum* spp. also chemically modify acid mine drainage. *Sphagnum* is an acid-producing genus so it has no appreciable impact on the pH of mine waters. It does have a significant effect on concentrations of ions such as sulfate, nitrate, iron, manganese, potassium, calcium, and sodium (Wieder and Lang, 1984). Kleinmann (1985) has shown that 1 kg (wet weight) of *Sphagnum recurvum* can remove up to 92% of the iron in an influent containing 50 mg/L iron at a pH of 3.8. An 80% reduction of total metal concentration by *Sphagnum* spp. has been documented (Gerber et al., 1985).

Efficient removal of iron and manganese by *Sphagnum* occurs via cation exchange (Tarleton et al., 1984; Wieder and Lang, 1984; Gerber et al., 1985; Kleinmann et al., 1983). Organically bound iron is the principle form of iron in *Sphagnum* peat. Saturation of available exchange sites in the moss is thought to be prevented by the presence of *T. ferrooxidans*. This microbial species regenerates cation exchange sites in the moss by oxidizing iron, causing its precipitation (Gerber et al., 1985). Manganese removal is not as efficient by *Sphagnum* as is iron removal because, of the two, iron is absorbed preferentially. Differences in the outer shell electron configuration of the two metals is thought to be the factor responsible for preferential iron adsorption. Iron's outer shell configuration allows it to interact better with the moss; therefore, the presence of iron inhibits the adsorption of manganese (Gerber et al., 1985).

12.2.6 Sediment Control

The high sediment loads typical of acid mine drainage can be controlled with a sediment pond prior to the wetland system. Accumulated sediments could be cleaned from the pond without disturbing the wetland itself, thereby increasing the lifetime of the wetland (Brooks, 1984). A sediment pond could also be used to control discharge into the wetland, thus ensuring that there is water in the wetland year round. A weeklong pretreatment period in a small aeration pond with a continuous alum feed was found to reduce BOD levels by 60% and suspended sediments by 70% without significant sludge accumulation, thereby eliminating the need for periodic cleaning (Wile et al., 1985). Seasonal harvesting of plant biomass will also reduce the rate of sediment accumulation (Kadlec, 1985).

12.2.7 Morphometry

The geometric configuration of a wetland is important to its reliability and treatment ability. A serpentine shape outperformed a rectangular shape in reducing contaminant concentrations (Wile et al., 1985). Designs should move away from uniform basin perimeters having consistent slopes and depths to those with a more natural, variable shape. Creating diverse environments is more aesthetically pleasing, provides better wildlife habitat, and enhances macrophyte establishment (Brooks, 1984; U.S. EPA, 1985). Islands within a wetland increase spatial and habitat diversity and act to impede water flow (Brooks, 1984; Mitsch et al., 1985).

12.2.8 Seasonality

A major concern with using wetlands to ameliorate acid mine drainage in north temperate latitudes is seasonal variations of the biological systems.

The vegetation's dormant season affects the capacity of a wetland to treat mine water. *Typha* spp. wetlands are especially subject to seasonal efficiency owing to seasonal dieback. The efficiency of metal removal by a *Sphagnum* wetland, however, continues at rates in the winter comparable to rates found in the summer. *T. ferrooxidans* is also active in the nongrowing season and so continues to remove metal cations from the moss, liberating cation exchange sites (McHerron, 1985).

12.2.9   Regulatory Issues

Regulatory issues related to use of wetlands, both natural and man-made, for treatment of acid mine drainage may inhibit the widespread use of this technology. Questions arise as to whether wetlands can meet final effluent standards on a consistent basis and whether or not they can be maintained to sustain vegetation and so retain biological control of the system. The liability of the mining companies concerned with cases of drowning is also an impediment to the construction of artificial wetlands (Byron, 1985). Permission to mine is more easily obtained with post-mining land use plans that mimic pre-mining conditions on the site. Alternative land use plans must be proved to be a "higher and better use" of the land (Gleich, 1985).

12.3   CASE STUDY

12.3.1   Site Description

A wetland being investigated for its role in reducing the effects of acid mine drainage is located in Coshocton County, in eastern Ohio. The created marsh covers an area of 0.22 ha and was installed in November, 1985. It is a system of three wetland areas or "cells" separated by small mixing pools of open water (see Figures 12.1 and 12.2). Prior to the installation of the wetland, the mine effluent flowed into an open stream and directly to an aeration pond. Mine drainage from an abandoned underground mine enters the system at a rate of approximately 380 L/min (550 m^3/day), causing a dry weather residence time of approximately 0.7 days (assuming 0.175 m water depth). Vegetation is presently dominated by *Typha latifolia*, the common cattail, although some diversification of the plant community, particularly with rice cut-grass (*Leersia oryzoides*) and duckweed (*Lemna minor*), is occurring.

12.3.2   Research Questions and Methods

From an ecological engineering point of view, two aspects of the wetland are under scrutiny. One is the effectiveness of wetlands to improve the quality of water draining from abandoned and active mine lands. The second is the impact of the drainage on the productivity of the vegetation, as well

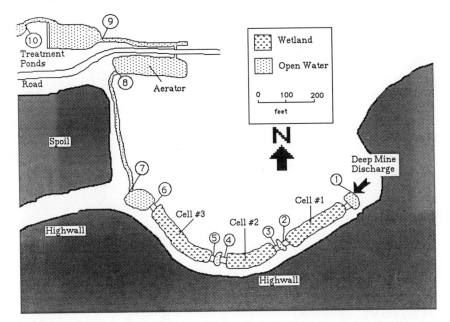

Figure 12.1 Site map of experimental wetlands being used to treat acid mine drainage, Coshocton county, Ohio.

as any successional changes that occur within the ecological communities. Water quality in the wetland has been monitored to determine what reductions are occurring in the concentrations of contaminants typically found in mine drainage, such as iron, manganese, sulfates, turbidity, alkalinity pH, and conductivity. Vegetation has been sampled to determine both the chemical makeup of the plant tissues and the primary productivity of the system, thus giving an indication of its fitness. Successional changes in the plant community have also been monitored.

12.3.3 Changes in Iron Concentrations

Prior to wetland construction, the concentration of iron in the mine effluent was reduced by approximately 33% as it flowed in an open stream to the aeration pond (see Figure 12.1). Immediately after the wetland was installed the average total iron concentration was reduced by 35%, representing an increase in treatment efficiency of 2%. The percent reduction in the wetland proper (to the end of cell three) averaged 23% (Stark et al., 1988).

Over the first growing season, the efficiency increased dramatically to a high of 59.5% in the wetland, and a drop of 81% in iron levels was reached by the time the water reached the aeration pond. This can be attributed to both an increase in the retention time of the water in the wetland, and its contact with the vegetation. During the winter of 1986/1987 the percent re-

Figure 12.2 Photograph of experimental wetland site.

duction averaged 49.8% in the wetland. This is a substantial increase in treatment as compared to the winter of 1885/1986, in part due to the senesced vegetation, which provided the means for low-velocity sheet flow, hastening the settling of ferric hydroxides. Over the summer of 1987, the system's second growing season, the efficiency of the wetland rose to a high of 63.4% and averaged 51.6%.

In September of 1987, substantial changes were made in the design of the wetland in an effort to improve its performance. This new arrangement effectively halves the hydrologic loading to cells 1 and 2; cell 3 still receives the total flow of water. Although any results showing improvement in treatment efficiency will not be conclusive until further study has been done, over the months September, 1987 to January, 1988, an average 58.1% reduction occurred between stations 1 and 6, a 6.5% increase in efficiency over the previous growing season and an 8.3% increase over the previous winter. A 94.0% reduction occurred between stations 1 and 10. For the first time cell 1 showed the highest iron reduction efficiency at 43.4% over the months September, 1987–January, 1988. The decreased loading rate is most likely responsible for the increase in efficiency.

There is a steady decline in the concentration of total iron in the mine water as it passes through the wetland cells (Figures 12.3 and 12.4). Cell 3 (the cell farthest from the mine effluent) consistently showed the highest rate of reduction in pollutant concentrations. Over the first growing season,

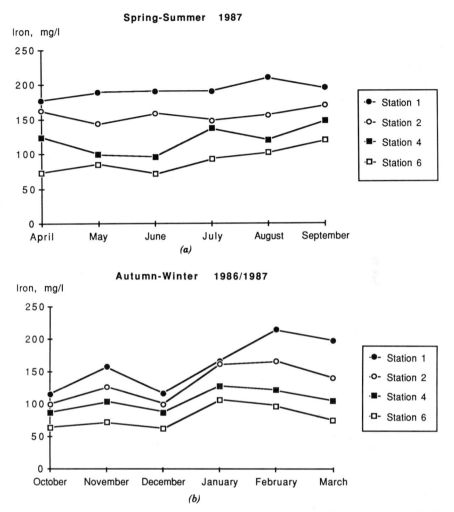

Figure 12.3 Seasonal patterns of iron at influent (Station 1) and effluent from wetland cells (Stations 2, 4, and 6) for (a) autumn–winter, and (b) spring–summer of 1986–1987. Station numbers refer to map in Figure 12.1.

cell 3 averaged a 35% reduction in iron, whereas the iron concentration in cells 1 and 2 fell only 22 and 20%, respectively (Stark et al., 1988). During the second growing season the average drop in cell 3 was 26%, compared to 16% in cell 1 and 20% in cell 2. This phenomenon can be partially explained by the high vegetation density found in cell 3; it has consistently had the highest vegetation cover and density, as well as the highest species diversity. This cell, which is the most ecologically healthy and stable, acts as the best pollutant processing system. This observation is supported in other studies, which show that high efficiencies of nutrient and metal uptake are coincident

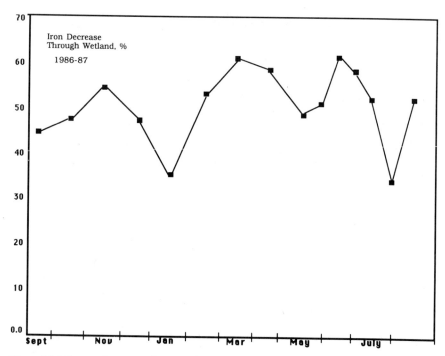

Figure 12.4 Seasonal pattern of percent iron reduction in experimental wetland through entire wetland.

with long detention times and a high level of contact with vegetation (Tilton and Kadlec, 1979; Brooks, 1984).

12.3.4   Other Water Chemistry

Table 12.2 shows the average inflow and outflow levels and the percent reduction for each water quality parameter in the experimental wetland. Manganese concentrations and turbidity are the only parameters that are not reduced by the wetland system. Manganese concentrations fluctuate greatly between sampling stations and tend to increase over the distance of the wetland. At all times, however, manganese levels remain below the legal effluent limit of 4.00 mg/L. The turbidity of the water increases in the wetland as ferric hydroxide precipitates form in the water column. Conductivity levels remain high and tend to mirror the levels of sulfate throughout the wetland. Little sulfate reduction is occurring because of prohibitively high redox potentials. pH levels show a gradual trend to circumneutral between stations 1 and 6.

12.3.5   Plant Uptake of Chemicals

Levels of iron in the vegetation seem to be correlated with the concentrations of iron held in the water in the different cells (Table 12.3). The foliar iron

Table 12.2 Changes in Water Quality Through Experimental Wetland Site in Coshocton County, Ohio

Parameter	Influent[a] (Station 1)	Effluent[a] (Station 6)	Percent Reduction
pH	6.35 ± 0.13 (27)	6.5 ± 0.18 (27)	—
Conductivity, μmho/cm	2201 ± 272 (29)	1978 ± 27 (29)	10.1
Fe, mg/L	163.7 ± 40 (28)	81.8 ± 18 (28)	50
Mn, mg/L	1.89 ± 0.57 (28)	2.14 ± 0.29 (28)	− 13.2
SO_4^{2-}, mg/L	1,377 ± 349 (25)	1,282 ± 263 (25)	6.9
Turbidity, FTU	109.9 ± 33 (20)	140.0 ± 96 (20)	− 27.4
alkalinity, mg/L as $CaCO_3$	99.7 ± 32 (28)	36.8 ± 17 (28)	63.1
acidity, mg/L as $CaCO_3$	102.3 ± 36 (28)	53.5 ± 31 (28)	47.7

[a] Concentrations are averages for year ± standard deviation (no. of samples).

Table 12.3 Chemical Analysis of Plant Material from Experimental Wetland and from Nearby Control Site

Location	Fe (mg per g tissue)		Mn (mg per g tissue)	
	Aboveground	Belowground	Aboveground	Belowground
Experimental Wetland				
PEAK BIOMASS (1986)				
Cell 1	0.560	—	0.267	—
Cell 2	0.203	—	0.291	—
Cell 3	0.172	—	0.238	—
PEAK BIOMASS (1987)				
Cell 1	0.491	2.135	0.143	0.063
Cell 2	0.238	1.414	0.190	0.169
Cell 3	0.276	1.630	0.212	0.069
WINTER (1987)				
Cell 1	0.380	0.265	0.340	0.166
Cell 2	0.396	0.461	0.234	0.088
Cell 3	0.667	0.763	0.425	0.120
Control Wetland				
	0.051	—	0.550	—

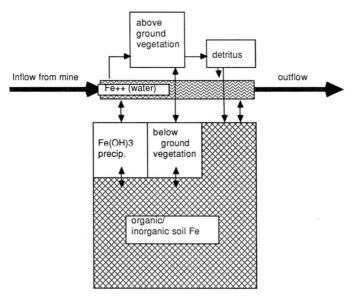

Figure 12.5 Conceptual model of iron budget for wetland receiving acid mine drainage.

levels were considerably higher in the plants taken from cell 1 than in those from cells 2 and 3. Foliar concentrations of iron were higher than concentrations of manganese in the cattail tissue in all three cells, a finding contrary to that of Mayer and Gorham (1951), who found that plants grown on acidic soils had a higher content of manganese than iron. Foliar concentrations in plants from cells 2 and 3 following the first frost show an increase in the levels of both iron and manganese, most notably in plants from cell 3. This is a finding similar to that of Davis and van der Valk (1978), who reported higher levels of iron in plant tissues after senescence than before in an Iowa marsh not affected by AMD. The accumulation of iron in the senesced tissue suggests that leaching may not cause an increase in the concentrations of metals in the effluent water. Iron concentrations in vegetation in the experimental wetland cells were significantly higher than concentrations in vegetation in a nearby unaffected "control" wetland. This indicates some luxury uptake of iron by the *Typha* in the experimental wetland.

12.3.6 A Conceptual Model and Preliminary Mass Balance Model

The most chronic problem at this site is the high level of iron in the mine water. Data from our study have been used to develop models of iron for the wetland (Figures 12.5 and 12.6). The conceptual model in Figure 12.5 was developed to illustrate the relationship among hydrology, iron loadings, vegetation growth and uptake, and sedimentation processes. The model allows the system to be viewed as a whole, providing clues to such questions

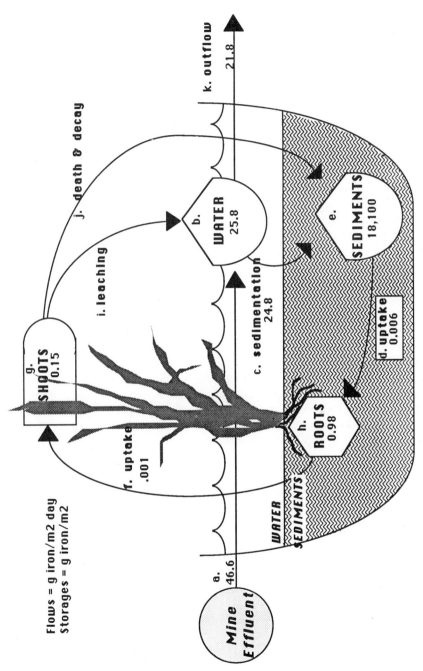

Figure 12.6 Mass balance model for the experimental wetland. Flows are expressed as g Fe per square meter per day and storages as g Fe per square meter (calculations of values in Fennessy, 1988).

247

as, what is the influence of hydrology on the iron removal capacity of the wetland? What is the role of the vegetation in Fe removal? And, what will be the effect of iron buildup on the "aging" process of the wetland?

As illustrated in Figure 12.6, the vegetation itself in the wetland treatment system does not appear to act as a major sink for iron in the mine water. Although the foliar iron concentrations indicate some luxury uptake of iron by the vegetation, the total amount of iron taken in annually in the biomass is insignificant when compared to the yearly flow through the system. Cell 1 shows the highest concentrations of iron in the *Typha* tissue as well as the highest levels of iron in the water. The processes involved in the removal of iron from the mine water appear to be primarily physical/chemical ones such as sedimentation and precipitation of iron hydroxides. This is consistent with findings by Gersberg et al. (1986), who concluded that removal of suspended sediments by wetlands was due almost entirely to physical processes (sedimentation, filtration). We conclude, as have others (Nichols, 1983; Gersberg et al., 1986), that vegetative accumulation of pollutants alone cannot account for the high pollutant removal rates that many wetlands demonstrate.

The sediments are the major sink of iron in the experimental wetlands. Although the microbiology of nonacidic acid mine drainage is not well understood (Lundgren et al., 1972), Olsen et al. (1979) report that sulfate-reducing bacteria were common in both mine waters and sediments of nonacidic water originating in a strip mine. Metal sulfide precipitation that results from microbiologically mediated sulfate reduction in wetland sediments can provide a valuable means for removal of trace metals from mine drainage.

Hydrologic loading rates to the entire wetland average approximately 2,500 m^3/ha-day (25 cm/day). This is much higher than the recommended loading rate of 200 m^3/ha-day (2 cm/day) for wetlands receiving wastewaters (Wile et al., 1985). Loading rates of this magnitude lessen the ability of the wetland to achieve high pollutant removal efficiencies. Efforts to increase the wetland area in order to reduce hydrologic loading would result in increased treatment levels. We believe the hydrologic regime of a wetland that is to be used as a water pollutant processing system is ultimately the most critical design criteria in determining the treatment capability of the wetland.

12.4   CONCLUSIONS

Our research demonstrates ecological engineering with wetlands as a natural reclamation alternative to costly chemical treatments presently in use. This approach also provides a means of creating new wetlands to replace the many which have been lost in Ohio and elsewhere in the midwestern United States. These will serve to help reestablish areas disturbed by mining while possibly providing new fish and wildlife habitats and, at minimum, improving water quality in downstream systems. Based on current information, wetlands appear to be effective treatment systems for the renovation of mine drainage. Their ability to reduce and, to some extent, to process contaminant loads provides a unique opportunity for the establishment of natural, self-perpetuating pollutant processing systems. From an engineering standpoint they are "self-designing" systems, needing little or no management following their construction. The limitations of wetlands' ability to treat polluted waters are sorely under-researched. Hesitations to utilize this technology are caused primarily by the paucity of data on their aging process and their long-term processing capacity. Studies of the long-term effects on wetlands receiving acid mine drainage are needed to validate further their use as pollution control systems.

ACKNOWLEDGMENTS

We acknowledge the support of American Electric Power Corporation for allowing us access to and research use of the experimental wetland site. We also appreciate the cooperation of the previous owner of the site, Peabody Coal Company and particularly that of Brad Wills and Early Murphy. Field and laboratory work was assisted by Brian Reeder, Julie Cronk, and Vanessa Steigerwald. The REAL Laboratory of the Ohio Agricultural Research and Development Center (OARDC) of Ohio State University provided plant analyses. Portions of salaries and research support were provided by state and federal funds appropriated to the Ohio Agricultural Research and Development Center, The Ohio State University. Manuscript number 60–88.

REFERENCES

Ackman, T. and R. L. Kleinmann. 1985. In-line aeration and treatment of acid mine drainage: Performance and preliminary design criteria. In Control of acid mine drainage. Proceedings of a technology transfer seminar. Bureau of Mines Information Circular no. 9027. U.S. Department of the Interior, pp. 53–61.

Bosserman, R. W., and P. L. Hill. 1985. Community ecology of three wetland ecosystems impacted by acid mine drainage. In *Wetlands and Water Management on Mined Lands*. Proceedings of a Conference 23–24 October. The Pennsylvania State University, University Park, PA, pp. 287–302.

Boto, K. G., and W. H. Patrick, Jr. 1979. Role of wetlands in the removal of suspended sediments. In P. E. Greeson, J. R. Clark, and J. E. Clark, Eds., *Wetland Functions and Values: The State of Our Understanding*, Proceedings of a National Symposium on Wetlands. Lake Buena Vista, Florida. American Water Resources Assoc. Tech. Publ. TPS 79-2. Minneapolis, MN, pp. 479–489.

Brooks, R. P. 1984. Optimal designs for restored wetlands. In J. E. Burris, Ed., Treatment of Mine Drainage by Wetlands. Contribution no. 264. Department of Biology, Pennsylvania State University, University Park, PA, pp. 19–29.

Burris, J. E., D. W. Gerber, and L. E. McHerron. 1984. Removal of iron and manganese from water by *Sphagnum* moss. In J. E. Burris, Ed., Treatment of mine drainage by wetlands. Contribution no. 264. Department of Biology, Pennsylvania State University, University Park, PA, pp. 1–14.

Byron, G. J. 1985. Man-made wetlands as a post mining use: regulatory issues and conflicts. In *Wetlands and Water Management on Mined Lands*. Proceedings of a Conference 23–24 Oct. The Pennsylvania State University, University Park, PA, pp. 181–183.

Cappenberg, T. E. 1974. Interrelations between sulfate-reducing and methane-producing bacteria in bottom deposits of a freshwater lake. I. Field observations. *Antonie van Leeuwenhoek J. Microbiol. Serol. 40:285–295.*

Curtis, W. R. 1977. Effect of strip mining on water quality in small streams in eastern Kentucky, 1967–1975. Report 1977-703-078/38. U.S. Government Printing Office, Washington, DC.

Davis, C. B. and A. G. van der Valk. 1978. Litter decomposition in prairie glacial marshes. In R. E. Good, D. F. Whigham, and R. L. Simpson, Eds., *Freshwater Wetlands: Ecological Processes and Management Potential*. Academic, NY, pp. 99–114.

Dills, G. and D. Rogers, Jr. 1974. Macroinvertebrate community structure as an indicator of acid mine pollution. *Environ. Pollut. 6:239–262.*

Down, C. G. and J. Stocks. 1977. *Environmental Impact of Mining*. Applied Science Publishers, London, 371 pp.

Dykyjova, D. and J. Kvet. 1978. *Pond Littoral Ecosystems: Structure and Functioning*. Springer-Verlag. New York.

Erikson, P. M., R. L. Kleinmann, and S. J. Onysko. 1985. Control of acid mine drainage by application of bactericidal materials. In Control of acid mine drainage. Proceedings of a technology transfer seminar. Bureau of mines information circular no. 9027. U.S. Department of the Interior, pp. 25–35.

Everhart, H. W., A. W. Eipper, and W. D. Youngs. 1975. *Principles of Fishery Science*. Cornell University Press, Ithaca, NY, 288 pp.

Fennessy, M. S. 1988. Reclamation of acid mine drainage using a created wetland: Exploring ecological treatment systems. M. S. thesis. The Ohio State University, Columbus, OH.

Gambrell, R. P. and W. H. Patrick, Jr. 1978. Chemical and microbiological properties of anaerobic soils and sediments. In D. D. Hook and R. M. M. Crawford, Eds, *Plant Life in Anaerobic Environments*. Ann Arbor, Mich.

Gerber, D. W., J. E. Burris, and R. W. Stone. 1985. Removal of iron and manganese ions by a *Sphagnum* moss system. In *Wetlands and Water Management on Mined Lands*. Proc. of a Conference 23–24 Oct. The Pennsylvania State University, University Park, PA, pp. 365–372.

Gersberg, R. M., B. V. Elkina, S. R. Lyon, and C. R. Goldman. 1986. Role of aquatic plants in wastewater treatment by artificial wetlands. *Water Res. 20:363–368.*

Gilbin, A. E. 1985. Comparisons of the processing of elements by ecosystems: Metals. In P. J. Godfrey, E. R. Kaynor, S. Pelczarski, and J. Benforado, Eds., *Ecological Considerations in Wetlands Treatment of Municipal Wastewaters*, Van Nostrand Reinhold, New York, pp. 158–179.

Girts, M. and R. Kleinmann. 1986. Construction wetlands for treatment of mine water. Proc. Society of Mining Engineers 7–10 Sept. St. Louis, MO.

Gleich, G. J. 1985. Why don't coal companies build wetlands? In *Wetlands and Water Management on Mined Lands*. Proc. of a Conference 23–24 Oct. The Pennsylvania State University, University Park, PA, pp. 191–194.

Gotoh, S. and W. H. Patrick, Jr. 1974. Transformation of iron in a waterlogged soil as influenced by redox potential and pH. *Soil Sci Soc. Amer. Proc. 38:*66–70.

Hill, P. L. 1983. Wetland-stream ecosystems of the western Kentucky coalfield: Environmental disturbance and the shaping of aquatic community structure. Ph.D. dissertation. University of Louisville, Louisville, Kentucky.

Jaynes, D. B., A. S. Rogowski, and H. B. Pionke. 1984. Acid mine drainage from reclaimed coal strip mines. I. Model description. *Water Resources Res. 20:*233–242.

Jørgensen, S. E. and I. Johnsen. 1981. *Principles of Environmental Science and Technology*. Elsevier, Amsterdam, 516 pp.

Joost, R. E., F. J. Olsen, and J. H. Jones. 1987. Revegetation and minesoil development of coal refuse amended with sewage sludge and limestone. *J. Environ. Qual. 16:*65–68.

Kadlec, R. H. 1985. Aging phenomena in wastewater wetlands. In P. J. Godfrey, E. R. Kaynor, S. Pelczarski, and J. Benforado, Eds, *Ecological Considerations in Wetlands Treatment of Municipal Wastewaters*. Van Nostrand Reinhold, New York, pp. 338–347.

Kleinmann, R. L. 1985. Treatment of acid mine water by wetlands. In Control of Acid Mine Drainage. Proceedings of a technology transfer seminar. Bureau of mines information circular no. 9027. U.S. Department of the Interior, pp. 48–52.

Kleinmann, R. L., T. O. Tiernan, J. G. Solch and R. L. Harris. 1983. A low cost, low maintenance treatment system for acid mine drainage using Sphagnum moss and limestone. In Symposium on Surface Mining, Hydrology, Sedimentology and Reclamation. 27 November–2 December, University of Kentucky, Lexington, Kentucky, pp. 241–244.

Letterman, R. D. and W. J. Mitsch. 1978. Impact of mine drainage on a mountain stream in Pennsylvania. *Environ. Pollut. 17:*53–73.

Lundgren, D. G. 1975. Microbial problems in strip mine areas: relationship to the metabolism of *Thiobacillus ferrooxidans*. *Ohio J. Sci. 75:*280–287.

Lundgren, D. G., J. Vestal, and F. R. Tabilta. 1972. The microbiology of mine drainage pollution. In R. Mitchell, Ed., *Water Pollution Microbiology*. Wiley, New York, pp. 69–88.

Maltby, E. 1987. Soils science base for freshwater wetlands mitigation in the Northeast United States. In J. S. Larson and C. Neill, Eds., Mitigation, Freshwater Wetland Alterations in the Glaciated Northeastern United States: An Assessment of the Science Base. Publication No. 87-1, The Environmental Institute, University of Massachusetts, Amherst, pp. 17–52.

Mayer, A. M. and E. Gorham. 1951. The iron and manganese contents of plants present in the natural vegetation of the English Lake District. *Ann. Bot. N.S. 15:*247–263.

McHerron, L. E. 1985. The seasonal effectiveness on a Sphagnum wetland in removing iron and manganese from mine drainage. In *Wetlands and Water Management on Mined Lands*. Proc. of a Conference 23–24 Oct. The Pennsylvania State University, University Park, PA, pp. 385–386.

Mitsch, W. J., and J. G. Gosselink. 1986. *Wetlands*. Van Nostrand Reinhold, New York. 537 pp.

Mitsch, W. J., M. A. Cardamone, J. R. Taylor, and P. L. Hill, Jr. 1985. Wetlands and water quality management in the eastern interior coal basin. In *Wetlands and Water Management on Mined Lands*. Proc. of a Conference 23–24 Oct. The Pennsylvania State University, University Park, PA, pp. 121–137.

Nawrot, J. R. 1985. Wetland development on coal mine slurry impoundments: Principals, planning and practices. In *Wetlands and Water Management on Mined Lands*. Proc. of a Conference 23–24 Oct. The Pennsylvania State University, University Park, PA, pp. 173–179.

Nawrot, J. R., and S. C. Yaich. 1982. Wetland development and potential of coal mine tailings basins. *Wetlands* 2:179–190.

Nichols, S. 1983. Capacity of natural wetlands to remove nutrients from wastewater. *J. Water Pollut. Control Fed.* 55:495–505.

Olson, G., S. C. Turbak, and G. A. McFetters. 1979. Impact of western coal mining II. Microbial studies. *Water Res.* 13:1033–1041.

Seidel, K. 1976. Macrophytes and water purification. In J. Tourbier and R. W. Pierson, Jr., Eds., *Biological Control of Water Pollution*. University of Pennsylvania Press, Philadelphia, PA, pp. 109–122.

Singer, P. C., and W. Stumm. 1970. Oxygenation of ferrous iron. U.S. Department of Interior, Fed. Water Qual. Admin., Water Pollut. Cont. Res. Series Rept. 14010-06/69.

Snyder, C. D., and E. C. Aharrah. 1985. The *Typha* community: A positive influence on mine drainage and mine restoration. In *Wetlands and Water Management on Mined Lands*. Proc. of a Conference 23–24 Oct. The Pennsylvania State University, University Park, PA, pp. 187–188.

Stark, L. R., R. L. Kolbash, H. J. Webster, S. E. Stevens, Jr., K. A. Dionis, and E. R. Murphy. 1988. The Simco #4 wetland: biological patterns and performance of a wetland receiving mine drainage. In Mine Drainage and Surface Mine Reclamation: Mine Water and Mine Waste. Bureau of Mines Information Circular 9183, Pittsburgh, PA, pp. 332–344.

Tarleton, A. L., G. E. Lang, and R. K. Wieder. 1984. Removal of iron from acid mine drainage by Sphagnum peat: results from experimental laboratory microcosms. In Symposium on Surface Mining, Hydrology, Sedimentation and Reclamation, University of Kentucky, Lexington, pp. 413–420.

Tilton, D. L. and R. H. Kadlec. 1979. The utilization of a freshwater wetland for nutrient removal from secondarily treated waste-water effluent. *J. Environ. Qual.* 8:328–334.

Topper, K. F., and B. R. Sabey. 1986. Sewage sludge as a coal mine spoil amendment for revegetation in Colorado. *J. Environ. Qual.* 15:44–49.

U.S. Environmental Protection Agency. 1985. Freshwater Wetlands for Wastewater Management Handbook. EPA Region 4, Atlanta, Ga. EPA 904/9-85-135.

Waltzlaf, G. R. 1985. Comparative tests to remove manganese from acid mind drainage. In Control of acid mine drainage. Proceedings of a technology transfer seminar. Bureau of mines information circular no. 9027. U.S. Department of the Interior, pp. 41–47.

Warburton, D. B., W. B. Klimstra, and J. R. Nawrot. 1985. Aquatic macrophyte propagation and planting practices for wetland establishment. In Wetlands and Water Management on Mined Lands. Proc. of a Conference 23–24 Oct. The Pennsylvania State University, University Park, PA, pp. 139–152.

Wheeler, B. D., M. M. Al-Farraj, and R. D. Cook. 1985. Iron toxicity to plant in base-rich fen wetlands: comparative effects on the distribution and growth of *Epilobium hirsutum* L. and *Juncus subnodulosus* Schrank. *New Phytol. 100:*653–669.

Wieder, R. K., and G. E. Lang. 1982. Modification of acid mine drainage in a freshwater wetland. In Proceedings of the Symposium on Wetlands of the Unglaciated Appalachian Region. 26–28 May. West Virginia University, Morgantown, pp. 43–53.

Wieder, R. K., and G. E. Lang. 1984. Influence of wetlands and coal mining on stream chemistry. *Water, Air Soil Pollut. 23:*381–396.

Wieder, R. K., G. E. Lang, and A. E. Whitehouse. 1985. Metal removal in *sphagnum* dominated wetlands: Experience with a man-made wetland system. In *Wetlands and Water Management on Mined Lands*. Proc. of a Conference 23–24 Oct. The Pennsylvania State University, University Park, PA, p. 353–364.

Wile, I., G. Miller, and S. Black. 1985. Design and use of artificial wetlands. In P. J. Godfrey, E. R. Kaynor, S. Pelczarski, and J. Benforado, Eds., *Ecological Considerations in Wetlands Treatment of Municipal Wastewaters*. Van Nostrand Reinhold, New York, pp. 26–37.

Wolverton, B. C., R. M. Barlow, and R. C. McDonald. 1976. Application of vascular aquatic plants for pollution removal, energy, and food production in a biological system. In J. Tourbier and R. W. Pierson, Jr., Eds. *Biological Control of Water Pollution*. University of Pennsylvania Press, Philadelphia, PA, pp. 141–150.

Yeasted, J. G. and R. Shane. 1976. pH profiles in a river with multiple acid loads. *J. Water Pollut. Control Fed. 48:*91–106.

13

ECOLOGICAL ENGINEERING OF COASTLINES WITH SALT-MARSH PLANTATIONS

Chung-Hsin Chung

Institute of Spartina and Tidal Land Studies, Biology Department, Nanjing University, Nanjing, China

13.1   INTRODUCTION

13.1.1   Coastline Problems

Most people are unaware of the very important problems of increasing atmospheric carbon dioxide and rising sea level in the next century. Titus (1986a,b) concluded that a 1.11°C warming by the year 2065 cannot be avoided and estimated that the sea level is likely to rise 1.22–2.134 m by 2100, but that a rise as low as 30.48–15.24 cm or as high as 3.658 m cannot be ruled out. Table 13.1 illustrates many of the anticipated problems due to sea level rise on the coastline, including flooding of coastal plains, aggravation of catastrophic storm tides, intrusion of salt water, and erosion of tideland (Yang, personal communication, 1987).

13.1.2   Management Objectives

There are several management objectives involved in the protection of the coastline. They include (1) effective protection of the coastline, the prerequisite for all other objectives; (2) transformation of tideland to fertile and firm soil for agricultural and industrial development—this requires transforming soft marsh soil to a more compact texture to increase porosity; (3) increasing marshland area and its elevation; (4) creating and extending communities of primary producers, especially macrophytes for colonizing barren salt flats and embanked saline soil, after which diversity is stressed to gain homeostasis.

Based on ecological principles, different portions of the coastline are utilized in various ways to work out a development plan that coordinates different interests.

13.1.3   Ecological Engineering Principles

According to the introduction to this book, ecological engineering is the design of human society with its natural environment for the benefit of both. Ecological engineering of coastlines, conforming to this definition, may mean the design of a coastline human society with its environment for the benefit of both. Ecological engineering of coastlines has its basis in the scientific theories of general, estuarine, marine, and salt-marsh ecology that have been developed over the past 80 years. Application of these principles to solve coastline problems will sustain existence of both human society and its environment. We started from two ecological principles: (1) the capability of plant communities to modify their environment and (2) the existence of plant communities that can withstand harsh conditions, except for the most extreme ones. We selected *Spartina anglica* C. E. Hubbard to achieve or partially achieve the following goals: stabilization of coastline; acceleration of accretion for reclamation or other purposes; uses as green manure, animal

Table 13.1 Coastline Problems, Including Sea Level Rise

PROBLEMS DUE TO WATER ACTION

1. Heavy rainfall causing inundation of land
2. Intrusion of salt water by overflowing of tides
3. Drought leading to shortage of water to flush silt to sea and resulting sedimentation on river bed
4. High wave energy inducing erosion, etc.
5. Problems due to soil factors
6. Dredging operations including dredge spoil flushed back to sea and river with resulting high cost
7. Salinization of embanked land
8. Intensive evaporation, concentrating dissolved salts of soil water
9. Slow salinization and desalinization of clay soil
10. Desalinization of soil water lagging desalinization of soil
11. Wind blowing sand landward, causing need for reseeding of crops
12. Leaking of saltern
13. Low organic matter
14. Much slower composition of organic matter in clay than in sandy soils
15. Digging sand or shell sand for building purposes
16. Mud flats too soft for walking
17. Impossible or difficult to cultivate embanked tideland owing to barrenness, salinity, and soil poverty even after being fallow a long time

PROBLEMS OF RESOURCES

1. Shortage of timber, firewood, and fuel to be supplied from outside
2. Overfishing by domestic and foreign fishing boats, causing abrupt decline in total and per capita catch, such as in China
3. Possible extinction of common fish such as yellow croaker (*Pseudoscinena crocca*) and small yellow croaker (*P. polyactis*)
4. Pollution making the coastline a reservoir of pollutants, leding to ecosystem fragility
5. Reduced survival of many species of invertebrates and toxic food chain established
6. No or very little use of tidal, wind, and solar energy
7. No or few macrophytes as "machines" to harness solar energy and for elemental recycling

PROBLEMS OF LABOR FORCE

1. Labor repairing earthen sea walls as at Qidong in Jiangsu, China
2. All-out defensive activities against high tides, storms, and typhoons
3. Repair of breaches of sea bank
4. Labor wasted as a result of improper planning of coastline use
5. Dredging and digging operations
6. Embankment at very low elevations leading to financial loss

PROBLEMS OF COASTAL MORPHOLOGY

1. Subsidence in some coastlines in addition to sedimentation, erosion, and siltation

fodder, fish feed, and fuel; increasing production of invertebrates; and partial control of waterway siltation and of pollution.

There are advantages and disadvantages to ecological engineering and of coastal engineering, and situations in which methods from each are most appropriate. Our biological measures are inexpensive, simple, easy to operate, and lasting, but they are unable to cope with high wave energy situations. Coastal engineering is effective in such situations, where salt-marsh vegetation cannot be created. However, it is expensive and time- and labor-consuming. Salt-marsh vegetation, on the other hand, is a renewable resource, so it can be used again and again.

13.2 HISTORY AND RECENT RESEARCH

13.2.1 History of *Spartina anglica*

Mobberley's (1956) monograph on the taxonomy and distribution of the genus *Spartina* is indispensable. K. G. Boston (1980) gave a good, brief historical account as follows: "*Spartina anglica* C. E. Hubbard is an amphidiploid form of *S. townsendii* (*sensu lato*), which arose in Southampton water *circa* 1870 as a result of hybridization between English *S. maritima* and the accidentally introduced American species *S. alterniflora*. Early collections including the 1870-type specimen were of a sterile F1 hybrid, now named *Spartina X townsendii* H. & J. Groves. The fertile *S. anglica,* which seeds profusely, is believed to have appeared in about 1890, when the new *Spartina* first began to spread rapidly by natural dispersal along the south coast of England. *S. anglica* is now the common form—of an estimated 12,036 ha of *S. townsendii* (*s.l.*) in Great Britain, and 20 ha are of sterile *S. x townsendii.*" Other interesting historical accounts of *Spartina anglica* are those by Hubbard (1965) and Marchant (1967). The review by Goodman et al. (1969) and Ranwell's (1972) book *Ecology of Salt Marshes and Sand Dunes* are also recommended.

13.2.2 Recent Research on *Spartina anglica*

Our basic research of *Spartina anglica* is summarized here to assist with an understanding of its biological characteristics in relation to its application to ecological engineering.

Seed Morphology in Relation to Germination. Difficulties of seed germination of *Spartina anglica* make production of uniform-age seedlings infeasible. Experimental results showed normally developed seeds with high percentages of germination, those with endosperms larger than embryos next in germination, and seeds with poorly developed embryos or with soft embryos and endosperms incapable of germination at all. Only 22.2% germination was observed in seeds with milky-white endosperms, even though seedlings were unable to survive. Morphological and anatomical characteristics enabled one to distinguish the presence or absence of germinability and the degree of germination to a certain extent. Zhou and Chung (1985) tackled this problem to pave the way for other studies.

Phenomena of plants of shorter stature after *Spartina*'s introduction to China were repeatedly reported, until Hubbard (1969) put forward an explanation of stature increase with increased daylength. Our latitudes and day lengths are lower and shorter, respectively, than those of England.

Anatomical Studies. Wang et al. (1979) confirmed the findings of Long et al. (1975) on leaf anatomy. Sung and Dou (1982), in their study of culm and leaf sheaths, found a circle of large, well-developed air passages about three cells below the epidermis in the cross-section of its culm. Vascular bundles were seen to be arranged in three circles. Adaxial and abaxial surfaces of the leaf sheath differ strikingly. Stomata were found on both surfaces. Salt glands occur only on the abaxial surface. Wang and Dou (1985) discovered root hairs on secondary adventitious roots and side roots, but their numbers were less than on plants grown in petri dishes. The presence of air was shown to play a key role. Zhou et al. (1982) found no salt glands on stems. The most efficient type of salt gland in the Gramineae examined comprised merely two cells—cap and basal cells. In both, dense cytoplasms, large nucleoli, numerous mitochondria, and a few other organelles were discernible. Clear plasmodesmata were perceived in walls connecting basal and epidermal cells and in those connecting basal and cap cells. Both optical and electron microscopy were employed. Jiang and Huang (1982) devised a better method of preservation of membranous systems, nucleoli, and lipid droplets with digallic acid-treated transmission electron microscopic specimens of plant materials.

Cytogenetics. Chromosomes of *S. anglica* were re-counted by means of colchicine, cellulase, and pectinase treatment under the optical microscope and found to be 116, instead of 122 and up as reported by Marchant (1968), using an electron microscope. Fang et al. (1982) reported 116 after repeated examinations.

A slightly improved Bernard's method was tested and found to have advantages. Impurities are eliminated without using organic solvents or enzymes and high-speed centrifuging is not required, creating a simple and effective method to purify DNA. A saving of many steps in Marmer's method was claimed by Lü and Lu (1985).

Chen and Duan (1985) reported success in transferring *S. anglica* DNA to rice with an increase of contents of protein and 16 amino acids, whereas other morphological characters were observed to be reduced in size or in number. Three DNA transfer plants designated D_1 (descendants) evolved, and continuous propagation of seven generations without any segregation has been noted.

Genecological Studies. A four-year study clarified the issue of reported tall forms. These grew to similar heights under similar experimental conditions in our botanical garden except for one case of slight persistence (Zhuo, unpublished). We may consider them as ecophenes of the same ancestors of the Essex plants, and time has been too short to differentiate them genetically.

However, three introduced populations from Essex, Poole Harbour, and Lancashire have been tested to be ecotypes, although no distinguishing morphological characters can be immediately recognized. Nevertheless, in stress conditions, changes emerged in the ultrastructure of leaf cells, proline accumulation, and two isozyme bands. The primary differences existed in stress ecophysiology, hereditary substances, and hence their regulation. Zhoun and Chung (1987) determined the Lancashire ecotype as low-temperature, salt-tolerant; the Essex ecotype as low-temperature, low-salt-tolerant; and the Poole Harbour ecotype as high-temperature, salt-tolerant.

Plant Physiological and Ecological Studies. For a long time, research on anatomy, absorption of ions, and abstraction of DNA was handicapped by the plant's peculiar, sporadic way of germination. Zhou and Chung (1985) succeeded in solving this difficult problem by an easy method of puncturing seeds. Percentages of germination as high as 91.7%, against 25% for the control, were recorded. Zheng et al. (1985) worked out a method of reducing atmospheric pressure to 0.4 and 0.8 with profit. One to three atmospheres had no appreciable effects, but four actually decreased germination.

Probably the first experiment of alkali tolerance by *S. anglica* in China was performed by Ou et al. (1982). Culm height, number of leaves, and plant dry weight of study plants grown in a 500–2000 ppm sodium carbonate nutrient solution (pH. 9.5) exceeded those of a control. The best performance was found at 1000 ppm, doubling the weight of that at 4000 ppm, its lethal concentration. Its maximum tolerable concentration was observed to be 3000 ppm. Later Ou et al. (1985) found chlorophyll content in fresh leaves of plants grown in alkaline soils versus controls to be 1.267 versus 1.080 mg/ g. Moreover, contents of both chlorophyll a and b exceeded those in the control. These results demonstrated no unfavorable effects to the formation of chlorophyll of plants in alkaline soil. Contents of protein in roots, stems, and leaves of 100-mg mixed dried samples were 1.51 versus 1.49 mg for the control. Similar results were also found for K^+, Na^+, and Cl^-. Higher amounts of Na^+, Cl^-, and Ca^{2+} excretion by plants in alkaline soils over the control and lower content of K^+ excreted were considered beneficial to *S. anglica*. This plant, grown in alkaline conditions, was shown to decrease pH and conductivity of nutrient solutions.

Crude protein of leaves was found to exceed that of shoots (stem plus leaf) in samples from Qidong county: 13–19% versus 9–13%. During the period of emergent lush growth (May and June), protein was more abundant than in the fruiting period. Carotene content was noted to vary with protein, but there was no appreciable seasonal variation in crude fat. Leaf crude cellulose was less than in the shoot, with its maximum in July. Higher calcium content in leaves ranged from 0.6 to 0.8%. Calcium in the shoots ranged from 0.3–0.4% in June–September to 0.65% in the fruiting period. Phosphorus in leaves fluctuated little, and was generally 0.2–0.3%. Total ash accumulated as high as 20–25%. Among 18 amino acids, glutamic acid was the most abundant and histidine the least. Considerable amounts of 10 essential amino acids were analyzed, with 6 of them, including phenylalanine and valeric acid, exceeding in amount those of representative forage grasses in the western world. Lu and Jiang (1981) analyzed samples collected over a 10-month period.

Chung and Qin (1983, 1985) wanted to find out if *S. anglica* absorbed and accumulated mercury, as does *S. alterniflora*. The amount of mercury absorbed from water culture of different concentrations was determined to be 10–56 times the initial quantity absorbed by the shoots and 250–2500 times by the roots. The most efficient absorption by tillers was almost half of the original in 1 ppm concentration after 4 weeks. The absorption curve is a positive relative-component curve, so this plant has been chosen as an agent to remove pollutants, especially mercury.

Wang et al. (1985) extended this line of research with four radioisotopes: ^{137}Cs, ^{90}Sr, ^{115m}Cd, and ^{85}Zn. Different concentration factors in different organs were detected with concentration factors in roots greater than in leaves, which in turn were greater than in stems. Root concentration factors were usually 300, but the maximum could be 600. Different radioisotopes had different concentration factors in different organs of this plant, with ^{90}Sr having the greatest, 2–10 times those of other radioisotopes. Concentration factors increased progressively in different organs. Most of them reached their maxima in 40 days and then declined, except for ^{85}Zn, which remained unchanged, and ^{90}Sr, which kept increasing after 40 days.

Stewart and Lee (1974) emphasized proline in relation to salt tolerance, arousing great interest here. Zhang et al. (1985) found much less proline in their plants than in those in England, but aspartic acid, glutamic acid, alanine, and glycine were more abundant. Essential amino acids were present to a much greater extent in leaves than in roots. Other workers confirmed a correlation between external salt concentration and proline production. A newly devised gas chromatographic method for measuring free proline by Yuan et al. (1981) has been recognized to be superior to colorimetric methods, because the color complex derived from indigo is reduced and proline is very unstable, light sensitive, and liable to experimental errors. Lu and Jiang (1983) improved Paquin's method of proline extraction with efficiency and a stable product.

Tissue culture success using the apex of *S. anglica* was first reported by

Li and Zhang (1985). Ten days after inoculation of stem tips in test tubes, differentiation of juvenile leaves began and after 20 days plantlets 2.0–2.8 cm in size appeared and elongated. It took half a month to form two to four juvenile roots 1–2 cm long after transferring the plantlets to rooting media. Rapid growth during 1 month led to roots 4–6 cm long. Another group of juvenile plants grew very slowly at 6–8°C, yet survival rates of 100% were amazing.

Microbiological Studies. A preliminary study by Yu and Cao (1983) had the following results. The greatest number of total bacteria, including nitrogen-fixing bacteria, was in the rhizosphere soil of *S. anglica*, the next highest in the nonrhizosphere soil of *S. anglica*, and the lowest in the control soil. The greatest number of total bacteria occurred in May, with less variation than in the control. There were changes of dominant types with *Micrococcus*, *Staphylococcus*, *Corynebacteria*, and *Pseudomonas* in the beginning of the growth phase of *S. anglica* and *Bacillus*, *Pseudomonas*, *Actinobacter*, and *Alkaligenes* later. Spore-forming bacteria were found as dominant types in the control soil and non-rhizosphere soil of *S. anglica* from beginning to end. They were rarely found in the rhizosphere of *S. anglica* in its beginning growth phase, but increased in later phases. A 10-fold increase of rhizosphere nitrogen-fixing bacteria over the control provides a convincing explanation of soil fertility of *Spartina* polders.

Zhou et al. (1985) carried out a numerical taxonomic study on the Gram-negative bacteria isolated from the plant surface of *S. anglica* by means of a TRS-80 microcomputer, with 133 morphological, physiological, and biochemical characters. Dendrograms were prepared to express phenon clustering. *Pseudomonas alkaligenes*, and 23 unidentified strains of *Pseudomonas* were found to be dominant. Though nitrogen-fixing capacity characters scattered into many phenons, this capacity still was found to occur more frequently in dominant strains.

Zoological Studies. The objectives of this 2-month survey in Qidong County, Jiangsu were to collect basic data on animals of salt flats and *Spartina* marshes and to clarify whether antagonism exists between *Spartina* plantations and animal survival, especially commercially important species. Among the latter, *Meretrix meretrix*, *Cylina sinensis*, and *Mactra veneriformis* were seldom collected. This was not entirely due to the presence of *Spartina*, because *Meretrix meretrix* generally occurs in sandy flats of 50–90% sand and silt. Its optimal growth has been observed in 60–80% sandy and silty soil. Tong et al. (1985) expressed the view that further study was needed.

13.3   CASE STUDY: WENLING, ZHEJIANG

This section reports our first comprehensive treatment of a 22-year case study from 1964 to 1986 in Wenling, on the east coast of China (Figure 13.1).

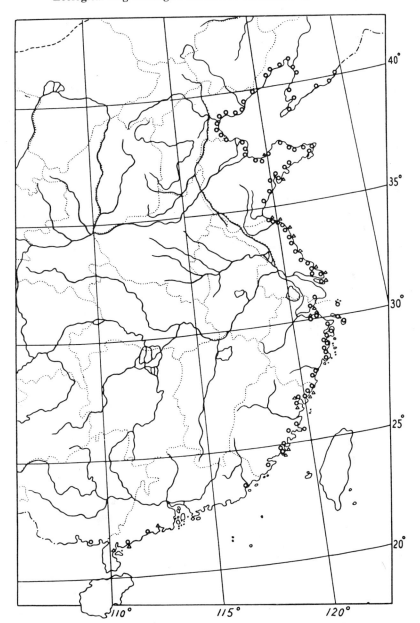

Figure 13.1 The distribution of *Spartina* on the coast of China. Circles show *Spartina anglica;* triangles show *Spartina alterniflora* (Chung, 1985; reprinted by permission of Gebrüder Borntraeger, Berlin Stuttgart).

The study is an ecological engineering project transforming the tidelands to farmlands and citrus groves, thereby improving the human society and its natural environment for the benefit of both. Preparatory work prior to trial plantings or even original thinking since the 1930s are not discussed at length here, but data gathered all through these years are synthesized. The published data are drawn from a special issue on advances in *Spartina* research of the Journal of Nanjing University, 1985. The unpublished data were collected in my fieldwork, in meetings, in visits and discussion, through surveys and correspondence, and so on.

13.3.1 General Description and Site Location

Our original site was located in a tideland 150 m east of a sea wall, Dongpian Farm, in the most northeastern part of the county. Wenling lies about 28°N along the East Sea of China. A tidal creek flows northwest to southeast north of the site. Three hundred meters eastward lies the Jingqing River. Our planting site was surrounded by a sea wall landward and a series of rocky islands seaward. In fact, the site was enclosed by more than three sides, more favorable than a planting site at Sloedam, the Netherlands. Waves and winds were greatly lessened by this particular topography.

Climatic Conditions. Based upon three-year records (1965–1967), the following data were useful. January mean temperature ranged from 5.1° to 8.0°C with the absolute minimum −5.0°C. The August mean ranged from 27.4° to 28.9°C, with the absolute maximum 36.4°C. Annual precipitation fluctuated from 1074 to 1698.9 mm, with 146–178 rainy days per year. There were 234–255 frostless days, with the earliest frost day on October 29, 1966 and the latest frost on March 24, 1967.

Soil Conditions. Analytic results of soils samples collected on October, 1965 (the top 10 cm of soil) were as follows: heavy clay soil with soluble salts 1.8182%, organic matter 0.8575%, total nitrogen content 0.0756%, and total phosphorus 0.0359%.

Tidal Submergence at Different Elevations. At elevations less than 3 m above sea level in July, August, and September no case of nonsubmergence was observed. During these three months consecutive days of nonsubmergence were recorded: 2 days at 3.5 m and 5 days at 4 m.

13.3.2 Problem in Detail

Increasing the amount of arable land for an increasing population has been of prime importance in China. Vast areas of land have been formed along a prograding coast in this country, as witnessed by names of inland villages such as Long Island and, in a neighboring county, Peach Island. The famous philosopher of the Sung dynasty Chu Xi (1130–1200 AD) studied tidal waves near Long Island for the construction of sea banks, but now that site is about 30 km from the coast. In 1979, only 0.033 ha of arable land was owned by each citizen of Wenling! The agricultural authority of Taizhou Prefecture told me of their preference to gain arable land from the sea rather than to cultivate seafood on the tideland.

Erosion of Tideland and Sea Walls. The three sides of the Wenling coastline total 160 km. Even though progradation takes place in the long run, erosion also has been taking place. For example, about one-third of the sea banks of Dongpian Farm were destroyed even though they were protected by islands. As a result of this, a new portion of sea wall had to be built.

Poor drainage of inland water to the sea leads to inundation of the land and siltation of river beds. Digging of mud and spreading it along the sides of the river and dredging operations by boats were tried, but the ultimate solution relied upon building a new dam at mouth of the Jinqing River.

Saline Soil Problem. High precipitation has not been effective in leaching salts from the soil to the water table. In the days before *Spartina anglica* plantings, embanked soil needed a fallowing of 3 years, followed by sweet potato cultivation for 2 years and then rice culture. Poor soil structure resulted from using newly embanked salt flats as well as from overapplication of chemical fertilizers. Increasingly sticky soil structure inadvertently occurred.

Soft Mudflats. People collecting bivalves, mollusks, and mussels sometimes sank into very soft mud and died. In 1966 I sank thigh deep and found it impossible to move my legs from the mire. Two colleagues of mine saved me at last.

Shortage of Fuel. I saw even very short grass being cut for fuel inside the sea wall.

Poor Quality of Animal Fodder and Its Shortage. *Arundo donax,* a robust Mediterranean tall grass, was planted on the slopes of earthen sea walls for protection but never succeeded because of grazing by cattle.

13.3.3 Original Thinking and Preparatory Work

Since 1934 I have been immensely interested in reading the *Spartina* literature and had hoped someday to be able to establish plantations for the benefit of human society. Visits to Holland, England, and Scotland were made in 1964, although in my early visit to Ireland in 1935 I had seen a *Spartina* plantation without knowing it at that time. I asked my friends to visit J. A. Jørgensen, who helped us to collect living plants in Højer *Spartina* marsh. Later I learned from Bird that the marshy areas of Denmark are one of the few places of this country where a real accretion occurs, but he concludes that, in general, recent coastal recession predominates in Denmark (Bird, 1974).

In Holland I visited Van Schreven and Beeftink with a better understanding of this new hybrid form. I saw Beveland and Walcheren united by *Spartina* marsh, and natural dispersal of seeds in both Zeeland and Friesland strengthened my belief in the promise of *Spartina* plantings. Beeftink and the International Institute of Reclamation and Improvement presented more very useful literature. In addition, C. E. Hubbard in England and Bryce in Scotland offered good opportunities for learning, advice, and encouragement. At Poole Harbour Ranwell and J. C. E. Hubbard showed me the natural, rapid spread of *Spartina* in a favorable environment. Ranwell and Morley showed me Steart Flat, Bridgewater Bay, and Somerset, a site of former erosion. Lambert drove me to Hythe, the original site of natural hybridization, and Lymington. Essex River Board colleagues took me to see more *Spartina* marshes in the Blackwater Estuary and the Essex coastline. Swan took me to Wolverton, Norfolk, where we looked from a distance at the Wash, and to Kingslynn. More literature, including that from the Herbarium at the British Museum, has been enlightening to me all these years. I am very grateful to all these scientists who helped me in various ways.

13.3.4   Planning, Design, and Construction

In order to solve the problems listed in Section 13.3.2, I selected only halophytic macrophytes that provided different mechanisms of salt tolerance for survival on tideland and that had large biomass. Candidates for consideration had to be primary producers; accelerators of sediment particles; mud and sand binders; ameliorators of saline soil; natural buffers to shore erosion; machines to harness solar energy and energy flow; systems of absorbing, accumulating, and recycling biogeochemical elements; soil conditioners making soft clay mud compact and hard soil of moderate texture; and animal feed and fuel. Other attributes included rapid propagation of large number of individuals with less expenditure, fewer labor days and materials, and resistance to submergence and anaerobic conditions.

Chinese marsh plants such as *Suaeda salsa,* an annual, and *Aeluropus littoralis* var. *sinensis* were considered not up to standard, so selection was narrowed to *Spartina anglica.* Introduction, propagation, stress experiments, riverside trial plantings, and seashore trial plantings on small and large scales were planned to proceed in this order.

Introduction. Only *Spartina anglica,* which was named in 1968 by Hubbard (1968), was imported on account of its lower-elevation habitat, more vigorous growth, and wider seaward distribution (Bryce, personal communication, 1963). The sterile male species was suppressed year after year by the stress resistance, biomass, and aggressiveness, in addition to the above properties, of the amphidiploid fruiting species. Four batches in total were imported in 1963 and 1964. The first batch from Mundon, Essex had 35 individuals, among which only 21 were living and the other 14, with most of their roots cut, were incapable of growth. These 21 individuals were destined to be the ancestors of *S. anglica* plants of almost all Chinese plantations! The second batch was Højer plants, collected from a marsh of 100 ha that had developed from 0.5 ha in the early 1930s. The Xinyang Agricultural Experimental Station, Sheyang, Jiangsu undertook experimental work on these two consignments. The third batch consisted of both seeds and plants, also from Mundon, Essex, with only the former being in good shape. Finally, Ranwell sent 18 plants from Poole Harbour, among which 15 were used for student theses and only three were propagated for planting on the lowest elevations just above Jinquin River. This fringe vegetation is still existing there. The third and fourth batches were our experimental materials.

After a very long time, hope was realized in 1963. Just prior to importation I was asked whether success could be achieved or not. My prediction was positive because I felt there must be some places in our study area similar to *Spartina*'s native ecological conditions.

Chung (1982) described three methods of propagation in some detail. Here some new data are used to supplement former accounts.

Propagation from Seeds. Because of postal delivery delays, we found plants rotting and becoming soft owing to high room temperatures during about 2 months in the post office. We had no fresh grass to experiment with but we did have 507 seeds. We started our adventure in mid-February. First of all, fungi and molds on seed coats were washed away before disinfection. Most of the seeds, except those immediately used, were stored in a refrigerator. Germination tests were carried out five times by selecting plump seeds and spreading them on moist filter paper in petri dishes. After germination, water and soil cultures followed. Goodman's (1960) method of using Knop's solution was not successful for us because high greenhouse temperatures promoted prolific growth of green algae that suffocated seedling growth. Poor growth in the soil was due to the presence of *Fusarium* and to the leaching of water through the drainage holes of the pots. With better soil and with the filling of the holes with cement to maintain a saturated state of soil, these problems were solved.

Germination of seeds also took place in a refrigerator. From a total of 507 seeds, 157 germinated, a rate of 30.97%. Forty-four seedlings ultimately survived, a rate of 8.7%. We discovered that germination of seeds on ears was higher than that of fallen seeds. After-ripening may enter as a factor in the later case.

Tillering started only after formation of five or six true leaves, which occurred in early June, 1964. From then on, an increment rate like that of

Table 13.2 Number of New Tillers Increased per Month in Open (June–October) and in Greenhouse (November–April)

Month	New Plants (increase/month)
1964	
June	244
July	767
August	2096
September	3617
October	3939
November	2561
December	4143
1965	
January	3670
February	2710
March	3140
April	3670

compound interest resulted, as shown in Table 13.2. More than 10,000 individuals were propagated in the open (Figure 13.2) and more than that were produced in greenhouses with an average air temperature exceeding 20°C and a maximum of 30°C. A total of 30,601 individuals were counted on April 26, 1965. Tillers also included rhizomes growing above soil surface. Because it was hard to distinguish the latter, the general term tillers has been employed for convenience.

Seedlings in pots occupied the full soil surface after 1 month (June). They had to be transplanted from time to time into large glazed vessels 70 cm in diameter filled with very rich pond mud, treated with salt water, manure, and fertilizers. There was a 694.5-fold increase of the original 44 individuals in less than a year, excluding the period of germination of seeds to the beginning of tillering.

This method took a long time to reach the stage of tillering; it is recommended only for new introductions when transporting fresh material is not possible. Its advantage was immediately recognized to use for selection purposes, on account of segregation producing individual plants all different from one another.

Propagation of Sprigs. Prior to the June tillering in 1964, there were no plants in our botanical garden possessing the power of vegetative multiplication. I brought six individuals from Zhoushan back to make a trial test in May, 1964. Planting them in a large glazed earthen vessel and finally trans-

Figure 13.2 More than 10,000 individuals of *Spartina anglica* were propagated in the open from 44 seedlings just prior to removal to the greenhouse in early November, 1964 (photograph on October 30, 1964).

planting them to eight vessels were necessary for full growth of tillers in April, 1965; these totaled 3607 after 11 months and 11 days. Immediate tillering after planting has been observed to be advantageous and has been widely recommended. In fact, in planting with a high percentage of transplants, care was taken to keep roots moist by watering during transport, as in the case of transport from Sheyang, Jiangsu to Tianjin in 1965.

Propagation from Rhizome. Our second experimental site was on the tide-influenced river flats of River Chientang, Farm of River Chientang. Because other planting materials were not available then, we brought more than 200 individuals in two pots from Xinyang Agricultural Experimental Station, Sheyang, Jiangsu to our second station, south of the river at Hangzhou, Zhejiang. Rhizomes left over after planting in the tidal creek were collected and planted in one pot. This was done for two reasons, namely, to propagate more plants and to find out whether they were capable of producing shoots and roots. Four days after being set out in early May, one segment of rhizome began to function first with root, followed by stem and leaf formation. Fourteen individuals were counted 82 days after setting (Figure 13.3*a*), 152 after 184 days (Figure 13.3*b*), 320 after 375 days, 26,000 after 500 days, 400,000 after 700 days, 1,450,000 after 790 days, and 9,100,000 after 850 days. Actual counts were made of plants growing in pots and es-

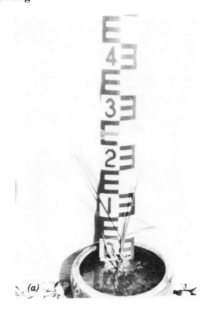

Figure 13.3 Propagation of *Spartina anglica* from rhizomes; (*a*) 14 individuals were counted 82 days after setting; (*b*) 152 individuals were counted 184 days of setting.

Figure 13.4 *Spartina anglica* individuals transplanted from pot to a 0.013-ha paddy in May, 1965 in Wenling, Zhejiang.

timates were based on the number of grasses in quadrats for plants growing in paddies (Figure 13.4). This method has never been used by others because of time and labor requirements. Only because of urgent need in 1964 did we resort to this.

Propagation in Paddies. For large-scale propagation work, paddies were very useful in Wenling. Chinese farmers know how to construct rice paddies. To avoid soil leaching, water has been considered to be a very important prerequisite for propagation work with paddies as nurseries.

Stress Experiments. There have been many hurdles for this new introduction to cope with. Our continental climate has very cold winters and very hot summers. Polar continental air masses prevail in winter and warm currents from the tropics do not flow close to the mainland to any great extent. Thus winter temperatures in the area from 40°53'N to a little north of 20°N along the Chinese mainland coast are lower than those of the same latitudes elsewhere. Cold waves in January, 1955 reached the South China Sea and south of the equator. Tropical air masses dominate in the summer, so summer temperatures are generally higher than those of the same latitudes elsewhere. If one contrasts these conditions with the favorable climate of the British Isles, with temperate conditions at high latitudes (50–60°N) and modest extremes owing to the prevailing winds toward shore from off the Gulf

Stream, one can see that *Spartina* had harsh conditions to overcome in China. The mean winter temperature is 4°C at sea level. In August, the mean temperature at Essex is 16°C. Snow cover is of brief duration, with 15 days of snow in the lower Thames valley. In China, it scarcely snows south of 25°N, but in extreme winters snow falls even at Guangzhou (23°N). The January zero isotherm lies in high latitudes in Europe, but ours is at middle latitudes, with the coastal zero isotherm just north of 34°N. Other stresses such as typhoons generating heavy precipitation, high tides, and tremendous wave energy have also been thought of as being fatal to salt-marsh propagation (Chung, 1983).

Temperature Stress Experiments. We did not worry about low temperature stress around our latitudes of eastern China, because we had learned from safe overwintering of *Spartina* in Sheyang in its first winter in China that its absolute minimum was −11.3°C. Sheyang's latitude is around 33°N. Our first station in Nanjing (32°00′N) and a second one south of Hangzhou (30°19′N) were no problem for overwintering. We had still to get a true picture of how this newly introduced species would adapt to our summer, because the plants were grown in the summer of 1963 in shade especially made to guarantee their growth. Three times a day we recorded air temperatures of grass clumps to determine the temperatures actually affecting the grasses. It happened to be an ideal year for this test, for the maximum temperature reached 39°C, the highest recorded there in 30 years. Maxima of the air temperature of grass clumps from July 10 to 17 ranged from 40.5 to 42.0°C at 2 pm. With a sufficient water supply, *S. anglica* was not only capable of withstanding this crucial test, but also kept on tillering every day. From the beginning of July to mid-September, peaks of increment of new tillers appeared at 33–35°C, the maximum air temperature in the grass clumps. Twenty-three percent of new tillers emerged during three winter months of 1964–1965 in Nanjing, with its absolute minimum at −5°C and even with frozen ice sheets at times. During the same season in the Farm of River Chientang, sand bar rhizomes persisted to elongated belowground, as had been reported in England.

Drought Stress Experiments. A day's trial of withholding water from plants in glazed jars showed signs of danger, so it was discontinued the next day. Another experiment with tidal creek plants with only five periods of submergence between June 29 to July 18 on rainless days deviated from the approximately normal daily pattern. Water content in the soil was determined to be only 60%, and loss of water from grass was accelerated. Watering by manual labor managed to prevent plant death on account of coarse grains of silt causing rapid leaching downward.

Submergence Stress Experiments. A period of dormant or partially dormant submergence far exceeding that of normal submergence of tideland (i.e., more than 3 days) did not affect *Spartina* performance in its next growing season.

Accretion Stress Experiments. Moderate accretion of sediments did no harm to the grass, but high accretion suddenly wiped out the entire plantation on the Chientang River in late August, 1966. In addition, we excavated plants in Rudong County, Jiangsu buried as deep as 80 cm after a period of 3 years of accretion. In other words, this plant can survive an average rate of accretion of 26.67 cm/year. This was discovered in the 1970s, not along with the other experiments, which were performed prior to establishment of *S. anglica* plantations.

Establishment of S. anglica on a Riverbank. A source of planting materials from Sheyang numbering more than 200 individuals was available. Planting took place on April 30 and on May 2, 1964 in a tidal creek in the Farm of River Chientang on the south bank of the river. On June 15 some plants were washed away and 60 individuals were planted in glazed jars for propagation and stress experiments. Advantages of the riverside site were more favorable conditions with fewer days of gales and less wave scour than the seashore. On October 11–12, 871 individuals were transplanted from the tidal creek to the riverbank, where they were regularly inundated by tidal water. A total of 305,814 transplantings were counted at the end of December, 1965. A *Spartina* plantation of 0.2 ha seemed to be a success in 1966, but it was suffocated by a super-accretion of 1 m of sediments in late August! Plantings on an eroded tidal flat with little grass did not prevent crumbling soil blocks from falling from microcliffs.

Establishment of Coastal Spartina Plantations. Two sources of planting materials were available including 5248 individuals that were shipped from the Farm of River Chientang to Wenling from November 15 to December 2. Glazed jars with grass were stored in the soil to prevent freezing. At the end of December, 5924 individuals were counted, and on May 12, 1965, a total of 8538 was recorded.

Three clumps of grass were immediately planted at each elevation on intertidal zones for comparison with native marsh grass as control. Overwintering was successful. After 1 year and 7 months, a 310.44-fold increase of plants at 3.0 m, 105.3 fold at 3.5 m, and 8.09 fold at 4.0 m were observed. In contrast, native species increased only 2.4-, 5-, and 5-fold at the respective elevations. Native people were convinced by this experiment. The other source from Nanjing totaled 33,208 individuals. These were shipped during the period April 30 to May 11, 1965 to Wenling.

Objectives of the small-scale experiments were (1) to gain experience for the forthcoming establishment of large plantations and (2) to establish a tideland nursery for genetic selection of better forms, because every seedling differs from each other owing to segregation. Unfortunately the latter objective was not achieved owing to reclamation in 1969. That was the end of our third batch of introduced plants, namely, the second from Essex.

Transplanting was completed in mid-May, 1965. Four months later the total number of plants amounted to 218,496. The widest clump diameter was measured as 93 cm, with 340 tillers and 40 cm high. Six months later 420,120 were counted. A complete plantation was formed through coalescence of clumps after 1½ years for 1-m spacing and 2½ years for 2-m spacing.

Two-year experiments to determine the optimal planting season in Wenling found that the first half-year plantings outperformed the second half-year plantings. Plantings twice a month were continued. After 1 year multiplication rates ranged from 34.54-fold (planting date March 30) to 77-fold (planting date June 15) for first half-year plantings. Corresponding figures for the second half-year ranged from 8.83-fold (November 15 planting) to 22.33-fold (July 15). April, May, and June plantings achieved better results than all other months, with the highest rate in June and the next highest in May.

The 3-m elevation was found to be optimal for transplanting, with poor performance at 3.4 and 2.5 m and still worse at 4.0 and 2.3 m. At sites below 2 m no plants survived owing to too long a submergence and too short an exposure. We found a 136.4-fold increase of plants at 3 m, 88-fold at 3.4 m, and 13.76-fold at 2.5 m. Death was inevitable at 2.0 m after 140 days of survival, at 1.5 m after 110 days, and at 1.0 m after only 80 days. The optimal elevation range was 2.7–3.5 m. Less planting material was used in favorable sites than otherwise, for wide spacing meant economy of labor and time.

Establishment of Large Spartina Plantations. Sources of planting materials were derived from a segment of rhizome and four individuals out of the 60 grown in glazed jars in the Farm of River Chientang. Plants propagated from rhizomes in pots also were shipped to Wenling along with other jars, reaching their destination in early December, 1964. Up to May 17, 1965, the total number of plants amounted to 320. They were transplanted to a 0.013-ha rice paddy, and a second transplanting into a larger paddy of 0.297 ha became necessary because of complete cover of the former (Figure 13.5a). In late August, 1966, from the data of sample quadrats, about 9,100,000 plant individuals were calculated (Figure 13.5b). We had not anticipated that descendants of a segment of rhizome would be the pioneers of the largest plantation in China in the 1960s in just 2 years and 4 months.

Descendants of four individual plants were propagated in 1.067 ha of newly constructed paddies. Manure was applied as a base dressing, with irrigation and smoothing of the soil surface before planting taking place in early April, 1965. Two to three individuals in a sprig were set at 5–7 cm in depth. Water was supplied all year round to keep the soil moist even in winter. In the growing season, a water layer no higher than 3 cm was desirable. Application of nitrogen more than of phosphorus and potassium was appropriate. During the summer solstice, dressing was for lush growth, whereas prior to transplanting from the nursery, fertilization was to accel-

Figure 13.5 Creating a large-scale *Spartina anglica* plantation (*a*) 3 months after setting of sprigs (photograph taken in December, 1966); (*b*) after 15 months' growth (photograph taken in December, 1967). Photographs taken by R. Z. Zhuo (Chung, 1982; reprinted with permission from *Creation and Restoration of Coastal Plant Communities*, copyright, CRC Press, Boca Raton, FL). Nine million, one hundred thousand individuals propagated in 0.297-ha paddies were transplanted from those in the 0.013-ha paddy shown in Figure 13.4.

erate growth. Pest and weed control and other forms of management were also performed.

The first transplanting in September, 1966 on 78 ha, the second in the second half of April, 1967 on 65 ha, and the third in the summer of 1968 on more than 133 ha totaled more than 276 ha. In 1969 this plantation was not yet well established when I visited there. In 1972 four communes began to embank, and embankment was completed in December, 1973. Soil profiles and stakes both indicated 80 cm of accretion within 7 years, conforming to permissible elevations for embankment. By comparison, it took 24 years for Sloedam to empolder, from 1924 to 1948.

Optimal planting time was April and May, with survival rates as high as 98%, the next highest in September and October with 80%, and the lowest in July and August, with merely 30%. In 1979, transplantings in January and February also were successful, with survival rates of 90%. Average rates of increase of plant individuals generally approximated 30-fold, with higher rates of 50–60-fold. The temperature initiating growth was observed to be 12°C, 20°C was the optimum for transplanting, and 25–30°C was best for the most rapid propagation. Special personnel were commissioned to inspect and protect newly planted clumps from trespassing and harvesting.

13.3.5   Effects on Coastal Morphology

Accretion and Reclamation. Accretion by *S. anglica* was first observed when a clone spread and became the site of low mounds, rising above the general mud flat level up to 19 cm in 2½ years (Chung, 1985). Commune members witnessed the sudden change of muddy tidal water into clear water owing to obstruction by the *S. anglica* plantation or clumps. I had not anticipated this rapid accretion from such water, which flows from a medium-size river north of Wenling. In the first three years of plantings, accretion rates were lower than those in the years after formation of closed stands. There were stems, roots, rhizomes, and leaves that had not undergone decomposition in the excavated profile. This was certainly proof of a rise of the soil surface of 80 cm within 7 years by *Spartina* accretion. This has been viewed as a process of collection and reuse of sediment particles suspended in tidal water. Otherwise the particles are merely transported from one spot to another, with a waste of tidal energy and with no contributions to humans or to ecosystems. Moreover, the suspended material prevents penetration of solar energy into seawater for phytoplankton use, thus decreasing production of fish and other animals in the sea. It is known that reclamation promotes accretion, which in turn shortens the time intervals between reclamations.

The following data may give some idea of the actual situation in our site. The marsh surface rose 10–15 cm 14 months after the coalescence of clumps over that of barren unvegetated soil surface of tidal flats. Comparative measurements of stakes showed an average rise of 14.6 cm in a vigorous

Spartina zone and an average rise of 5 cm in a poor growth zone over control surfaces from September, 1966 to October, 1967. An increase of 66–68 cm over the control after 4 years was measured by Dongfang Commune. A lessening of wave energy and a slowing down of currents were proved by spectacular events of changing coastal morphology.

Reclamation. After completion of an embankment of the Union Polder in November 1973, nine breaches were inflicted by an August typhoon accompanied by high tides. In October 1974 they were filled, enclosing an area of 333 ha. In 1975, *Suaeda solida* was sown over 276 ha of polder land to remove soil salts. Robust growth was constructive, whereas sparse, slim, and weak growth was conspicuous on barren unvegetated soil. Shoot production of 9000 kg/ha for the marsh soil contrasted with 2250 kg/ha for the barren soil. In 1976 *Sesbania canabina* was sown over 320 ha for amelioration of saline soil, because this popular legume has been used for summer green manure in saline coastal land. Production of stalks and straws was 7500 versus 525 kg/ha for the barren soil. Total seed yield amounted to 125,000 kg. A range of 225–375 kg/ha was harvested, but there was seldom a seed harvest for the control.

Crop Production. In the autumn of 1976, 320 ha of barley were sown and 200,000 kg were harvested. Average yield was 900–1125 kg/ha, with a maximum of 2250 kg. Barley on barren tideland soil generally was observed to show no growth, with best performance yielding 225–375 kg/ha. Barley sown on 306.67 ha in 1977 produced 599,500 kg in 1978. On more than 213 ha, its average yield was 960 kg/ha, with a maximum approaching 4500 kg/ha and a minimum 1875 kg/ha. A survey of 93.33 ha revealed a striking contrast with an average yield of 937.5 kg/ha, with a maximum near 1500 kg/ha and a minimum of 150–225 kg/ha. The sown area of barley in 1978 was 25.67 ha and a harvest in 1979 totaled 548,500 kg. In addition to the above, total productions of the following from 1976 to 1979 were also realized: plant oil 4500 kg; *Sesbania* seeds 405,000 kg; *Sesbania* bark 415,000 kg; stalks and straw for fuel 13,500,000 kg; asparagus 10,500 kg; tomatoes 50,000 kg; and cereals 1,444,000 kg.

Citrus Production. Since 1979, citrus groves have been developed on 149 ha, of which 115 ha is in production. During several years of observation, growth and production of these groves far outstripped those grown in barren tideland soils. Observations indicate sharp differences of growth under similar conditions in all aspects. In 1982, citrus trees in the control soil were noted to have short stature, small canopy, poor growth form, and slight chlorosis of autumn leaves. Those on *Spartina* marsh soils, in contrast, were characterized by high stature, large crown, good growth form, and essentially no chlorosis. A 1986 survey discovered that marshland citrus trees were 28.0 cm higher, 28.6 cm wider in crown diameter, 3.3 cm larger in trunk circumference, and 22.7 cm thicker in chlorophyll layers.

The citrus trees first fruited in 1985. Random selection was made of 10 trees of *Citrus unshiu* in *Spartina* marsh soil and control soil. Weights of fruits of the former turned out to be 157.2 versus 31.75 kg for the control, a 4.9-fold increase. A 1986 survey of 10 trees from each habitat produced 2165 fruits versus 982. Based upon the standard of 12 fruits/kg, the estimated production is 180 kg versus 82 kg. The marshland fruits were larger in size, more uniform in shape, and sweeter than the fruits from the control site. Several other *Spartina* plantations have since been reclaimed profitably.

Coastal Stabilization. We did not pay much attention to stabilization problems until the breaches caused by a 1973 typhoon. After that, it was the people's initiative to plant *S. anglica* at the toes of the sea wall and on the barren salt flats seaward. To gain a new polder has been their hope. Although stems and leaves take care of waves and currents, we must concur in Boston's opinion of the desirability of "a dense network of roots and rhizomes which helps to stabilize the intertidal flat throughout the year," as cited by Guilcher (1981). The numerous roots increase year after year and accumulate in the soil, but the new growth of shoots decays after their death. So the former exceed the latter in biomass by several times, as high as five times. Soil particles are bound by roots firm enough to withstand the scour of waves. Measurements of accretion rates of *Spartina* on the coast of China were compiled by Chung (1985); these show its stabilizing effect on substrate under normal wave conditions (Table 13.3).

13.3.6   Effects on Soils of Intertidal and Supratidal Zones

Amelioration of Saline Soils. Soil analyses from year to year offered convincing evidence of the superiority of *Spartina* marsh soil over barren control. Representative results are shown in Table 13.4. Samples collected in April 1974 in the Union Polder showed more organic matter in the 0–20-cm and 20–40-cm strata of *Spartina* soil. The marsh soil of citrus mounds contained 52.8% more organic matter than the control in 1985. More available iron and soluble iron were consistently determined in the marsh soil. Although NaCl increased as high as 40.2% over the control after embankment but before cultivation, rapid desalinization reduced NaCl to 0.021% in the 0–20 cm layer in April 1980. Even during the rainy season 0.152% was still left in the control soil. Salinity was also lower in the 20–40- and 40–60-cm profiles. Higher pH values likewise decreased after cultivation.

Soil structure, as shown in Table 13.4, was improved up to 1980, with higher percentages of granulation (with diameter larger than 0.25 mm) in the marsh soil. Lower bulk density was determined in the marsh soil. Less water content was found in a lower layer (20–25 cm), but the reverse was true in

Table 13.3 Accretion Rates Under *Spartina* Cultivation on the Coast of China

Province	Site	Period (years)	Total Accretion (cm)	Mean Annual Accretion (cm)
Liaoning	Qinxi	10	29.0	2.9
Hebei	Haikou[a]	1	1.6	1.6
Hebei	Haikou[b]	2	8.0	4.0
Shandong	Showguan	4	50.0 (74.0)[f]	12.5 (18.5)
Shandong	Yexian	9	30.0	3.3
Jiangsu	Sheyang[c]	1	7.0	7.0
Jiangsu	Sheyang[d]	1	19.0	19.0
Jiangsu	Qidong	4	93.0	23.3
Zhejiang	Union Polder Wenling	1, 2[e]	14.6	12.6
Zhejiang	Union Polder Dongfang	7	80.0	11.4
Fujian	Cuanzhou	3	20.0	6.7

Source: Chung, 1985.
[a] Spartina cover 30%.
[b] Spartina cover 90%.
[c] Low-density sward.
[d] High-density sward.
[e] 1 year, 2 months.
[f] Numbers in parentheses indicate maximum accretion rates.

an upper layer (5–10 cm). Higher total porosity and higher air content were also measured in the marsh soil.

Fertility of *Spartina* marsh soil prior to cultivation was characterized by the three following merits: an increase of aeration, mitigation and even elimination of resalinization, and great potential of organic matter accumulation. After plowing and sun-baking, large blocks of soil were easy to break. Many crumbs of 2–4 cm diameter formed by root action were observed. No dispersion due to absorbing water occurred even in a downpour, nor did the soil become sticky. Good aeration still was maintained, a very favorable condition for the growth of crop plants. Owing to a winter drought of 1975, there was a serious rise of salt in the barren control soil, determined in early January 1976 to be 0.635% of NaCl in the 0–5 cm layer. On the other hand, a 0.36% maximum (rare case) was noted in the *Spartina* marsh soil. In a 4-year-old *Spartina* marsh in Bayi Polder, about 15 kg of roots and rhizomes was present under a soil surface of 1 m², or 150,000 kg/ha (Chung et al., 1985).

Other results obtained in the above polder during and after cultivation were high stress tolerance, high production with a saving of fertilizers in the first year of reclamation, earlier plantings of crops of medium salt tolerance, formation of highly productive arable land with stability in a relatively short

Table 13.4 Comparison of Soil Analyses of *Spartina* Marsh Soils and Barren Control Salt Flats in Union Polder, China

Soil Analysis	*Spartina* Marsh Soil	Barren Control Soil
ORGANIC MATTER, %[a]		
0–20 cm	1.52	1.07
20–40 cm	1.52	1.12
SALINITY, %[b]		
0–20 cm	0.021	0.152
20–40 cm	0.059	0.196
40–60 cm	0.143	0.241
SOIL PARTICLE DIAMETER >0.25 mm, %		
0–15	30	13.8[c]
15–25 cm	28	5.4[d]
BULK DENSITY, g/cm³		
5–10 cm	1.26	1.38
20–25 cm	1.30	1.34
WATER CONTENT, %		
5–10 cm	34.8	31.1
20–25 cm	36.3	38.5
POROSITY, %		
5–10 cm	52.4	48.4
20–25 cm	51.1	49.7
AIR CONTENT, %		
5–10 cm	8.5	5.5
20–25 cm	3.9	1.9

[a] Measured April, 1974.
[b] Measured April, 1980.
[c] 0–20 cm.
[d] 20–30 cm.

time, rapid desalinization after 2½ years, a better developed soil profile, and favorable changes of physicochemical properties.

The question as to whether embanked barren unvegetated salt flats could be converted by *Spartina* plantings or not was satisfactorily answered by experiments from May 1973 to May 1975 in Gouqing Polder, Meiou Brigade, Lunyian Commune, Yuhuan County. That site had been a part of Wenling. Salt contents were determined to be 3 parts per thousand (ppt) for the *Spartina* marsh versus 5 ppt for the control. A 76% net gain of oil rapeseeds was realized over those grown in control soil that had been improved by Indian *Vigna* and *Sesbania canabina*. An area of 0.056 ha was used for this experiment.

Use as Green Manure. The first trial was carried out by young farmers of an inland commune; coastal farmers were unwilling to make this trial on account of the high salt content in *Spartina* plant bodies. The plots were rice paddies of 1.33 ha. Six thousand kilograms of fresh grass was applied per hectare; an increase of 1,026.75 kg/ha of the late crop of rice over control stimulated other farmers to try this technique, also with success. Net increase ranged from 750 to 1,125 kg/ha over control in inland communes. Coastal farmers lagged in this application not only because of the high salt content of *Spartina*, but also because of the difficulties of decomposing its heavy coating of wax and cutin. Intense heat from the end of July to mid-August led to water temperatures of around 45°C, or even 50°C, and helped to decompose the *Spartina* rapidly; 53.6% of the dry matter decomposed in the first 9 days and 60.3% in 15 days. People were surprised by this phenomenon.

Further experiments investigated the increase of the salt content in the soil after application of *Spartina*. Growth of rice seedlings was previously believed to be harmed by this. However, 15,000 kg/ha of fresh grass dressing increased the salinity only 0.02–0.04% over that of the control. We conclude that salt injury would not occur in the inland area of Wenling, in contrast to the general opinion regarding addition of *Spartina* to old marsh soils. Coastal communes also increased their late rice crop 452–677 kg/ha after as many trials as the inland farmers had conducted.

In 1973, prior to completion of an embankment, Hunghe Commune harvested 2,900,000 kg of fresh *Spartina* for basal dressing of late rice. The increase in yield over 1972 averaged 18.3% with a maximum increase of 23.2%. Its fertility was assessed to be equivalent to urea, with a ratio of 100 kg:1 kg, and better than that of an equal quantity of stable manure. Inland farmers preferred using *Spartina* to chemical fertilizers. In 1973 I asked coastal farmers why they preferred it; the answers included its longer effective period, improvement of soil structure instead of the deterioration that chemical fertilizers cause, and few weeds surviving owing to odorous gas released during its decomposition. From 1968 to 1969 it was used as a manure to increase the late rice crop, and estimates of total harvest and average increase of yield were 7.7 and 20 million kg, respectively. Our investigation in 1984 of the area of *Spartina* plantations in Wenling totaled 8.965.4 ha, 75% of which was used as green manure, with a net gain of the late crop of rice estimated to exceed 2,570,000 kg since 1978.

A comparison of *S. anglica* with *Medicago hispida* revealed lower nitrogen content, 0.323 versus 0.400%, but more phosphorus and potassium, all in fresh weight: 0.149 versus 0.080% and 0.315 versus 0.240%, respectively. Field observations on comparative rates of becoming effective after absorption by the second crop of rice reported fresh *Spartina* being 5–7 days later than fresh *Sesbania* and rotted manure.

13.3.7   Effects on Animals

Nesting and Feeding Ground of Migratory Birds, Waterfowl, and Domestic Fowl. Prior to creation of the community, the Chinese wild geese had used wheat fields as their feeding ground, but they began to shift to the newly planted vegetation as both nesting and feeding grounds. People told us the geese neglected their old feeding place entirely thereafter. Swans and other waterfowl also did the same.

Spartina Marsh Pastureland. As early as 1970, Goushantow Brigade, Guanou Commune began grazing cattle and pigs on a *Spartina* marsh every day, resulting in the rapid growth and fattening of the animals. The advantages of *Spartina* were then recognized.

As Animal Fodder. In 1968 I suggested to farm workers of Dongpian Farm that they harvest more than 10,000 kg of fresh grass to feed cattle and pigs. Xinhe District exceeded all districts in their county in implementing their plan of pig farming because of using more *Spartina* as fodder. Rabbits, sheep, and goats also thrived on it. Different processing forms, such as ground powder or syrup of fresh grass, all produced desirable effects. In neighboring Yuhuan County, people began to feed *Spartina* to domestic geese, reaching specifications for export in 70 days.

Nereid and Crab Production. Boring for nereids for export destroyed 600 ha of *Spartina* plantation in 1981 and that is why the planned polder in front of Union Polder has not been progressing as rapidly as expected. The nereids found in the *Spartina* marsh were larger, heavier, and more abundant than in barren flats, which may be explained by food chain relationships. I witnessed fibrous portions of *Spartina* being extruded from the anus of a nereid once in Qidong, Jiangsu. Increases of crabs since *Spartina* plantings have also been noticed.

13.3.8   Effects on Humans

Economic Value

Total economic returns from agricultural produce from 1977 to 1980 amounted to 2,500,000 yuan, but investment for empoldering Union Polder amounted only to 960,000 yuan, not including the cost of labor.

The plantings of citrus trees on *Spartina* marsh since 1978 have been helpful in demonstrating to people *Spartina*'s rapid effect, lower investment, and greater profit. A nursery cost per sapling of 0.44 yuan versus 1.13 yuan for the control resulted in a profit of 5530 yuan for 1.4 ha. Then in March 1980 transplanting to mounds took place and in 1985 fruit production began. An early seedless variety averaged 7.56 kg/tree for the marshland, or 2.405 kg/tree more than the control. Based on a price of 1.6 yuan/kg, an increase in income of 3,232 yuan/ha was attained, a 46.6% raise. A late seedless variety weighed 7.08 kg/tree for the marshland, exceeding the control by 2.505 kg/tree. A net surplus per hectare amounted to 3,366 yuan more than the control, or a 54.8% increase. There is a promising prospect for even greater production after the first few years.

Social Value

Spartina may be used as a fuel and, a new development, for marsh gas production. In the 1960s, when *Spartina* plantations were not well established, I saw people collecting grass for fuel. In 1973 I saw farmers transporting bunches of grass home, the grass being *Spartina anglica*. In our 1984 investigation, we learned that 20% of total *Spartina* grass production was devoted to this important need of the people there.

Marsh Gas Production. This is an account of what happened in Zhenhai Commune's social and family life. Three hundred and forty tankfuls of marsh gas were collected during two months of autumn 1978 and more than 130 were put up in 1979. There was still a shortage of fuel at that time. Ninety-five percent of the families in 10 teams had marsh gas tanks, enabling them to cook, boil water, and illuminate with methane. Four advantages were unanimously appreciated by the people:

1. A new solution of the fuel problem on the farm by saving fuel and reducing expenditures was described as a revolution in the history of fuel. The families no longer cooked by burning rice straw or used kerosene lamps for light. For a family of five a 6–8 m^3 tank, with good management served the purpose of daily cooking and illumination the year round. For cooking three meals a day for a year and lighting a kerosene lamp for 10 months, each family would have burned 4.5 kg rice straw. This means that an entire commune (provided 70% of families have built tanks) would save 4,960,000 kg rice straw, 210,000 kg coal, and 55,000 kg kerosene. In other words, an annual saving of 221,000 yuan could be achieved. It took less than a year to pay back the tank investment.

2. An increase of organic manure for agricultural development was realized. After this operation, rice straw was saved and used as padding material in pig pens. This in turn enhanced the source of organic fertilizers, because stable manure, weeds, garbage, and sewage sludge were all used as raw materials generating marsh gas. Moreover, raw manure was transformed into well rotted farmyard manure. Large-scale disposal and change of raw materials twice a year supplied 10,000 kg methane manure equivalent

to 3000 kg standard manure, for an increase of 125,000 kg ammonium carbonate. This resulted in the elimination of 800 sinks of manure decomposition, with an area of 0.4 ha freed for other purposes.

3. The women's labor force was liberated. Women, no longer occupied with heavy, tedious and complex housework, have been able to occupy themselves with sewing and other household affairs. If one saves 1.5 hours daily, more than 200,000 days per year would be saved in an entire commune.

4. Family and community sanitation was greatly improved. In former times cooking in smoky kitchens not only was unsanitary and unsafe, but people also suffered suffocation from smoke and dust. Thereafter there has been no air pollution or ditch slush, and all weeds have been cleared up to the tanks, leading to decreased propagation of mosquitoes and houseflies. Methane light has also been useful for studying and working at night. All agree its brightness surpasses that of a 40-candle electric lamp or eight kerosene lamps, with better quality.

Ecological Value. We may cite one example of a citrus grove on *Spartina* marsh soil with better growth of normally vigorous trees, essentially no freezing injury, and very few fallen leaves and dried up branches. Production of standard transplants exceeded the control by 36.6% and growth was taller by 28.9 cm. For a single plant the air-dried weight was found to be 3.59 times the control.

Papermaking. Dongfang Commune was the first commune that succeeded in making unbleached brown paper from *Spartina*. A series of processes including fermentation, pulp making, collecting, and finishing were developed. A lower cost than any other raw materials and rate of yield surpassing rice straw were found. Users remarked on the strong tensile strength, good quality, and low price of the paper and its suitability for packing. Production began in 1968. However, the Wenling people and we rather prefer to prolong the *Spartina* food chain, for papermaking does not recycle materials.

13.3.9   A Systems Model

What happened since *Spartina anglica* changed the interrelationships between ecosystem and humans in Wenling is shown as a qualitative model in Figure 13.6. Of course this model needs much improvement. Although we determined the caloric value from a Jiangsu sample to be 4570 cal/g dry weight, a quantitative model is not ready, for lack of detailed studies of food chains and human usage.

13.4   CONCLUSIONS

Several conclusions may be drawn from the above ecological engineering research:

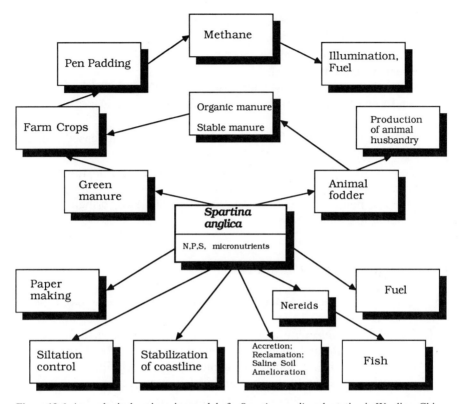

Figure 13.6 An ecological engineering model of a *Spartina anglica* plantation in Wenling, China.

1. Man-made emergent vegetation ecosystems have been demonstrated to be more efficient in energy flow and material recycling than natural ecosystems in coastal China.

2. Ecological engineering in enclosed intertidal land, with *Spartina anglica* playing the main role, has been demonstrated to be successful in accretion and reclamation of land with an advance of 4–5 years of land use.

3. Plantings have been beneficial for coastline stabilization in normal wave conditions by stilling currents, dissipating wave energy, and causing suspending particles to fall.

4. Saline soil amelioration has been especially successful, with the creation of *Spartina* marshland improving the structure of the soil as well as its chemical properties.

5. People in Wenling and neighboring counties have been fond of using *S. anglica* as a summer green manure. Because formerly they lacked summer green manure for the late crop of rice, the importance of *Spar-*

tina has grown since 1968. Even in recent years, 75% of its fresh grass has still been used in this way.

6. *Spartina* marsh pastureland was appreciated by Wenling people for its effect in fattening animals. Use as animal fodder and feed was tried with cattle, pigs, rabbits, sheep, and goats. Xinghe District has been the foremost in pig farming in Wenling, giving evidence of the value of marsh grass.

7. Enormous quantities of nereids of larger size, and increased production of other animals have been decumented.

8. The marsh's efficiency for harnessing solar energy, transforming materials, recycling elements, and activating as well as creating ecosystems all point to a positive ecological equilibrium, bringing mutual benefits to both humans and the environment.

13.5   FUTURE RESEARCH NEEDS

There are many subjects needing further study or beginning exploration:

1. The complex interplay of various biological and physical factors involved in accretion and erosion should be explored. Relationships between accretion and animals and algae ought to be clarified. Prevention of erosion by animal growth is an important project.

2. Stabilization of coastlines with plantings of *Spartina* species by means of other protective devices and in conjunction with coastal engineering works ought to be tried.

3. Amelioration of saline and alkaline soils in dry and cold inland areas with improved varieties of plant breeding will be very useful in developing those areas.

4. Uses as green manure, animal fodder, fish feed, and as raw materials generating marsh gas require more study.

5. Large-scale planting experiments should be performed to prevent siltation of river beds and navigational channels.

6. The role of *Spartina* in solving problems of pollution and of dredge spoil ought to be explored.

7. Last but not least, there is urgent need for study of the use of *Spartina* plantings in coping with sea-level rise in the twenty-first century.

REFERENCES

Bird, E. C. F. 1974. Coastal changes in Denmark during the past two centuries. Geol. Inst., Aarhus Univ., Denmark. 20 pp. (mimeographed).

Boston, K. G. (1980) A note on *Spartina anglica* as an agent of change on marshy shorelines. Unpublished manuscript of Department of Geography, Melbourne State College.

Chen, S. B. and X. L. Duan. 1985. Character changes induced by transfer of DNA into rice. *Scientia agricultura Sinica* (3):6–9 (in Chinese).

Chung, C. H. 1982. Low marshes, China. (Chapter 6) In R. R. Lewis, Ed., *Creation and Restoration of Coastal Plant Communities*. CRC Press, Boca Raton, FL, pp. 131–145.

Chung, C. H. 1983. Geographical distribution of *Spartina anglica* Hubbard in China. *Bull. Mar. Sci. 33*(3):753–758.

Chung, C. H. 1985. The effects of introduced *Spartina* grass on coastal morphology in China. *Z. Geomorph. N. F. Suppl. Bd. 57*:169–174.

Chung, C. H. and P. Qin. 1983. An inquiry into the mercury absorption of *Spartina anglica* and its environmental purification effect. *Marine Science (Qingdao), 12*:6–11. Abstract: 1985. *J. Nanjing Univ. Res. Adv. Spartina*, p. 328 (in Chinese with English summaries).

Chung, C. H. and P. Qin. 1985. Absorption of mercury by *Spartina anglica* and its purification effect. *Coll-Oceanic Works (Tianjin, in English)*. (1):24–53.

Chung, C. H. (Zhong Chongxin), R. Z. Zhuo, H. B. Zhou, G. H. Ye, C. H. Hu, B. Y. Yin, L. Z. Jin, and X. L. Pan. 1985. Experiments of tidal plantings of *Spartina anglica* Hubbard and effects of saline soils amelioration in China. *J. Nanjing Univ. Res. Adv. Spartina*, pp. 44–82 (in Chinese with English summary).

Fang, T. S., J. L. Ho, R. L. Zhou, I. X. Dai, T. S. Ma, and G. H. Lin. 1982. Chromosomes numbers of *Spartina anglica* and *S. patens. J. Shandong Coll. Oceanology. 12*(1):65–68 (in Chinese with English summary).

Goodman, P. J. 1960. Investigations into 'Die-back' in *Spartina townsendii* agg. II The morphlogical structure and composition of the Lyminton Sward. *J. Ecol. 40*:711–724.

Goodman, P. J., E. M. Braybrooks, C. J. Marchant, and J. M. Lambert. 1969. 4. *Spartina townsendii* H. & J. Groves sensu lato in Biological Flora of the British Isles. *J. Ecol. 57*:298–313.

Guilcher, A. 1981. Shoreline changes in coastal salt marshes and mangrove swamps (mangals) within the past century. In E. C. F. Bird and K. Koike, Eds., *Coastal Dynamics and Scientific Sites*. Komazawn University, Tokyo, pp. 31–33.

Hubbard, C. E. 1968. *Grasses,* 2nd edition, Penguin, Harmsworth.

Hubbard, J. C. E. 1965. *Spartina* marshes in southern England, VI. Pattern and Invasion in Poole Harbour. *J. Ecol. 57*:759–804.

Jiang, H. X. and J. S. Huang. 1982. The preliminary observation on the digallic acid-treated transmission electron microscopic specimens of *Spartina anglica* C. E. Hubbard. *J. Cell Biol. (Shanghai)*, *4*(2):15–16. Abstract: 1985. *J. Nanjing Univ. Res. Adv. Spartina*. pp. 334–335 (in Chinese with English summary).

Li, M. W. and B. Zhang. 1985. Tissue culture of stem apex of *Spartina anglica* Hubbard. *J. Nanjing Univ. Res. Adv. Spartina*. pp. 185–188 (in Chinese with English summary).

Long, S. F., L. D. Incoll, and H. W. Woolhouse. 1975. C^4 photosynthesis in plants from cool temperate regions with particular reference to *Spartina townsendii*. *Nature 257*:622–624.

Lu, B. S. and F. X. Jiang. 1981. Studies on the nutritive value of *Spartina anglica*. *J. Nanjing Univ.* (*Nat. Sci.*) (4):531–536. Abstract: 1985. *J. Nanjing Univ. Res. Adv. Spartina*, pp. 342–343 (in Chinese with English summary).

Lu, B. S. and F. X. Jiang. 1983. The colorimetric determination of free proline in plant tissue. *J. Nanjing U.* (*Nat. Sci.*) (4):694–700. Abstract: 1986. *J. Nanjing Univ. Res. Adv. Spartina*, p. 350.

Lü, H. M. and B. S. Lu. 1985. Purification and identification of DNA in seedlings of *Spartina anglica* Hubbard. *J. Nanjing Univ. Res. Adv. Spartina*, p. 351.

Marchant, C. J. 1967. Evolution in *Spartina* (*Gramineae*). I. The history and morphology of the genus in Britain. *J. Linn. Soc.* (*Bot.*) *60:*1–24.

Marchant, C. J. 1968. Evolution in *Spartina* (*Gramineae*). II. Chromosomes, basic relationships and the problem of *Spartina Townsendii* agg. *J. Linn. Soc.* (*Bot.*) *60:*381–409.

Mobberley, D. G. 1956. Taxomony and distribution of the genus *Spartina*. *Iowa State Coll. J. Sci. 30:*471–574.

Ou, H. C., T. Z. Fu, R. S. Que, and C. H. Chung. 1982. A preliminary study of tolerance to alkalinity and salinity by seedlings of *Spartina anglica* Hubbard. *Plant physiol. Comm.* (1):36–38. Abstract: 1985. *J. Nanjing Univ. Res. Adv. Spartina*, pp. 338–339.

Ou, H. C., T. Z. Fu, A. T. Zhou, B. Yin, C. H. Chang, and R. S. Que. 1985. Biological basis of longtime establishment of *Spartina anglica* in alkaline soil. *J. Nanjing Univ. Res. Adv. Spartina*, pp. 179–184.

Ranwell, D. S. 1972. *Ecology of Salt Marshes and Sand Dunes*. Chapman & Hall, London.

Stewart, G. R. and J. A. Lee. 1974. The role of proline accumulation in halophytes. *Planta* (*Berl*), *120:*279–289.

Sung, R. J. and R. L. Dou. 1982. Biological characteristics of *Spartina anglica*. II. The structure of the culm and the leaf sheath. *J. Nanjing Univ.* (*Nat. Sci.*), (1)111–116. Abstract: 1985. *J. Nanjing Univ. Res. Adv. Spartina*, p. 331.

Titus, J. 1986. *Can we delay a Greenhouse Warming?* U.S. Government Printing Office, Washington, DC.

Titus, J. 1986. *Protecting future sea level Rise: Methodology, Estimates to the year 2100, and Research Needs*. U.S. Government Printing Office, Washington, DC.

Tong, Y. R., W. X. Meng, and Q. Xu. 1985. A preliminary survey of animals in *Spartina anglica* marsh. *J. Nanjing Univ. Res. Adv. Spartina*, pp. 133–140.

Wang, R. J. and R. L. Dou. 1985. Biological characteristics of *Spartina anglica*. III. Anatomy of root system. *J. Nanjing Univ. Res. Adv. Spartina*, pp. 149–158.

Wang, H. J., H. B. Zhou, R. J. Suug, H. X. Jiang, J. Chen, and R. L. Dou. 1979. Biological characteristics of *Spartina anglica*. I. Preliminary observation on the structure of the leaf. *J. Nanjing Univ.* (*Natural Science*), (3):45–52.

Wang, B. K., A. H. Zhou, X. R. Wang, D. Zhou, Z. R. Zhang, and D. M. Pan. 1985. Accumulation and distribution of radioisotopes ^{127}Cs, ^{90}Sr, ^{115m}Cd and ^{65}Zn in *Spartina anglica*. *J. Nanjing Univ. Res. Adv. Spartina*, pp. 124–132.

Yuan, Y. S., C. H. Wang, Y. Ting, C. G. Zhang, and H. S. Zhou. 1981. Gas chromatographic analysis of proline free in *Spartina anglica* Hubbard. *J. Nanjing Univ. (Natural Science)*. (1):265–268. Abstract: 1985. *J. Nanjing Univ. Res. Adv. Spartina*, pp. 348–349.

Yu, W. H. and Y. C. Cao. 1981. Preliminary studies of the rhizosphere bacteria of *Spartina anglica* Hubbard. *J. Nanjing Univ. (Natural Science)*. (3):365–370. Abstract: 1985. *J. Nanjing Univ. Res. Adv. Spartina*, pp. 329–330.

Zhang, C. G., H. S. Zhou, C. H. Wang, and Y. S. Yuan. 1985. Preliminary analysis on amino acid pool and total amino acids of *Spartina anglica* Hubbard. *J. Nanjing Univ. Res. Spartina*, pp. 206–211.

Zheng, J. S., R. X. Ye, H. C. Ou, T. Z. Fu, and Z. R. Zhang. 1985. The effect of atmospheric pressure to seed germination of *Spartina anglica*. *J. Nanjing Univ. Res. Adv. Spartina*, pp. 171–179.

Zhou, H. B. and C. H. Chung. 1985. An inquiry to seed germination of *Spartina anglica*. I. Seed morphology in relation to germination. *J. Nanjing Univ. Res. Adv. Spartina*, pp. 159–163.

Zhou, H. B. and C. H. Chung. 1985. An inquiry on seed germination of *Spartina anglica* II. The relationship between puncture on seeds and germination. *J. Nanjing Univ. Res. Spartina*, pp. 166–170.

Zhou, H. B., H. X. Jiang, and R. L. Dou. 1982. Morphology of salt gland of *Spartina anglica* Hubbard. *Acta Botanica Sinica, 24* (2):115–119. Abstract: 1985. *J. Nanjing Univ. Res. Adv. Spartina*, pp. 332–333.

Zhou, H. L., Z. F. Wang, Y. C. Cao, and W. H. Yu. 1985. Numerical taxonomic studies on the Gram negative bacteria isolated from the plant surface of *Spartina anglica*. *J. Nanjing Univ. Res. Adv. Spartina*, pp. 141–148.

Zhuon, S. H. and C. H. Chung. 1987. An investigation on ecotypic differentiation of *Spartina anglica* Hubbard. *J. Ecol. (Shenyang)* 6(6):1–9 (in Chinese).

14

EXPERIMENTAL STUDY OF SELF-ORGANIZATION IN ESTUARINE PONDS

Howard T. Odum

Department of Environmental Engineering Sciences and Center for Wetlands, University of Florida, Gainesville, Florida

14.1 INTRODUCTION

All over the world high-nutrient wastewaters from developing economies on land are going into estuaries, where new ecosystems are developing by self-organizational processes, making use of the marvelous diversity of species seeded into the estuaries from the oceanic gene pool by the tides. The wastewaters are potentially a rich contribution to estuarine production. Managing these new interfaces between humanity and nature is a major opportunity for the new field of ecological engineering, which may be defined as managing the self-organizational process of ecosystems for mutually useful purpose. In 1967 we proposed that a conscious effort be made to guide the self-organization of marine ecosystems to become interface partners of the human economy that was increasingly releasing eutrophic wastewaters to the environment (Odum, 1967). Under National Science Foundation–Sea Grant auspices in 1968–1971, a study was made in marine ponds at Morehead City, North Carolina.

This chapter summarizes this experimental study of the self-organization in six estuarine ponds, three receiving secondarily treated municipal wastewaters mixing with salt water and three control ponds receiving tap water and salt water. The study was a team project involving 84 investigators. It was organized using systems ecological models and was funded as a test of ecological engineering as a means to develop wastewater utilization. The idea was to use technology less and the multiple complexity of nature more to turn a waste product into a beneficial ecosystem. A data report was recently published (Odum, 1985).

14.1.1 Hypothesis of Power-Maximizing Self-Organization

The maximum power principle suggests that self-organization of available species occurs by the reinforcement of those that develop contributions to other parts of a system. The resulting complex web of food chains, hierarchical relationships, recycling patterns, and population control mechanisms is in a design that draws more resources and uses them toward utilizing more resources and increasing efficiencies of use.

The hypothesis may be tested by setting up new conditions, supplying a mix of available species, and studying what develops. Theory predicts increased organization accompanied by increased total metabolism (photosynthetic production and respiration, which measures use). Studies of this type have already been done in microcosms, which increased diversity and metabolism with time (Odum and Hoskin, 1957; Beyers, 1963; Nixon, 1969; Kelly, 1971).

14.1.2 Questions About Self-Organization with Estuarine Wastewaters

Appropriate questions for ecological engineering of estuaries are whether the self-organization process occurs readily there with new conditions from wastewater influence and how much time is required. How does self-organization with high nutrients differ from that in more ordinary conditions? Would the new ecosystems developing under the new conditions have economic potentials for reducing waste treatment costs and increasing aquaculture yields?

14.2   STUDY PLAN AND METHODS

Ponds were constructed with pumping systems to provide similar conditions to three replications of wastewater ponds and three replications of control ponds. These ponds are shown in Figures 14.1 and 14.2. Measurements were made regularly of metabolism, nutrients and other chemical parameters, physical characteristics, and the living populations that developed. Studies dealt with daily, seasonal, and successional changes. It was a group project in systems ecology and ecological engineering. Quarterly discussions of new data were held so that investigators could relate their measurements to concurrent data on other aspects of the system. Models were used to organize the work, and an ecosystem model was developed to summarize the facts and relate to concepts of ecosystem operation. Simulation models were developed to test the consistency between ideas about mechanisms and the observed patterns with time. The details of treatments and what happened in these ponds over a 3-year period were recently published in a Sea Grant data report (Odum, 1985).

The main physical, chemical, and biological features of the emerging ecosystem were monitored, including structures, functions, and species. Experiments were conducted to elucidate mechanisms and rates. Nutrient-rich waste ponds and control ponds were compared and the role of high nutrients studied. Most of the work considered phenomena at two levels, one involving components and the other the role of the process in affecting the larger ecosystem. Aquaculture possibilities were considered. Before summer 1971, policy was against removing species or otherwise changing the ecosystem's self-organization process during the study. In July 1971, after 3 years of self-organization, the ponds were inventoried, including the seining out of all the fishes. Details of the many methods have been given by Odum (1985).

14.2.1   Experimental Ponds

Three ponds were constructed in 1968 in marshes adjacent to the Morehead City Sewage Plant on Calico Creek. The ponds were supplied with a mixture of estuarine water and treated sewage wastewaters. Three control ponds were constructed adjacent to the Institute of Marine Sciences at Morehead City and supplied with a mixture of estuarine water and tap water. Figure 14.1a shows the locations of the ponds. Figure 14.1b shows the arrangements

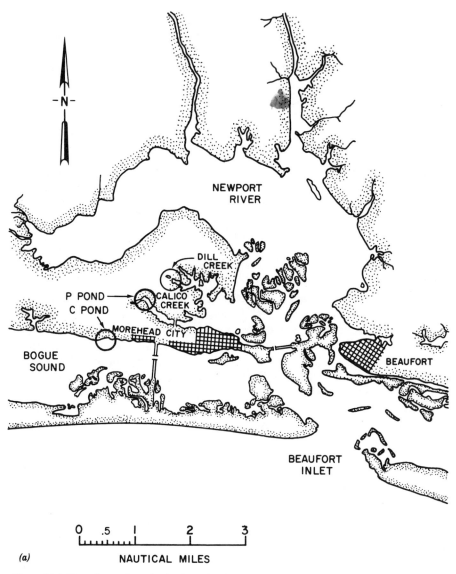

Figure 14.1 Experimental ponds. (*a*) Location in Morehead City, NC; (*b*) arrangement of ponds and water flows.

for water flow through mixing tanks so that replicate ponds received similar inputs.

Pond maps with details are given in Figure 14.2. Special substrates added for development of ecological subsystems are indicated, including patches of marsh grass planted along margins, a shell reef bar with live oysters on top, creosote-treated wood pilings for attachment studies, pier pilings

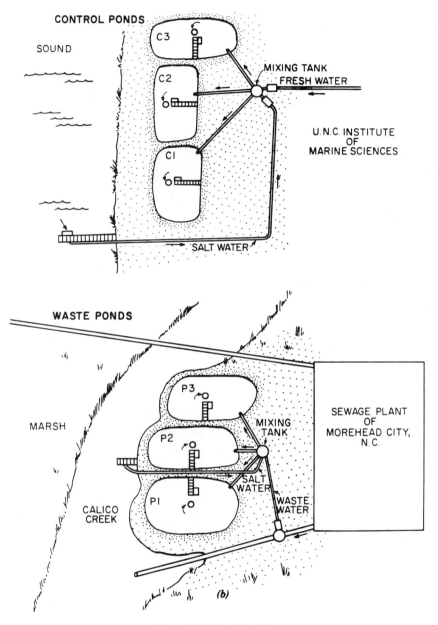

Figure 14.1 (*continued*)

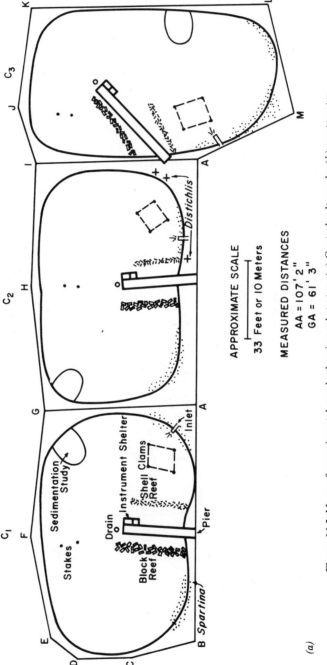

Figure 14.2 Maps of experimental ponds showing substrates. (*a*) Control salt ponds; (*b*) wastewater salt ponds.

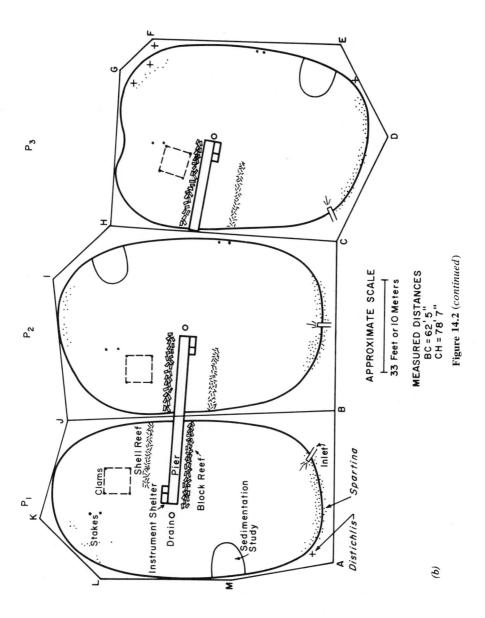

APPROXIMATE SCALE

33 Feet or 10 Meters

MEASURED DISTANCES
BC = 62' 5"
CH = 78' 7"

Figure 14.2 (*continued*)

(b)

297

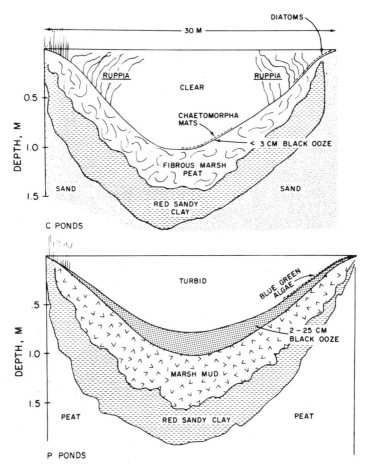

Figure 14.3 Cross-sectional pattern of the ponds in July 1971. C ponds = control ponds; P ponds = experimental ponds.

(treated with pentachlorophenol), masonry plates for fouling studies, rows of concrete blocks and rubble as a micro-reef, and shallow sandy beach margins supporting benthic algae. Ponds were about 1 m deep (mean depth 0.5 m). Maps with depth contours and hypsographic curves used to estimate volumes are given elsewhere (Odum, 1985). After basins were scooped out with a dragline, both sets were lined with about 41 cm of reddish clay to reduce permeability and give both sets of ponds the same substrate. Both bottoms were then floored with about 30 cm of black marsh mud from the Calico Creek marsh site. Figure 14.3 is a cross-sectional sketch of the ponds. Figure 14.4 contains views of the ponds just after construction. Later views after development of marsh grass and aquatic macrophytes in control ponds are shown in Figure 14.5.

Figure 14.4 Views of ponds at the start in 1968. (*a*) Wastewater ponds; (*b*) Control ponds.

Figure 14.5 Views of ponds after 3 years in 1971. (*a*) Wastewater pond, P2; (*b*) control ponds C2 and C1.

14.2.2 Initial Filling, Acid Period

During construction in June 1968, black marsh mud was piled up at the control ponds and exposed to oxidation by air for several weeks. The smell of hydrogen sulfide was strong. After the mud was spread on the pond bottoms, estuarine water was introduced with a portable pump July 8. The waters were soon found to be acid, pH 3. The first of these ponds (C1) received a mud slide and its standpipe was accidentally removed so that the acid water was immediately discharged. Ponds C2 and C3 remained acid until the continuous pumping system began October 31. Interesting phenomena during this period were studied. Differences observed later between C1 and the other two control ponds may have resulted from these different treatments initially.

(b)

Figure 14.5 (*continued*)

In the waste ponds, the marsh mud used for the floor was transferred without delay, and without much chance for oxidation. Acid conditions did not develop.

14.2.3   Water Flow and Salinity

Pumping of water into the ponds through mixing tanks began in September 1968. The experimental plan was to mix estuarine waters with wastewaters in the waste ponds and with tap water in the control ponds so as to maintain pond salinity in the mesohaline range, 15–25 ppt. The salinity of estuarine water available to the control ponds was steady so that inflows of estuarine water and tap water through the mixing tank maintained the desired salinity without much variation (Chestnut et al., 1985a).

However, the estuarine water available to the waste ponds was from Calico Creek, a small estuary with a wide range of salinity depending on rainfall. Consequently, in order to keep salinity in the mesohaline range, pumping of treated sewage had to be varied. The hours of pumping into waste ponds were recorded. Turnover was estimated from the flows divided by pond volumes and graphed (Odum, 1985). Approximate water turnovers for control ponds were 2.0 turnovers per month in pond C1 and 1.5 turnovers per month in ponds C2 and C3. In wastewater ponds, turnovers ranged from

0.5 to 3.0 times per month. Nutrient input was estimated by multiplying hours pumped times nutrient content of wastewater and estuarine water. The estuarine water pumped into the waste ponds was fairly high in nutrients because the intake was only 60 m from the sewage plant outfall (Figure 14.1b).

The salinity in the control ponds ranged from 12 to 26 ppt and similarly in wastewater ponds except that periods of lower salinity down to 4 ppt were observed for several weeks in the winters of 1968 and again in 1969. The variation over a 3-year period was not unlike that in many estuaries. The chronology of events affecting the ecosystems and detailed salinity records are included in the data report. After September 1971 management changed and the ponds were used for other purposes (Odum, 1985).

14.2.4 Multiple Seeding of Species

The continuous self-organization of estuarine ecosystems was facilitated by continuous seeding of species aided by tidal exchanges and migrations. In the ponds, some plankton species were found to enter through the pumps. Because Bogue Sound and Calico Creek are somewhat intermixed by tidal exchange, in the course of a week many of the same organisms were carried to the vicinity of the intake pipes of both sets of ponds. With exceptional storm tides, back flow introduced species through the standpipes, especially into waste pond P3.

To simulate nature and as an ecological engineering technique to minimize genetic limitations, multiple seeding of other species was carried out especially in the springs of 1969 and 1970. Species were systematically introduced by adding estuarine waters, sediments, fouling communities, reef materials, plankton from plankton net tows, larger larvae with coarse plankton nets, small fishes from seines, and so on. Seeding efforts were recorded in a logbook and tabulated (Odum, 1985).

Some of the added species were measured for growth, including Rangia clams, oysters, and shrimp.

14.3 RESULTS

Graphs and tables in the data report (Odum, 1985) are summarized as here.

14.3.1 Physical and Chemical Characteristics

Water Regime. Salinities varied because the salinity of the waters pumped in from the bay was varying. Salinity was 12–26 ppt in control ponds and 5–25 ppt in waste ponds. Evaporation ranged from 3 to 11 mm/day, highest in summer, when evaporation took most of the inflowing waters.

Temperature. Water temperatures ranged from 0°C on a few winter days after strong cold fronts to 35°C on calm, sunny, summer afternoons. The diurnal range of temperature, 1–6°C, was similar in all ponds, depending on daily insolation and wind conditions. Freezing and near-freezing caused fish kills several times and the ponds froze on January 9, 1970. With light winds, a slight vertical stratification was observed with a gradient of several degrees.

Turbidities and Blooms. The Secchi disk reading was about 50 cm in two control ponds (C1 and C2) during summer clearing in midwinter. Ponds C2 and C3 developed *Ruppia* beds, much of it floating at the surface by the end of the summer. Light penetration was greater in control pond C3. Pond C1, with slightly more phytoplankton turbidity and without *Ruppia,* maintained a more active reef of oysters and other filter feeders. In control ponds the particulate matter was about 100 mg/L, of which organic matter was about 10–20%.

The waste ponds had a Secchi disk reading of only 20 cm during winter with the dense *Monodus* populations, and about 40 cm during the summer with more diversified phytoplankton. The ponds cleared briefly in the first week of May when the *Monodus* bloom crashed, quickly replaced with summer plankton regime. In wastewater ponds, particulate matter was 200–300 mg/L with organic content 20–50%.

Oxygen. In the control ponds, diurnal ups and downs of oxygen with daytime photosynthesis and nighttime respiration were 2–5 ppm in control ponds depending on insolation (Odum, Hall, and Masarachia, 1985; Smith, 1971). With only a small daily range, oxygen concentrations did not fall below 4 ppm, and a higher diversity of adapted animals was present. In control ponds, oxygen levels were higher in winter when temperatures were lower. Diurnal range was less than in summer when more light was available. A typical diurnal record is given in Figure 14.6.

In the wastewater ponds, diurnal ups and downs were 5–18 ppm because of the much higher metabolism. In winter very high oxygen levels occurred with supersaturation because there was a net production by *Monodus.* In summer the wide range of daily variation operating at a high temperature causes very low oxygen levels at night, often near zero. Only animals with adaptations to low oxygen survived these conditions. Blue crabs came to shallows and bathed their gills with waters of the surface film. With light winds in summer afternoons, a slight vertical stratification was observed with less oxygen near the bottom.

pH. In control ponds pH values were usually between 7.5 and 8.5, rising with photosynthetic use of carbon dioxide during the day and declining at night. In wastewater ponds there was a wider range of pH values in the

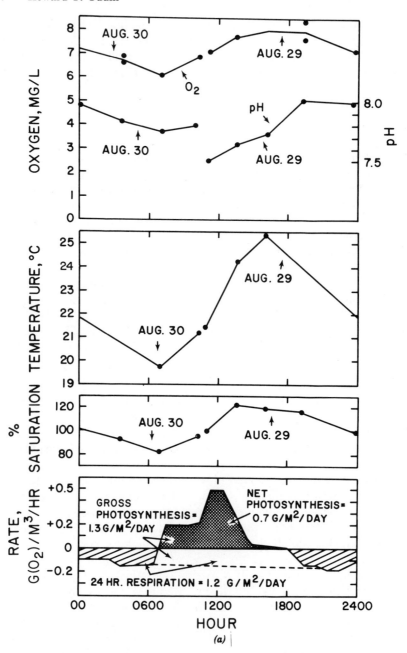

Figure 14.6 Typical diurnal oxygen curve observed in ponds Aug. 29–30, 1968 (Odum et al., 1985). (*a*) Pond P3, normal pH; (*b*) Pond C3, acid pH.

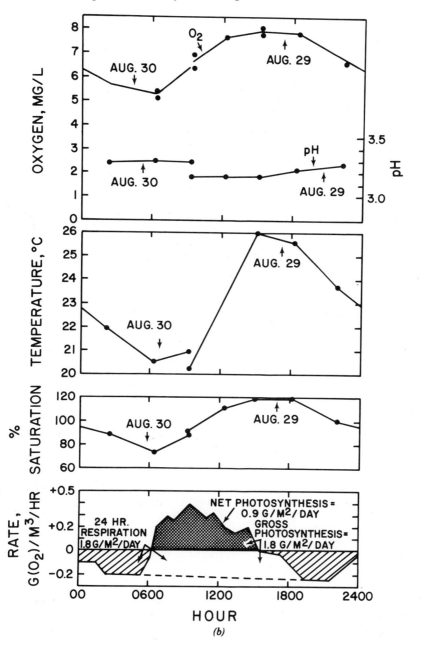

Figure 14.6 (*continued*)

diurnal cycle and seasonally. In winter with the *Monodus* bloom, pH ranged from 9 to 11, whereas summer values ranged from 7.3 to 8.6.

Carbonate Alkalinity. Carbonate alkalinity in control ponds ranged from 2.8 meq/L in winter to 4 meq/L in summer (Day, 1971, 1983). In wastewater ponds carbonate alkalinity ranged from 3.5 meq/L in summer to 0.5 meq/L in the *Monodus* bloom conditions of winter.

Carbon Dioxide. From alkalinity, pH, and direct measurement of partial pressures of carbon dioxide at the surface of the ponds, total inorganic carbon, free carbon dioxide, and total carbon were measured (Day, 1971, 1983), showing the diurnal variations due to metabolism. Because control ponds were receiving hard groundwaters, they were generally supersaturated with inorganic carbon dioxide, with outward diffusion. Wastewater ponds at higher pH were undersaturated with carbon dioxide, often with diffusion from the atmosphere into the ponds.

Organic Carbon. In control ponds dissolved and microparticulate organic carbon ranged from 4 mg/L in winter to about 11 mg/L at other seasons. In the wastewater ponds dissolved and microparticulate organic carbon ranged from 12 mg/L at the end of the summer rising to 67 mg/L at the end of winter just before the crash of the *Monodus* bloom (Day, 1971, 1983).

Total Inorganic Carbon. Total inorganic carbon in control ponds ranged from 35 mg/L in winter to 42 mg/L at other seasons. In wastewater ponds inorganic carbon ranged from 35 mg/L at end of summer to 5 mg/L during the winter *Monodus* regime.

Total Carbon. Total carbon in control ponds ranged from 42 to 62 mg/L, most of which was inorganic throughout the year. In wastewater ponds total carbon ranged from 35 mg/L in summer to 70 mg/L at end of the winter *Monodus* regime. In the wastewater ponds carbon shifted from a mainly organic state in winter bloom to a more equal organic–inorganic distribution in summer.

Bottom Organic Matter. After ecosystem development, the organic concentration in bottom ooze in the pond centers was higher in the waste ponds. The chemical oxygen demand was 35.7 mg/g in control ponds (mean of 10) and 102.0 mg/g in waste ponds (mean of seven samples). Loss on ignition was 95.9 mg/g in control ponds and 118.2 mg/g in waste ponds. In peripheral zones loss on ignition was 303 mg/g in control ponds (mean of 334 samples) and 280 mg/g in wastewater ponds (mean of 30 samples). However, interpretation of loss on ignition values in sediments with clay is difficult because of uncertainty about the water initially bound to the clay.

Sediment. More sand was in the periphery and more clay in deeper central part of the ponds. Clay minerals were kaolinite, illite, and a 14 Å hydroxy-interlayer mineral (LeFurgey and Ingram, 1985).

Phosphorus (Kuenzler, 1970; McKellar, 1971; Woods, 1985). In control ponds in spring and summer much of the phosphorus in the waters was in particulate fraction owing to phytoplankton activity. Particulate phosphorus was 0.5–2.2 mg-atom/m^3, the dissolved organic fraction was 0.2 mg-atom/m^3 or less, and the dissolved inorganic fraction was 0.1 mg-atom/m^3 or less. In the winter all fractions were 0.3 mg-atom/m^3 or less.

In wastewater ponds phosphorus concentrations were much higher, 20–70 mg-atom/m^3 as dissolved inorganic phosphorus, 3–50 mg-atom/m^3 as particulate phosphorus, highest in winter in *Monodus* populations, and dissolved organic phosphorus 1–5 mg-atom/m^3.

Nitrogen (Raps, 1973; Masarachia, 1985; Woods, 1985). In control ponds nitrogen was widely varying, less so in winter. Total nitrogen was 1–90 mg-atom/m^3; particulate nitrogen 1–73 mg-atom/m^3; dissolved nitrogen 1–55 mg-atom/m^3; nitrate 1.5 mg-atom/m^3 or less; nitrite 0.01–0.2 mg-atom/m^3; and ammonia 0.03 to 0.3 mg-atom/m^3.

In wastewater ponds nitrogen concentrations were much higher, less in summer. Most of the nitrogen was in particulate form. Total nitrogen and particulate nitrogen were 2–420 mg-atom/m^3, total dissolved nitrogen 2–100 mg-atom/m^3, ammonia 0.15–0.4 mg-atom/m^3, and nitrate and nitrite 0.01–0.09 mg-atom/m^3.

The rate of outward diffusion of ammonia was 5–45 mg nitrogen/m^2-day in wastewater ponds and 3–9 mg N/m^2-day in control ponds. The nitrogen-fixing potential was also much higher in wastewater ponds than in control ponds.

In the control ponds most of the inflowing nitrogen was accounted for in outflowing waters. In the wastewater ponds half of the inflowing nitrogen was in the outflowing waters, most of the rest being retained in animals and sediments.

14.3.2 Living Components

Phytoplankton and Phytobenthos. Populations of diatoms and other algae were counted and the taxa illustrated (Campbell, 1985). Biomass was about 0.033 g dry/m^2 in control ponds winter and summer. Biomass in wastewater ponds was about 7.6 g/m^2 with winter *Monodus* and 0.34 g/m^2 in summer. Phytoplankton diversity in summer was 5–30 species per 1000 individuals counted, higher in wastewater ponds. There were many successive blooms of various species during the self-organization period (Campbell, 1985; Kuenzler, 1985; May, 1985).

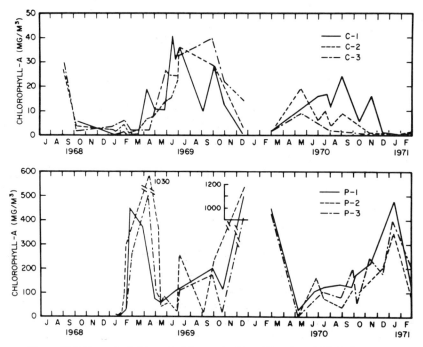

Figure 14.7 Record of chlorophyll in ponds. From Woods and McKellar (1985).

Monodus Bloom. Starting in the fall of the year and increasing during the winter there were enormous blooms of *Monodus* in the wastewater ponds, 10 million cells per milliliter. In the first week of May in three different years there was a crash in the population, the pond clearing for a few days as the bloom settled as bottom detritus, followed by a surge in normal diversified summer plankton diatoms (Campbell, 1985; Hommersand and Talbert, 1985).

A laboratory study was conducted of the physiology of *Monodus* (Kuenzler, 1985; Hommersand and Talbert, 1985), and growth rates were obtained for different conditions. Maximum growth rates were found in conditions of high pH, low temperature, and low light intensity. The species was adapted to varying salinity. It was oil filled at low temperature, but the oil was metabolized at higher temperatures, causing loss of buoyancy. These studies proved the special suitability of the species as a dominant capable of holding the ecosystem's store of biomass and nutrient content in the winter when animal populations were small.

Chlorophyll. Plankton chlorophyll in control ponds was low in winter and high in summer, as shown in Figure 14.7. In the wastewater ponds there were very remarkable dense *Monodus* blooms in winter with exceptionally high chlorophyll concentrations (Woods, 1985; Leeper and Woods, 1985).

Ruppia Beds. Dense beds of the underwater, estuarine flowering plant *Ruppia* developed in control ponds C2 and C3. By the end of the second year these beds filled much of the water space, with fronds floating at the surface. There was dieback in winter and regrowth in summer. Attempts to pull out and weigh all the *Ruppia* in June 1971 were only partly successful, yielding 425 kg wet weight in Pond C2 (760 g/m^2) and 165 kg from pond C3 (314 g/m^2). By midsummer 1972, beds had regrown and were more extensive. Pond C1 was several times planted with *Ruppia* but it did not grow, for reasons not understood. Turbidity was slightly higher there, water turnover time was 25% higher than in the other ponds, and snapping shrimp and oyster reef components prevailed more than in other ponds. In short, the plankton type ecosystem with filtering reefs in some way took resources away from the alternative *Ruppia*-based ecosystem.

Salt Marsh. After patches of *Spartina, Juncus,* and *Distichlis* were planted in each pond, *Spartina* marshes developed rapidly, soon fringing the ponds. Growth was extremely heavy in the wastewater ponds, even without a tide (see Figure 14.5). There was heavy flowering in summer. Other species patches grew only slightly (Marshall, 1985).

Marsh-shrouded pond margins developed populations of fiddler crabs soon after larvae were seen in the plankton of the inflowing waters. The spread of the *Spartina* was facilitated by the fiddler crab holes ahead of the marsh edge (154–320 burrows/m^2). *Littorina* populations did not develop in the pond marshes, although they occurred erratically in marshes around the ponds. A high diversity of microarthropods was found in the pond marsh fringe, 73 species per 1033 individuals in the wastewater pond marsh, and 88 species per 865 individuals in the control pond marsh (McMahan et al., 1972, 1985).

Bacteria. Direct counts of bacteria were made by Marsh (1985) and Rabin (1985). These were 1–8 million cells/ml in control ponds and 2–26 million cells/ml in wastewater ponds.

Much higher coliform concentrations, 200–1100 MPN/ml, were found in wastewater ponds than in control ponds (0.5–23 MPN/ml). Coliform counts in waste ponds were lower than in inflowing waters (May and Harvell, 1985). A survey of fungi found eight phycomycetes, six ascomycetes, and seven fungi imperfecti, with more species in control ponds than wastewater ponds (Rao and Koch, 1985).

Foraminifera. Biomass averaged 0.7 dry g/m^2 from 80,000 individuals/m^2 in control ponds, but 0.23 g/m^2 from 27,000 individuals/m^2 in wastewater ponds. More juveniles were in control ponds. Diversities were 22–24 species per 1000 individuals. *Elphidium clavatum* was most common in all ponds; *E. becarii, Miliammina fusca, Ammontium salsum,* and *E. tumidum* also occurred (LeFurgey, 1972; LeFurgey and St. Jean, 1976).

Table 14.1 Weight (kg) of the Most Abundant Macrofauna Seined Out of the Ponds in June, 1971

Species	Control Ponds			Waste Ponds		
	1	2	3	1	2	3
MOLLUSKS						
Crassostrea virginica	4.1	1.4	1.8	0.9	0.4	1.8
CRUSTACEANS						
Callinectes sapidus	0.7	0.1	0.1	5.4	3.1	2.0
Palaemonetes pugio	2.9	9.6	0.5	5.8	2.1	5.3
Alpheus heterochelis	1.3	P	P	0	0	0
FISHES						
Fundulus heteroclitus	P	0.2	0.6	2.4	3.7	4.7
Cyprinodon veriegatus	0	0	0	20.6	2.4	0
Paralichthys lethostigmus	2.7	2.4	2.3	0	0	0
P. albiguttus	0.7	0.8	1.1	0	0	0
Leiostomus xanthurus	0.3	1.5	0.7	0	0	0
Lagodon rhomboides	0.4	1.1	3.3	0	0	0
Total	13.1	17.1	10.4	35.1	11.7	13.8

Bottom Microzoa. The predominant bottom life comprised small worms, *Capitella, Leonereis,* and oligochaetes, 108 g wet biomass/m^2 in waste ponds and 39.7 g/m^2 wet biomass in control ponds. Meiofaunal density was 15,000–20,000 individuals/m^2, 98% nematodes (McMahan et al., 1985).

Zooplankton. Extensive studies of zooplankton showed populations of estuarine plankton species and larval stages entering the inflow pumps (McCrary and Jenner, 1985). Typical estuarine species prevailed. *Oithona* was dominant in the wastewater ponds in summer with 40–1720 thousand individuals/m^3. *Acartia tonsa* was a dominant in control ponds, but there were population surges and periods without the predominant species, especially in winter. Larger zooplankton included Paleomonetes, xanthid crabs, oyster larvae, barnacle larvae, and *Mysidopsis.*

Attached Animals. Mainly barnacles and a few oysters developed on ceramic fouling plates. *Limnoria* survived in transplanted boards that contained them initially. Transplanted oysters and Rangia clams survived, showing growth in control ponds, but did not survive in wastewater ponds (Chestnut et al., 1985).

Crabs and Shrimp. Grass shrimp, Paleomonetes, were a dominant species in all ponds (Table 14.1). Peneid shrimp were estimated at 0.47, 0.75, and 0.26 g dry/m² in ponds C1, C2, and C3. Blue crabs became a predominant carnivore in both sets of ponds (g dry/m²: C1, 2.5; C2, 1.4; C3, 1.3; P1, 2.3; P2, 1.8; P3, 1.5). Pond C1 had large populations of snapping shrimp, with a few in the other control ponds (Williams et al., 1985).

Fishes. Flounders, pinfish, and spots were the principal fishes in control ponds, whereas mullet and air-breathing top minnows (baitfish) were the principal members of the wastewater ponds (see Table 14.1). Diversity of fishes was higher in the control ponds (17.9, 14.8, and 12.2 species per thousand individuals) than in wastewater ponds (8.8, 8.1, and 5.9 species per thousand individuals). Mullet were found in wastewater pond P3, which had some backing up of estuarine water through the standpipe during exceptional tides, possibly seeding that pond more than the others (Schwartz and Hyle, 1985).

Ducks and Wading Birds. Ducks and wading birds of 17 species were attracted to the ponds, especially to the wastewater ponds, in the early morning when oxygen conditions were low and many animals were at the surface in the shallows. A semi-wild mallard brood were regular visitors. In a survey, 3.4 birds per hour of observation were found in the wastewater ponds but only 0.3 birds per hour in the control ponds (Spears and Williams, 1985).

Baitfish. Top minnows *Fundulus heteroclitus* and *Cyprinodon variegatus* were abundant in the wastewater ponds, in spite of large numbers taken by wading birds (Table 14.1). The ponds showed that baitfish could be raised in wastewater ponds with little effort. Screening out birds would reduce their predation and might be cost effective (Field et al., 1985).

14.3.3 Metabolism

Diurnal records of pH and oxygen were used to evaluate rates of photosynthesis and respiration. Figure 14.6 shows the graphic method of evaluating metabolic rates of change with corrections applied for oxygen exchange with the atmosphere. The gaseous exchange rate with the atmosphere was evaluated with floating dome experiments and found to be small.

During the initial 6 weeks when two ponds were acid a regular diurnal oxygen pattern with substantial photosynthesis was observed (Figure 14.6). At that time there were phytoplankton species normally found in acid mine waters (Odum, 1985; Campbell, 1985).

Therefore, an abbreviated procedure was used (Smith, 1971) that measured oxygen change between minimum value at dawn and approximate time of highest value at 1700 in the afternoon. Values were obtained on most days in the year. Several months' data are given in Figure 14.8 comparing the ups and downs of control ponds with the wide ranges in wastewater

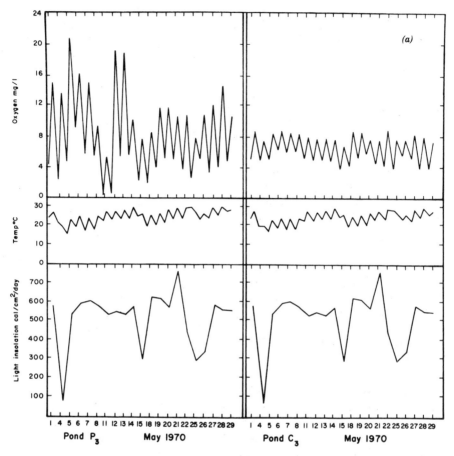

Figure 14.8 Comparison of oxygen, temperature, and insolation in four seasons (Smith, 1971).

ponds. Insolation and water temperatures are also shown. Then in Figure 14.9, net daytime photosynthesis is plotted upward and night respiration is plotted downward so that the distance between the two curves is an estimate of gross photosynthesis (sum of net production and night respiration). Salinity is also shown.

Production was intimately connected to respiration, with each a mirror image of the other. More photosynthesis makes the following night's respiration greater and vice versa. From day to day the magnitudes of both rose and fell with the available light and nutrient plankton oscillations from week to week. The monthly means for the six ponds are graphed in Figure 14.10. Production and consumption were much higher in wastewater ponds. As Table 14.2 shows, the ratio of organic production to sunlight input was much higher in nutrient rich ponds. Production was higher in summer in all ponds, but low in winter, in spite of the enormously dense *Monodus* bloom

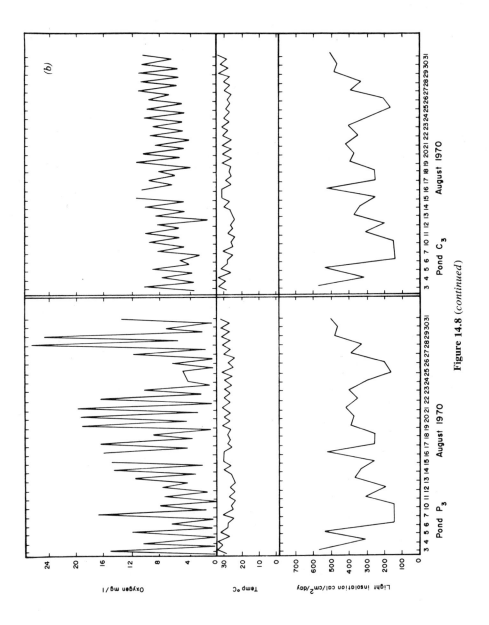

Figure 14.8 (*continued*)

313

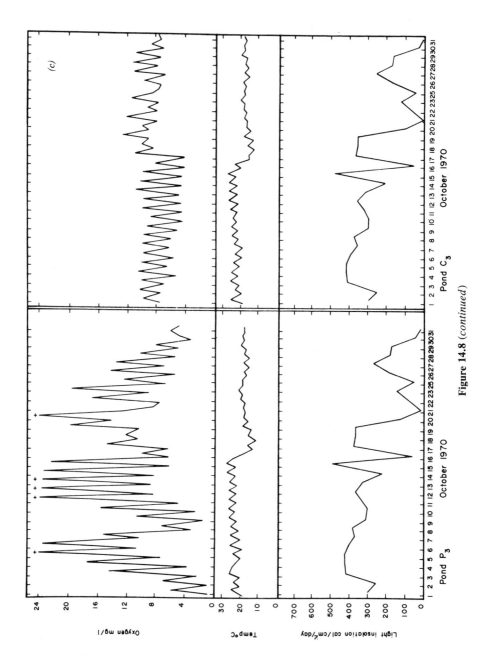

Figure 14.8 (*continued*)

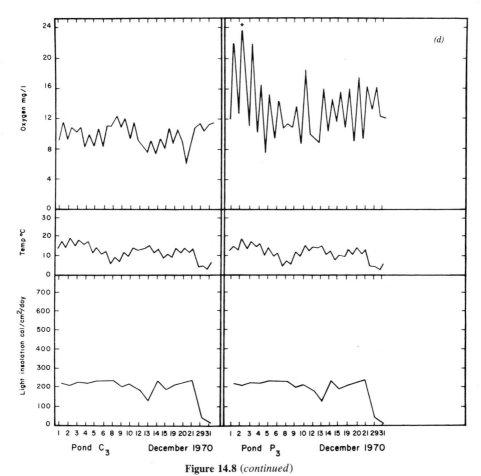

Figure 14.8 (*continued*)

and high chlorophyll concentrations in the wastewater ponds in winter. Metabolism estimated from free-water changes in pH and alkalinity were similar to that obtained from oxygen changes.

Table 14.3 summarizes the mean metabolism for two successive years during the self-organization process. The observation that performances consistently increased supports the hypothesis that self-organization tends to maximize power (consumption).

Estimates were also made of plankton and bottom metabolism using dark and light bottles, dark and light bell jars on the bottom, and dark and light plastic tubes with water and bottom cores. In control ponds more of the metabolism was in the plankton and the sum of bottom and plankton measurements was similar to the metabolism estimated with free-water methods. In wastewater ponds the bottom and plankton estimates were similar but the sum was much less than for the free water estimates.

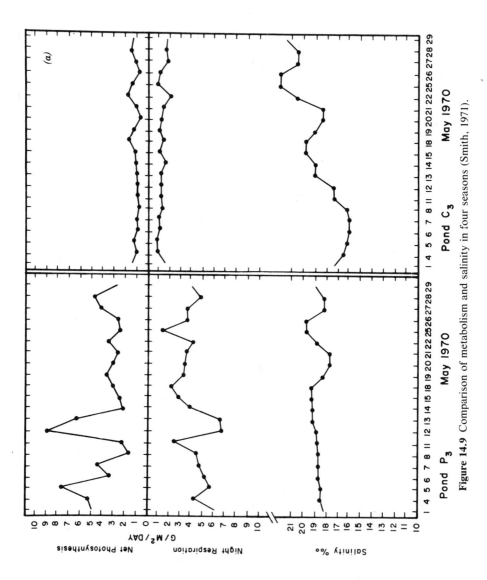

Figure 14.9 Comparison of metabolism and salinity in four seasons (Smith, 1971).

316

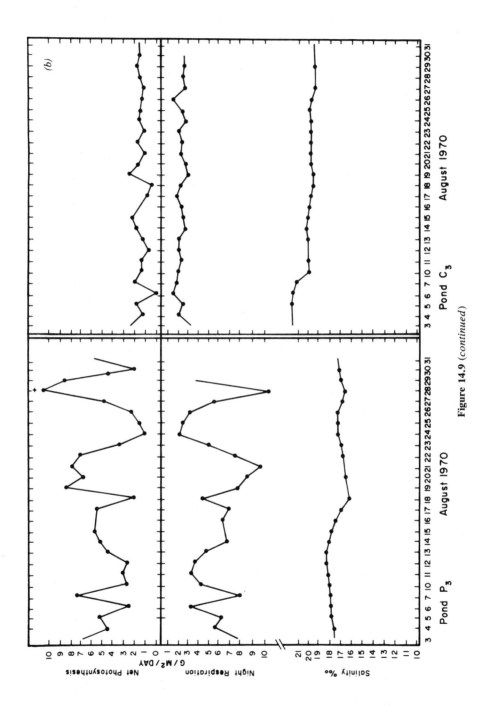

Figure 14.9 (continued)

317

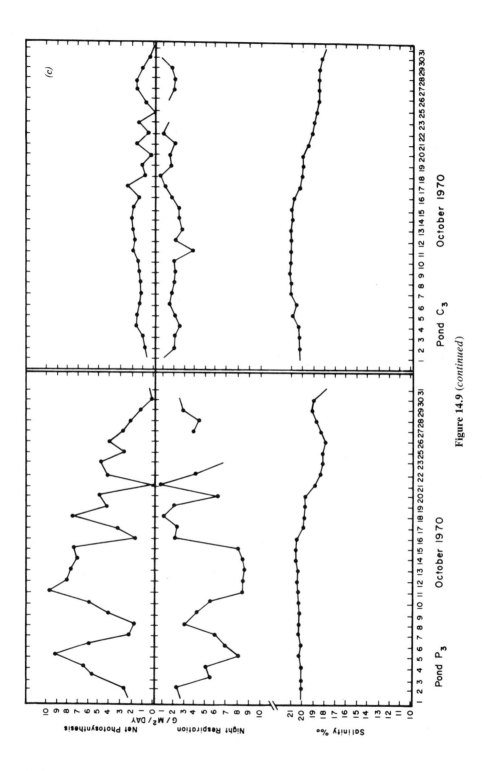

Figure 14.9 (*continued*)

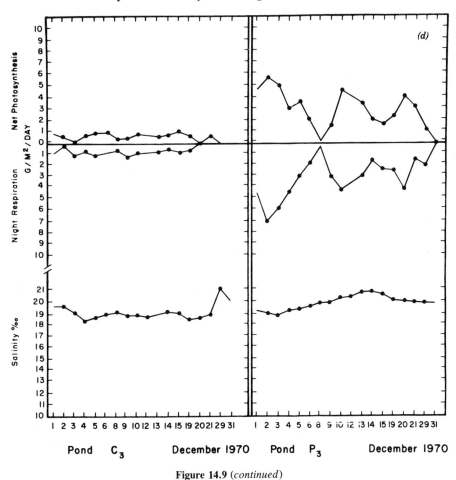

Figure 14.9 (*continued*)

Diurnal studies of phosphorus and radioactive uptake measurements were used to evaluate phosphorus metabolism in plankton (Kuenzler, 1970, 1985; McKellar, 1971, 1985). Turnover rates of uptake and release were higher in bottom than in plankton in waste ponds.

14.4 DISCUSSION

14.4.1 Summary of the Sequence of Events

After pond construction and filling, following an acid phase in control ponds, there was a period of buildup of larger components during which the smaller ones were variable and changing. A heavy fringe of *Spartina,* an enormous winter stock of *Monodus,* and a substantial population of killifish developed

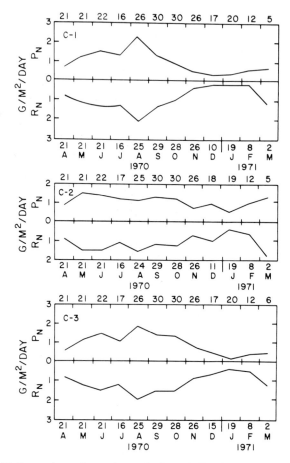

Figure 14.10 Comparison of annual record of metabolism in six ponds (Smith, 1971).

in the waste ponds, whereas a cover of *Ruppia* and a larger diversity of fishes developed in the control ponds. Variation in the smaller components was augmented by external variations in nutrient pumping, input salinities, and weather. Something of a repeating pattern had developed by the third year, after which pumping regimes were changed, causing ecosystem changes.

14.4.2   Climax Regime in the Waste Ponds

In the enriched ponds were heavy growth of fringing *Spartina* marsh, large populations of air-gulping, bait-sized minnows, blue crabs, and support for many aquatic birds. Organic matter was accumulating on the bottom. The eutrophic ecosystems here were like those so often observed in field studies, where the high nutrient zones act as net producers of organic matter that

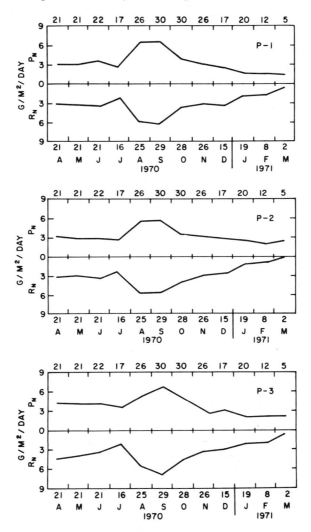

Figure 14.10 (*continued*)

flows out to support a longitudinal succession, enriching more typical oligotrophic ecosystems that are adjacent. The higher nutrient conditions produced much higher gross production but lower diversity, a characteristic that favors some utilization of net yields.

Figure 14.11 is a graphic summary of the seasonal sequence that emerged in the waste ponds and was repeated to varying degrees for three years until sewage input pumping was stopped. A winter *Monodus* regime developed in the fall, building up an enormous chlorophyll and plankton biomass, high pH, and oxygen in early spring. With a crash in early May, the bloom was sedimented with a surge of excess respiration followed by the onset of a

Table 14.2 Efficiency of Gross Photosynthesis for Control (C) and Wastewater (P) Ponds

Month	Mean Insolation (kcal/m²-day)	Mean Gross Production[a] (kcal/m²-day) C	P	Efficiency[b] C	P
1970					
April	4059	6.44	27.8	0.16	0.69
May	5018	10.40	26.8	0.21	0.53
June	5016	12.04	27.6	0.24	0.55
July	4375	9.40	19.7	0.21	0.45
August	3848	14.88	44.9	0.39	1.17
September	3985	10.80	49.4	0.27	1.24
October	2664	10.04	33.4	0.38	1.26
November	2223	5.88	25.2	0.27	1.13
December	1812	4.80	23.2	0.27	1.28
1971					
January	1450	2.80	14.7	0.19	1.01
February	2567	3.92	12.0	0.15	0.47
March	3643	4.84	11.4	0.13	0.31

Source: Smith, 1971.

[a] Mean oxygen production in g/m²-day times 4 kcal/g.
[b] Percent that mean oxygen production is of mean insolation. If half of the insolation is used as in wavelengths available for photosynthesis, efficiencies are twice these.

more photosynthetic regime of diverse phytoplankton, rise in microbial actions and temperature, and seasonal increases in the animals, followed by gradual decline in stocks and process with decline of sunlight and temperature in the fall.

14.4.3   Summarizing Ecosystem Overview

Ecosystem diagrams were made and discussed at the quarterly meetings of the project group as one of the communication tools for each scientist to relate his or her studies to those of others. Various sources, components, food relationships and other mechanisms believed to be important were included. When everything suggested by everyone was included, the ecosystem diagram became very complex.

To understand the self-organizational processes that developed estuarine ecosystems in the ponds, the results may be examined as a system. Figure 14.12 is an aggregated overview of the main components and processes. Any simplification of nature must also be an approximation with some arbitra-

Table 14.3 Annual Mean of Daily Metabolism (gO$_2$/m^2-day)

Pond	P net		R	
	1969–1970[a]	1970–1971[b]	1969–1970[a]	1970–1971[c]
C1	1.301	1.022	0.986	1.025
C2	1.274	1.111	0.822	1.125
C3	1.096	1.139	1.055	1.168
Mean	1.225	1.091	0.953	1.106
P1	1.945	3.559	1.521	3.793
P2	2.712	3.417	2.192	3.529
P3	3.123	3.927	3.384	4.084
Mean	2.594	3.634	3.364	3.802

Source: Smith, 1971.

[a] Odum, Hall, and Masarachia (1970): 127 days.
[b] 246 days.
[c] 226 days.

riness in what is shown and what is aggregated. Items included at one scale of view may become too small to stand out at the scale of next larger size. Figure 14.12 is a composite view that includes most of the categories identified in the research when viewed in the larger scale.

Figure 14.12 was drawn using energy systems language, in which symbols and pathways have precise energetic, mathematical, and hierarchical meaning (Odum, 1967, 1983). One may read the diagram by starting with the factors outside the boundaries such as sun and inflows of waters and follow pathways in, around, and through the system noting main components, production, consumption, and other process intersections. Hierarchical relationships are represented by positioning items large in quantity such as the sunlight on the left and items small in quantity but important in control actions such as fishes on the right.

Pond inflows and outflows are at the top, contributing to and draining from the pool of dissolved and suspended substances in the water. The biological web is shown below receiving and contributing to the pool. The dominants of both sets of ponds are included in the diagram, because these species were available to both. Of course the presence of high nutrients made a different set dominant from that in the oligotrophic ponds.

14.5 SIMULATION MODELS

Whether one's aggregated overview of a system is an accurate one or not depends on whether it is consistent with the mechanisms known to operate and the main events observed. A simulation model tests the consistency between ideas about relationships and observed events. In other words, a

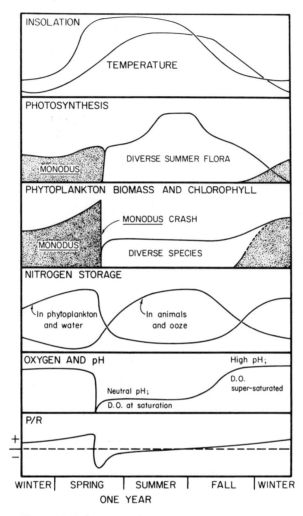

Figure 14.11 Summary of annual rhythm (Odum, 1985).

simulation can determine if the relationships represented in the overview ecosystem diagram (Figure 14.12) explain the events summarized in Figure 14.11.

Diurnal patterns were simulated with a basic production–consumption model, but simulation of seasonal and successional patterns required a more complex model that included temperature.

14.5.1   Simulation of Diurnal Variations

Analog computer simulations of the sharply varying diurnal process were previously published, including those for carbon dioxide and phosphorus

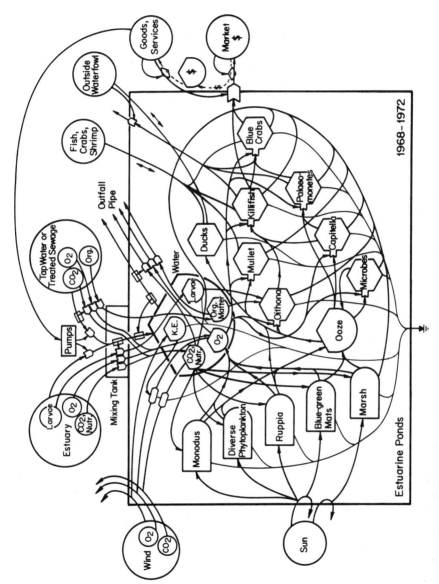

Figure 14.12 Summarizing diagram of the ecosystem that developed after self-organization (Odum, 1985).

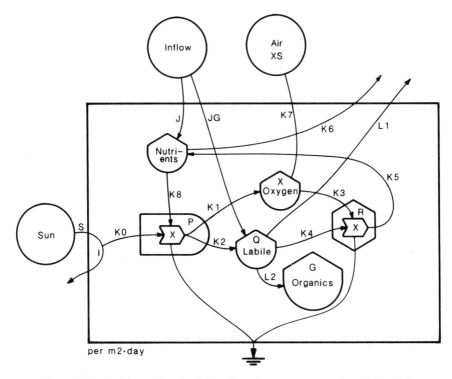

Figure 14.13 Model used for simulating diurnal oxygen patterns (see Table 14.4).

(Day et al., 1971; Day, 1971; McKellar, 1971, 1985). The model for these was similar to that for oxygen given here in Figure 14.13. Simulation of diurnal oxygen patterns in wastewater and control ponds generated the main patterns observed in the short term. The program in Table 14.4 is readily adapted to any aquatic ecosystem.

In the diurnal model in Figure 14.13, diurnal changes in light drive gross production, which generates oxygen and labile organic matter (metabolically available on a short range basis) and uses nutrients from the available pool. Respiration is a product function of organic matter and oxygen. Oxygen is exchanged with the air slowly in proportion to the gradient of oxygen partial pressure. Respiration regenerates nutrients available for gross photosynthetic production. In the course of a day some oxygen and organic matter are added with inflowing waters and exported with the overflow. The program is given in Table 14.4.

Simulation runs are given in Figure 14.14 for the high-nutrient ponds and for the low-nutrient ponds, the only difference in calibrations being the nutrient levels. Alternating sunny and cloudy days were simulated by varying

Table 14.4 Microcomputer Simulation Program for the Model in Figure 14.13

```
3 REM IBM PC
5 REM DIELPOND; E. C. and H. T. ODUM
7 CLS
8 REM GRAPHICS:
9 SCREEN 1,0
11 COLOR 0,1
15 LINE (0,0)-(320,180),3,B
18 LINE (0,30)-(320,30),3
20 LINE (0,60)-(320,60),3
25 LINE (0,100)-(320,100),3
27 LINE (0,120)-(320,120),1
30 LINE (0,140)-(320,140),3
100 REM INSOLATION ARRAY AT 1 HR INTERVALS, MEGAJOULES PER
  SQUARE METER
110 DIM A(24)
120 DATA 0, 0.3, 0.8, 1.6, 2.1, 2.3, 2.5, 2.6, 2.5, 2.3, 2.1,
  1.6, 0.8, 0.3, 0, 0, 0, 0, 0, 0, 0, 0, 0, 0, 0
130 FOR HR = 1 TO 24
140 READ A(HR)
150 NEXT
160 REM WIND ARRAY AT 1 HOUR TIME INTERVALS IN MILES PER HOUR
165 DIM B(24)
167 DATA 6.5, 6.7, 7.0, 7.1, 7.3, 7.6, 8.1, 8.2, 8.3, 8.4,
  8.4, 8.5, 8.4, 8.2, 8.0, 7.5, 7.3, 7.1, 6.9, 6.7, 6.5, 6.3,
  6.1, 6.0
170 FOR HR = 1 TO 24
175 READ B(HR)
180 NEXT
200 REM SCALING FACTORS
205 DT = .05
208 TO = .5
240 NO = .2
250 GO = 2
260 SO = .1
270 XO = 1.5
280 PO = .3
290 RO = .3
295 WO = .5
297 HR = 1
300 REM STARTING VALUES:
305 Z = .4
310 J = .1/24
315 JG = .625/24
320 XS = 6
330 X = 6
340 N = 3
350 Q = 100
```

Table 14.4 Microcomputer Simulation Program for the Model in Figure 14.13 (*continued*)

```
355 P = .2
360 R = .2
365 REM COEFFICIENTS
370 K0 = 19
380 K1 = 35!
390 K2 = 14
400 K3 = .001
410 K4 = .0004
420 K5 = .000022
430 K6 = .000833
450 K8 = .68
460 K9 = .01
470 L1 = .000166
475 L2 = .000417
500 REM PLOTTING TIME CURVES:
510 REM SUN IN RED:
520 LINE (T/T0,30)-(T/T0,(30-S/S0)),2
522 REM WIND IN WHITE
525 PSET (T/T0, 30-W/W0),3
530 REM NUTRIENTS IN WHITE:
540 PSET (T/T0, 60-N/N0),3
550 REM OXYGEN IN BLUE:
560 PSET (T/T0, 100-X/X0),1
563 PSET (T/T0,  60-G/G0),2
570 REM NET PRODUCTION RATE IN RED:
580 PSET (T/T0,120-(P-R)/P0),2
590 REM GROSS PRODUCTION RATE IN RED:
600 PSET (T/T0,180-P/P0),2
610 REM TOTAL RESPIRATION RATE IN BLUE:
620 PSET (T/T0,180 - R/R0),1
652 REM PERMANENT ORGANIC DEPOSITION IN RED
700 REM EQUATIONS
710 S = A(HR)
711 IF T<72 THEN S = .5* A(HR)
712 W = B(HR)
715 K7 = K9*W
720 IF S<0 THEN S = 0
730 I = S/(1 + K0*N)
740 DN = J - K8*I*N - K6*N +K5*X*Q
750 DQ = K2 *I*N - K4*X*Q +JG -L1*Q-L2*Q
760 DX = K1 *I*N - K3*X*Q +K7*(1-X/XS)
765 DG = L2*Q
770 REM NEXT VALUES
775 N = N +DN*DT/Z
780 Q = Q +DQ*DT
790 X = X +DX*DT/Z
```

Table 14.4 Microcomputer Simulation Program for the Model in Figure 14.13
(*continued*)

```
795 G  =  G  +DG*DT
800 P  =  K1*I*N
810 R  =  K3*X*Q
820 T  =  T+DT
825 HR  =  1+INT(T-24*D)
827 IF HR>24 THEN HR=1:D=D+1
830 REM GO BACK AND REPEAT:
840 IF T/TO < 320 GOTO 500
850 END
```

the light inputs on successive days. The diurnal oxygen curves obtained may be compared with data on oxygen graphed in Figures 14.6 and 14.8. The simulations generated the observed shapes and greater metabolism on and after the sunny days as well as the observed large oxygen amplitudes in high-nutrient ponds and the smaller oxygen amplitudes of the control ponds.

14.5.2 Simulation of Ecosystem Succession and Seasonal Patterns

Starting with some further aggregation of Figure 14.12, a simulation model of the ecosystems of the ponds was drawn (Figure 14.15). Next, a mathematical translation of the ecosystem model was written in the form of differential equations (Table 14.5). These relationships were included in a BASIC computer program given in the data report (Odum, 1985). Data were used to estimate flows and storages as given in Figure 14.16. Then coefficients were calibrated. Simulations were run generating graphs of pond components with time.

At first, simulation graphs had major differences from the observed details. For example, one preliminary model on an analog computer (Odum, 1972) generated broad features of production and respiration during successional buildup of organic matter, but did not generate characteristics of the *Monodus* regime. Temperature, not included at first, was found necessary to generate the *Monodus* behavior found by Hommersand's laboratory study. The version presented here was developed on a Compucolor microcomputer and later refined on an Apple microcomputer.

To be successful, an overview model should be able to generate behavior in both control and waste ponds, generating curves of one or the other when the nutrients are varied. Therefore the model was supplied with pathways for both regimes, thus representing multiple seeding. This was a way of introducing the self-organization observed in nature by which some pathways are reinforced and prevail over others.

The final version of the simulation model is given in energy language in Figure 14.15. By inspection, a reader may compare the flow in or out of a

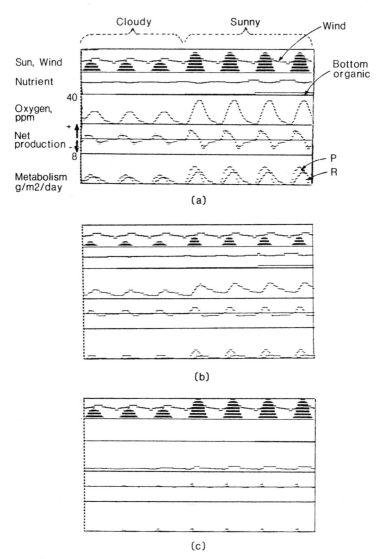

Figure 14.14 Results of simulating the diurnal oxygen model in Figure 14.13 with sunny and cloudy light conditions. (*a*) Wastewater ponds with summer regime; (*b*) wastewater ponds with winter regime and respiration coefficients K3, K4, and K5 reduced to 25%; (*c*) control ponds with summer regime.

storage tank with the storage number in the tank to get a turnover time. In this way the diagram helps visualize which compartments are rapid and which are slow.

The calibration of the model for the waste ponds (P ponds) used phosphorus for the nutrients labeled N and NØ in Figure 14.15. With high con-

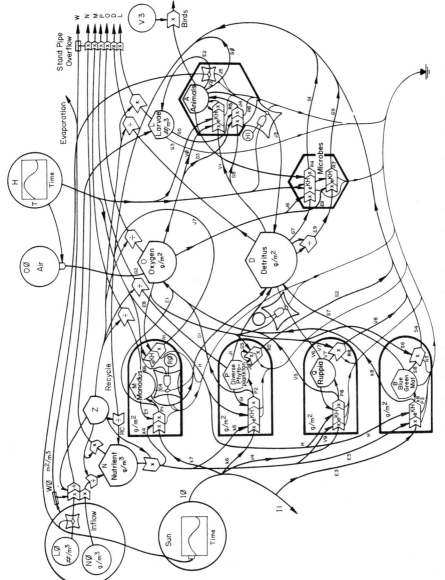

Figure 14.15 Model used for simulating seasonal and successional patterns (see also Table 14.5).

331

Table 14.5 Differential Equations of the Pond Model in Figure 14.15

$$\dot{N} = \frac{W_0 N_0}{Z} + R_e - \frac{k_4}{Z} INM - \frac{k_5}{Z} IN(e^{KH})P - V_9 IN(e^{KH})Q - E_5 I_1 N(e^{KH}) - N\frac{W}{Z}$$

$$\text{IF } H > R_0$$

$$\dot{M} = E_1 INM - \overbrace{E_4 M} - M(e^{KH})O - \frac{W_0 M}{Z}$$

$$\dot{P} = G_1 IB(e^{KH})P - J_1 P - J_0 P^2 O$$

$$\dot{Q} = V_5 IN(e^{KH}) - V_6 Q - V_7 QO$$

$$\dot{B} = k_8 I_1 N(e^{KH}) - S_8 BO - S_7 B$$

$$\text{IF } H > R_0 \qquad\qquad \text{IF } O > 0.2$$

$$\dot{D} = \overbrace{E_4 M} + J_1 P + \overbrace{V_6 Q} + S_7 B - E_9 D^2 O(e^{KH}) - G_7 DO(e^{KH})A$$

$$\text{IF } H > H_1$$

$$= + \overbrace{J_3 A} - V_1 DO(e^{KH})A - G_8 DOL$$

$$\dot{O} = G_2(O_0 - O) + E_1 I \frac{N}{Z} M + G_1 IN(e^{KH}) + k_8 I_1 N e^{KH} + V_5 IN(e^{KH})Q$$

$$- E_8 M(e^{KH})O - S_5 P^2 O - V_7 QO - S_6 BO - E_9 D^2 O(e^{KH})$$

$$- I_6 DO(e^{KH})A - V_0 DO(e^{KH})A - G_3 DOL$$

$$\text{IF } H > H_1 \qquad\qquad \text{IF } G_4 = 1$$

$$\dot{A} = S_9 DO(e^{KH})A + J_4 DOL - \overbrace{J_3 A} - J_5 A - \overbrace{V_2 A} - V_3 A$$

$$\text{IF } G_2 = 1$$

$$\dot{L} = \overbrace{E_2 A} + L_0 \frac{W_0}{Z} - S_3 L - G_5 DOL - L\frac{W}{Z}$$

$$R_c = J_7 M(e^{KH})O + S_0 A + S_1 DO(e^{KH})A + G_9 DO(e^{KH}) + S_2 P^2 O + S_4 BO + V_8 QO$$

Initial conditions indicated as N_0, W_0, etc. = N$\emptyset$, W$\emptyset$, etc.

centrations of nutrient inflow, simulation as shown in Figure 14.17a generated a *Monodus* bloom that dominates the winter, switching to diverse summer regimes in spring. The simulation did generate the pattern in Figure 14.11.

To run the program for the control pond conditions, six coefficients were substituted (Odum, 1985), reducing nutrient inflow. The calibration of the model for the control ponds (C ponds) used nitrogen for N and N$\emptyset$. As shown in Figure 14.17b, simulation did not develop *Monodus* blooms, but instead generated *Ruppia* beds in 3 years.

Some proof of system understanding was demonstrated by the ability of the model to generate the main features of the pond's annual cycle and succession by combining the components and mechanisms found by those studying ecosystem parts and processes.

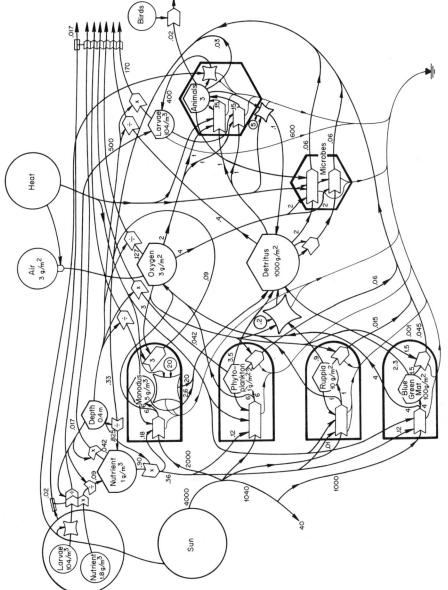

Figure 14.16 Storage and flows per day used to calibrate the seasonal–successional model in Figure 14.15.

333

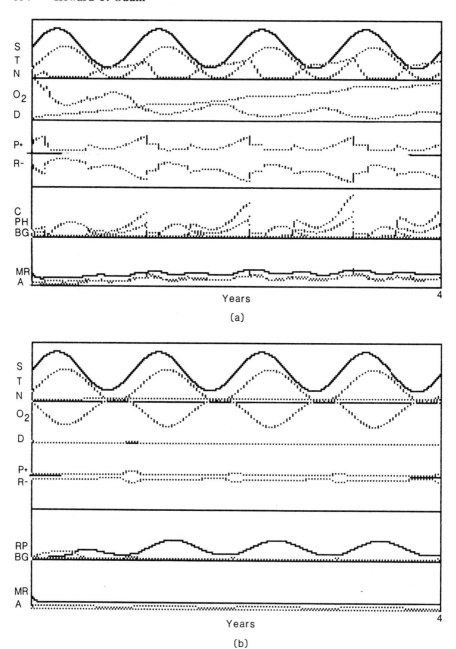

Figure 14.17 Results of simulation of the seasonal–successional model in Figure 14.15. S, sun; T, temperature; N, nutrients; O₂, oxygen; D, detritus; P⁺, net production; R⁻, respiration; C, chlorophyll; PH, phyloplankton; BG, blue greens; MR, microbes; A, animals; RP, *Ruppia*. (*a*) With high-nutrient inflows; (*b*) with low-nutrient inflows.

The model successfully representing pond events has the same kinds of design characteristics that have been enumerated already as emerging with self-organization. That self-organization was rapid in estuarine conditions when multiple seeding provided an excess of genetic variety shows how the tuning of ecosystems in estuaries is continual and automatic.

14.6 SUMMARIZING OBSERVATIONS

1. Estuarine ecosystems can develop main components in typical organization of food chains, nutrient cycles, and seasonal regimes in 3 years.

2. Blooms and surges of the smaller, high-turnover phytoplankton, microzoa, and bacteria probably represent variable seeding, competition, selection, waves of homeostasis in the coupling of production and respiration, and prey–predator oscillations, which collectively provide the choices and noise to facilitate a continuous process of adaptation and self-organization.

3. In one sense there was a succession of species when populations and patterns appeared briefly that did not continue later. Some of these intermediate components were different in each pond, representing varying initial conditions and input transients. The dominance of one species at one time in one pond through population interaction mechanisms may be the means for fine-tuning the maximum performances of the various guilds of the ecosystem.

4. Self-organization was accompanied by gradual buildup of storages of detritus, animal and plant biomass, productivity, total respiratory metabolism, and diversity. These may be the stabilizing, species-independent design criteria of successful ecosystems that are the end result of and system stabilizer of the noisy surges of populations.

5. Nutrient storages were developed in biomass in all ponds. In control pond C1, storages were in shellfish; in control ponds C2 and C3, storages were in macrophyte beds of *Ruppia* and associated grass shrimp; and in waste ponds, storage was in the ooze, in *Spartina* beds, and alternating seasonally between the winter *Monodus* stock and the summer animals.

6. Diversity in the ponds was less than in the comparable environments outside, except for microarthropods in the marsh grass, which were more diverse within the ponds.

7. Diversities of most groups of organisms were less in the eutrophic ponds than in the more oligotrophic control ponds, partly because of the extremes of oxygen, pH, and other variables.

8. Nontidal estuarine pond microcosms can develop many of the characteristics of oligotrophic and eutrophic estuaries, which may make them suitable for experimental estuarine study for many purposes.

9. Except during the initial filling time before sewage pumping started,

nitrogen was more limiting to photosynthesis than phosphorus, especially in waste ponds, as might be expected because a low nitrogen to phosphorus ratio is characteristic of treated sewage.

10. Eutrophic estuarine ponds, because of their oxygen fluctuations, develop a channeling of food chains into a few species, a useful ecological engineering aquaculture technique requiring no dollar management costs.

11. The high productivity of small fish and shrimp in estuarine eutrophic ponds has potential as an inexpensive means of supporting waterfowl and other wildlife. Or with suitable arrangement for outmigration in late summer, the ponds could enrich higher food chains outside.

12. The use of eutrophic estuarine ponds for baitfish aquaculture will require screening from waterfowl to maximize the net yield for economic purposes.

13. The estuarine ponds fertilized with waste had the typical characteristics of cultural eutrophy that are sometimes regarded as undesirable when they develop in natural waters (intensely green with plankton algae, low diversity, high and low extremes of oxygen and pH, missing game fish). Sometimes cultural eutrophy is regarded as pathological, unregulated, unstable, and undesirable. However, this systems study found a well organized ecosystem whose performance in generating overall production and protein was better than that of the control ponds or the normal surrounding estuaries.

14. The pond system was a fertile interface between wastewaters and public waters, a means to carry out and control tertiary treatment at low cost, while reducing and stabilizing effluent effects on outside public waters. A strip of such ecosystems between society (with its effluents from sheet sources and point sources) may be a useful design for a coastal zone that reduces open water eutrophication while utilizing the enrichment benefits. After fish were added there were no mosquitoes.

15. The ecological engineering technique of setting boundary conditions (inflows, turnover times, and controlling actions) and providing massive multiple seeding of species may generate rapidly new ecosystems for new conditions that are useful for the developing harmony between humanity and nature.

16. Self-organization of new ecosystems for new conditions occurs rapidly, utilizing available species. Development of ecosystems symbiotic with the human economy can be accelerated by the ecological engineering technique of multiple seeding.

ACKNOWLEDGMENT

The project "Optimum Ecological Designs for Estuarine Ecosystems in North Carolina" was supported by the Sea Grant Program of the National

Science Foundation in 1967 with NSF Grant GH-18 and transferred as Grant GH-03 when Sea Grant was moved to the National Oceanographic and Atmospheric Administration.

Experimental ponds were constructed with funds from the North Carolina Board of Science and Technology with grants No. 180 and 232. C. C. King prepared pond maps and legal description for a lease from Morehead City. The construction contract was with Howard Construction Co, Newport, NC. J. Lamb arranged pumping systems.

Principal investigators were H. T. Odum and A. C. Chestnut for 1967–1970, E. J. Kuenzler and A. C. Chestnut for 1970–1971, and C. M. Weiss for 1971–1972.

Eighty-four faculty and students of the University of North Carolina Marine Curriculum and Institute of Marine Science participated; their data and reports are detailed elsewhere (Odum, 1985). Participating were M. Adams, M. Beeston, A. R. Camp, P. H. Campbell, M. Canoy, A. C. Chestnut, C. Patterson, A. F. Chestnut, E. Danya, L. C. Davidson, F. E. Davis, J. Day, C. R. Dillon, R. E. Dowds, R. A. Farris, R. C. Field, R. S. Fox, C. Hall, J. R. Hall, A. Harvell, M. Hastings, P. Hebert, T. L. Herbert, M. Hommersand, Y. Huang, J. T. Hunter, R. Hyle, R. Ingram, C. Jenner, J. D. Johnson, J. Joyner, R. Klemm, R. L. Knight, W. Koch, E. J. Kuenzler, C. Lathrop, W. Laughinghouse, D. E. Leeper, A. LeFurgey, E. Lindgren, R. Mah, B. Marino, J. A. Marsh, D. E. Marshall, S. Masarachia, M. May, A. B. McCrary, H. N. McKellar, Jr., E. A. McMahan, N. Meith, D. Miller, J. Murray, B. Muse, D. Oakley, H. T. Odum, R. Outen, T. Owen, P. Parks, C. Patterson, A. Powell, A. N. Rabin, B. Rao, M. Raps, C. F. Rhyne, J. Richey, W. L. Rickards, R. J. M. Riedl, F. J. Schwartz, M. Smith, W. Smith, C. J. Spears, J. Staley, H. Stelljes, A. E. Stiven, J. St. Jean, D. Talbert, E. Walton, C. M. Weiss, A. B. Williams, W. J. Woods, and S. D. Wyman.

Initial data were issued in progress reports (Odum and Chestnut, 1970; Kuenzler and Chestnut, 1971a, 1971b; Kuenzler et al., 1973). Detailed documentation of data, contributions, and student reports are given in the data report (Odum, 1985).

REFERENCES

Beyers, R. J. 1963. The metabolism of twelve aquatic laboratory microecosystems. *Ecol. Monogr. 33*(4)281–306.

Campbell, P. H. 1985. In H. T. Odum, Ed., Self Organization of Ecosystems in Marine Ponds Receiving Treated Sewage. University of North Carolina Sea Grant Office, Publication #UNC-SG-85-04, N.C. State University, Raleigh, NC, pp. 105–109.

Chestnut, A. F., W. Laughinghouse, W. Smith, P. Parks, and R. Klemm. 1985a. In H. T. Odum, Ed., Self Organization of Ecosystems in Marine Ponds Receiving

Treated Sewage. University of North Carolina Sea Grant Office, Publication #UNC-SG-85-04, N.C. State University, Raleigh, NC, pp. 23–27.

Chestnut, A., A. Williams, B. Muse, M. Canoy, R. Dowd, and E. Lindgren. 1985b. In H. T. Odum, Ed., Self Organization of Ecosystems in Marine Ponds Receiving Treated Sewage. University of North Carolina Sea Grant Office, Publication #UNC-SG-85-04, N.C. State University, Raleigh, NC, pp. 71–74.

Day, J. W. 1971. The Carbon Metabolism of Estuarine Ponds Receiving Treated Sewage Wastes. Ph.D. dissertation. Curriculum in Marine Sciences, University of North Carolina, Chapel Hill, 128 pp.

Day, J. W. 1983. Carbon dynamics of estuarine ponds receiving treated sewage wastes. *Estuaries* 6:10–19.

Day, J. W., C. M. Weiss, and H. T. Odum. 1971. The carbon budget and total productivity of estuarine oxidation ponds receiving secondary sewage effluent. In Proc. Second International Symposium of Waste Treatment Lagoons. Nuclear Reactor Center, Lawrence, KS, pp. 100–103.

Field, R., M. Beeston, A. Williams, and F. J. Schwartz. 1985. In H. T. Odum, Ed., Self Organization of Ecosystems in Marine Ponds Receiving Treated Sewage. University of North Carolina Sea Grant Office, Publication #UNC-SG-85-04, North Carolina State University, Raleigh, NC, pp. 201–208.

Hommersand, M. and D. M. Talbert. 1985. In H. T. Odum, Ed., Self Organization of Ecosystems in Marine Ponds Receiving Treated Sewage. University of North Carolina Sea Grant Office, Publication #UNC-SG-85-04, North Carolina State University, Raleigh, NC, p. 119.

Kelly, R. A. 1971. The Effects of Fluctuating Temperature on the Metabolism of Laboratory Freshwater Microcosms. Ph.D. dissertation. Dept. of Zoology, University of North Carolina, Chapel Hill, 205 pp.

Kuenzler, E. J. 1970. Dissolved organic phosphorus excretion by marine phytoplankton. *J. Phycol.* 6:7–13.

Kuenzler, E. J. 1985. In H. T. Odum, Ed., Self Organization of Ecosystems in Marine Ponds Receiving Treated Sewage. University of North Carolina Sea Grant Office, Publication #UNC-SG-85-04, North Carolina State University, Raleigh, NC, pp. 88–94.

Leeper, D. E. and W. Woods. 1985. In H. T. Odum, Ed., Self Organization of Ecosystems in Marine Ponds Receiving Treated Sewage. University of North Carolina Sea Grant Office, Publication #UNC-SG-85-04, North Carolina University, Raleigh, NC, p. 55.

LeFurgey, A. 1972. Foraminifera in Estuarine Ponds Designed for Waste Control and Aquaculture. M.S. thesis. Curriculum in Marine Sciences, University of North Carolina, Chapel Hill, NC.

LeFurgey, A. and R. Ingram. 1985. In H. T. Odum, Ed., Self Organization of Ecosystems in Marine Ponds Receiving Treated Sewage. University of North Carolina Sea Grant Office, Publication #UNC-SG-85-04, N.C. State University, Raleigh, NC, p. 140.

LeFurgey, A. and J. St. Jean, Jr. 1976. Foraminifera in brackish water ponds designed for waste control and aquaculture studies in North Carolina. *J. Foraminiferal Res.* 65:274–294.

Marsh, J. A. 1985. In H. T. Odum, Ed., Self Organization of Ecosystems in Marine Ponds Receiving Treated Sewage. University of North Carolina Sea Grant Office, Publication #UNC-SG-85-04, North Carolina State University, Raleigh, NC, pp. 126–127.

Marshall, D. 1985. In H. T. Odum, Ed., Self Organization of Ecosystems in Marine Ponds Receiving Treated Sewage. University of North Carolina Sea Grant Office, Publication #UNC-SG-85-04, North Carolina State University, Raleigh, NC, p. 117.

Masarachia, S. 1985. In H. T. Odum, Ed., Self Organization of Ecosystems in Marine Ponds Receiving Treated Sewage. University of North Carolina Sea Grant Office, Publication #UNC-SG-85-04, North Carolina State University, Raleigh, NC, pp. 103–104.

May, M. S. 1985. In H. T. Odum, Ed., Self Organization of Ecosystems in Marine Ponds Receiving Treated Sewage. University of North Carolina Sea Grant Office, Publication #UNC-SG-85-04, North Carolina State University, Raleigh, NC, p. 169.

May, M. and A. Harvell. 1985. In H. T. Odum, Ed., Self Organization of Ecosystems in Marine Ponds Receiving Treated Sewage. University of North Carolina Sea Grant Office, Publication #UNC-SG-85-04, North Carolina State University, Raleigh, NC, pp. 128–130.

McCrary, A. and C. Jenner. 1985. In H. T. Odum, Ed., Self Organization of Ecosystems in Marine Ponds Receiving Treated Sewage. University of North Carolina Sea Grant Office, Publication #UNC-SG-85-04, North Carolina State University, Raleigh, NC, pp. 143–160.

McKellar, H. N. 1971. The Phosphorus System of Brackish Water Pond Ecosystems Exposed to Treated Sewage Waste. M.S. thesis. Curriculum in Marine Sciences, University of North Carolina, Chapel Hill.

McMahan, E. A., R. L. Knight, and A. R. Camp. 1972. A comparison of microarthropod populations in sewage-exposed and sewage-free *Spartina* salt marshes. *Environ. Entomol.* 1(2):244–252.

McMahan, E. A., R. L. Knight, and A. R. Camp. 1985. In H. T. Odum, Ed., Self Organization of Ecosystems in Marine Ponds Receiving Treated Sewage. University of North Carolina Sea Grant Office, Publication #UNC-SG-85-04, North Carolina State University, Raleigh, NC, pp. 165–168.

McMahan, E. A., A. Powell, and A. McCrary. 1985. In H. T. Odum, Ed., Self Organization of Ecosystems in Marine Ponds Receiving Treated Sewage. University of North Carolina Sea Grant Office, Publication #UNC-SG-85-04, North Carolina State University, Raleigh, NC, p. 141.

Nixon, S. W. 1969. Characteristics of Some Hypersaline Ecosystems. Ph.D. dissertation, Dept. of Botany, University of North Carolina, Chapel Hill.

Odum, H. T. 1967. Biological circuits and the marine systems of Texas. In T. A. Olsen and F. J. Burgess, Eds., *Pollution and Marine Ecology*. Wiley-Interscience, New York, pp. 99–157.

Odum, H. T. 1972. Chemical cycles with energy circuit models. In *Nobel Symposium No. 20,* Wiley, New York, pp. 223–257.

Odum, H. T., Ed. 1985. Self Organization of Ecosystems in Marine Ponds Receiving

Treated Sewage. University of North Carolina Sea Grant Office, Publication #UNC-SG-85-04, North Carolina State University, Raleigh, NC, 250 pp.

Odum, H. T., C. Hall, and S. Masarachia. 1985. In H. T. Odum, Ed., Self Organization of Ecosystems in Marine Ponds Receiving Treated Sewage, pp. 51–52. University of North Carolina Sea Grant Office, Publication #UNC-SG-85-04, North Carolina State University, Raleigh, NC, pp. 51–52.

Odum, H. T. and C. M. Hoskin. 1957. Metabolism of a laboratory stream microcosm. *Publ. Inst. Marine Sci. Univ. Texas 4:*115–133.

Rabin, A. N. 1985. In H. T. Odum, Ed., Self Organization of Ecosystems in Marine Ponds Receiving Treated Sewage. University of North Carolina Sea Grant Office, Publication #UNC-SG-85-04, North Carolina State University, Raleigh, NC, p. 126.

Rao, B. and W. Koch. 1985. In H. T. Odum, Ed., Self Organization of Ecosystems in Marine Ponds Receiving Treated Sewage. University of North Carolina Sea Grant Office, Publication #UNC-SG-85-04, North Carolina State University, Raleigh, NC, pp. 133–134.

Raps, M. 1973. The Effects of Treated Sewage Effluents on Nitrogen Fixation and Ammonia Diffusion. Thesis. Dept. of Environmental Science and Engineering, University of North Carolina, Chapel Hill, 116 pp.

Schwartz, F. J. and R. Hyle. 1985. In H. T. Odum, Ed., Self Organization of Ecosystems in Marine Ponds Receiving Treated Sewage. University of North Carolina Sea Grant Office, Publication #UNC-SG-85-04, North Carolina State University, Raleigh, NC, p. 206.

Smith, M. 1971. Productivity of Marine Ponds Receiving Treated Sewage Waste. M.A. thesis. Department of Zoology, University of North Carolina, Chapel Hill, NC.

Spears, D. J. and A. Williams. 1985. In H. T. Odum, Ed., Self Organization of Ecosystems in Marine Ponds Receiving Treated Sewage. University of North Carolina Sea Grant Office, Publication #UNC-SG-85-04, North Carolina State University, Raleigh, NC, pp. 187–189.

Williams, A., N. Beeston, R. Field, E. Walton, and R. Hyle. 1985. In H. T. Odum, Ed., Self Organization of Ecosystems in Marine Ponds Receiving Treated Sewage. University of North Carolina Sea Grant Office, Publication #UNC-SG-85-04, North Carolina State University, Raleigh, NC, pp. 173–180.

Woods, W. J. 1985. In H. T. Odum, Ed., Self Organization of Ecosystems in Marine Ponds Receiving Treated Sewage. University of North Carolina Sea Grant Office, Publication #UNC-SG-85-04, North Carolina State University, Raleigh, NC, pp. 87, 91–92.

15

CHANGES OF REDOX POTENTIAL IN AQUATIC ECOSYSTEMS

Sven Erik Jørgensen

Department of Environmental Chemistry, Institute of Chemistry AD, The Royal Danish School of Pharmacy, Copenhagen, Denmark

15.1 INTRODUCTION

The redox potential is one of the most crucial factors for aquatic ecosystems. Many chemical and biological processes are very dependent on the redox potential. Manipulation of the redox potential of aquatic ecosystems can often be used to create more favorable environments at a relatively low cost.

A redox process can be described by an equation of the general form:

$$\text{Reductant} \rightleftarrows \text{oxidant} + ne^- \qquad (15.1)$$

where n = number of electrons and e^- = electron. The corresponding reductant and oxidant form a redox pair. Electrons cannot exist alone in an

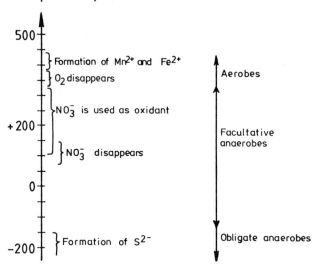

Figure 15.1 Redox potential relations and associated biological communities.

aquatic solution and two redox pairs are therefore necessary to realize a redox process, or transfer of electrons:

$$\text{reductant } 1 + \text{oxidant } 2 \rightleftarrows \text{oxidant } 1 + \text{reductant } 2 \qquad (15.2)$$

The term *redox potential* is introduced to indicate the tendency to transfer the electron. In general, the larger the positive magnitude of the potential, the greater the tendency the oxidant has to gain electrons. Similarly, the larger the negative magnitude of the potential, the stronger the reductant and the weaker the oxidant.

Redox potential is a relative measure that is standardized with reference to the hydrogen–proton couple:

$$H_2 \rightleftarrows 2H^+ + 2e^- \qquad (15.3)$$

The potential for this couple at 25°C with H^+ = 1 mol/L and H_2 at 1 atm pressure is defined as a potential of 0.000 V.

15.2   REDOX EFFECTS ON CHEMICAL AND BIOLOGICAL PROCESSES

Figure 15.1 shows a correlation of redox potential with reactions taking place in waters and sediments, including nitrification and denitrification. Some of the most important relationships between chemical and biological processes and the redox potential are given below.

1. The rate of nitrification depends highly on the redox potential, as can be seen from the chemical reaction scheme:

$$NH_4^+ + 2O_2 \rightleftarrows NO_3^- + H_2O + 2H^+ \qquad (15.4)$$

2. If the oxygen concentration is low, dentrification may occur:

$$nH^+ + nNO_3^- + n(CH) \rightleftarrows \frac{n}{2} N_2 + nH_2O + nCO_2 \qquad (15.5)$$

3. The oxidation state of sulfur is determined by the redox potential. Nitrate is able to oxidize sulfide, because

$$^2H^+ + 5HS^- + 8NO_3^- \rightleftarrows 5SO_4^{2-} + 4N_2 + 4H_2O \qquad (15.6)$$

4. The oxidation states of iron and manganese are related to the redox potential:

$$Fe^{3+} + e^- \rightleftarrows Fe^{2+} \qquad (15.7)$$

$$MnO_2 + 2e^- + 4H^+ \rightleftarrows Mn^{2+} + 2H_2O \qquad (15.8)$$

5. The redox potential determines the composition of the microorganisms culture (species and abundance).
6. A part of the phosphorus bound in the sediment may be released at lower redox potential owing to the following process:

$$2H_2O + FePO_4 + (CH) + 5HS^- \rightleftarrows 5FeS + CO_2 + 5HPO_4 + 5H^+$$

$$(15.9)$$

This implies that anaerobic conditions in the sediment may influence eutrophication significantly.

7. Species composition and diversity depend on the oxygen concentration, as can be seen in Figure 15.2.
8. Anaerobic decomposition of organic matter is different from aerobic decomposition. The complete oxidation under aerobic conditions can be expressed as

$$(CH_2) + \frac{3}{2} O_2 \rightarrow CO_2 + H_2O \qquad (15.10)$$

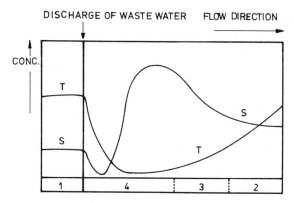

Figure 15.2 Patterns of pollution in a river showing number of higher organisms (S) and number of species (T) versus distance in running water. Classification with numbers in accordance with saprobic system (from Hynes, 1971).

and the partial decomposition of organic matter under anaerobic conditions can be represented by the following reaction scheme:

$$\text{organic matter} \rightarrow CO_2 + CH_4 + CH_3(CH_2)_n\,COOH \quad (15.11)$$

9. Many aquatic organisms utilize gills, whereby dissolved oxygen is passed from the water into the circulatory fluid of the organisms. Low oxygen concentrations in water lead to lower oxygen uptake. This can partly be compensated for by fish and other aquatic organisms by pumping the water more rapidly over the gills. Moderately reduced oxygen concentrations result in reduced physiological activity and insufficient muscle activity. Food consumption, growth, and swimming velocity will all decrease for fish at dissolved oxygen concentrations less than about 8 mg/L (Welch, 1980; see Figure 15.3) At slightly lower

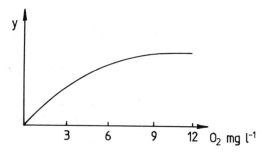

Figure 15.3 Activity of fish (food consumption, growth, or swimming activity) versus dissolved oxygen concentration. The relationship varies from species to species and can therefore be considered only very approximate.

Table 15.1 Limiting Oxygen Concentrations for Selected Organisms

Organism	Limiting O_2 concn. (mg/L)	Temp. (°C)
Brown trout	2.9	6–24
Coho salmon	2.0	16–24
Rainbow trout	3.7	11–20
Amphipods	0.7	—

oxygen concentrations, oxygen becomes the limiting factor, as shown in Table 15.1.

The redox potential is related to the oxygen concentration, as also indicated in Figure 15.1. The oxygen concentration is, however, a dynamic state variable. Oxygen is consumed by oxidation processes, and it is formed by the photosynthesis. Furthermore, oxygen can be exchanged with air. Reaeration is very important for the maintenance of a suitable oxygen concentration in many polluted rivers. The relations among consumption, formation, and exchange of oxygen are shown in a conceptual model in Figure 15.4.

The solubility of oxygen in water depends on the temperature and the salinity of water. High temperature means lower solubility, but the rate of oxygen consumption is higher at higher temperature. The relation between the rate constant and the temperature for oxidation of organic matter as well as nitrification may be formulated as follows:

$$K_T = K_{20} Q^{(T-20)} \tag{15.12}$$

where K_T and K_{20} are the rate constants at T°C and 20°C, respectively, and Q is a constant. These relations imply that the lowest oxygen concentrations are recorded in a warm summer before sunrise, for oxygen formation by photosynthesis depends on solar radiation (Figure 15.5).

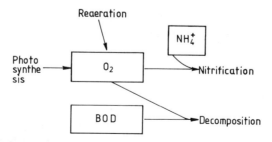

Figure 15.4 Conceptual diagram of a dissolved oxygen model.

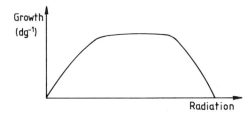

Figure 15.5 Growth of phytoplankton (photosynthesis) versus solar radiation.

A further complication of the oxygen exchange processes is caused by the formation of a thermocline or a halocline, which reduces considerably the rate of exchange between epilimnion and hypolimnion. Therefore anaerobic conditions are often observed in temperature lakes during the summer in the hypolimnion (Figure 15.6).

It is understandable that the problem of oxygen depletion is so complex that modeling has been used rather extensively to find a good management solution.

The redox conditions are further complicated by the influence of pH. Redox processes depend on pH, as can be seen from Equations 15.1–15.4. Hydrogen ions are formed or consumed by these processes and therefore a changed hydrogen ion concentration (changed pH) will change the concentration of the other components.

Figure 15.7 shows how the oxidation state of iron depends on the redox potential as well as pH. As seen, the redox potential that determines the transformation from one oxidation state to another depends on pH. The rate constants for oxidation processes also depend on pH. This is illustrated in Figure 15.8, where the oxidation rate of sulfide versus pH is shown.

15.3   APPLICATION OF ECOTECHNOLOGICAL METHODS

The redox potential of water is increased most easily by use of artificial aeration, which may be used in streams or lakes. The direct effect is, of

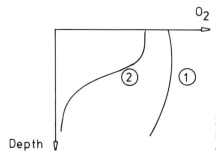

Figure 15.6 Dissolved oxygen profiles in a stratified lake. (1) Winter condition; (2) summer condition.

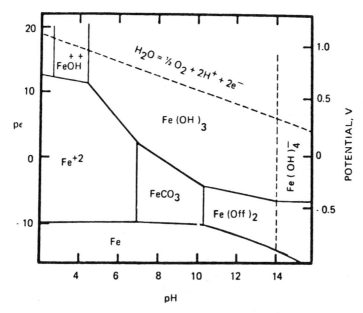

Figure 15.7 pε–pH diagram for iron. Total concentration is 2×10^{-3}M and ion activity is 10^{-5} Reproduced by permission from W. Stumm and J. J. Morgan, Aquatic Chemistry, Wiley-Interscience, New York, 1970, p. 533.

course, a higher oxygen concentration, but in addition the aeration may also affect the sediment–water exchange processes, because the release of phosphorus from sediments is affected by the redox potential.

The most effective method of solving the problem of oxygen depletion in streams is by treatment of the wastewater. The biological treatment should preferably take place at wastewater plants, not in the streams. However, if

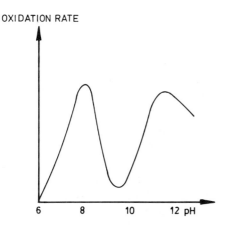

Figure 15.8 Sulfide oxidation rate as a function of pH (Chen and Morris, 1979).

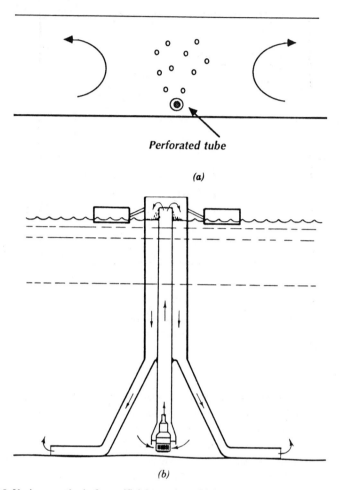

Perforated tube

(a)

(b)

Figure 15.9 Various methods for artificial aeration of lakes: (*a*) pressurized air in perforated tubes; (*b*) pumping hypolimnetic water with submerged pump; (*c*) pumping hyppolimnetic water with surface outboard motor; (*d*) aeration directly into hypolimnion.

the streams have an accumulation of organic sludge, aeration may be needed to accelerate the decomposition (oxidation) of the sludge.

For lakes the problem is more complex. First of all, the oxygen depletion is most often caused by eutrophication, not by discharge of biological oxygen demand. Secondly, the high concentration of nutrients cannot be reduced rapidly, owing to the long retention time of water in lakes. Therefore artificial aeration has been used more for lakes than for streams and it has been used mainly to prevent anaerobic conditions at the sediment surface.

Four different aeration methods have been applied. They are shown in Figure 15.9. Table 15.2 gives the characteristic features of the methods.

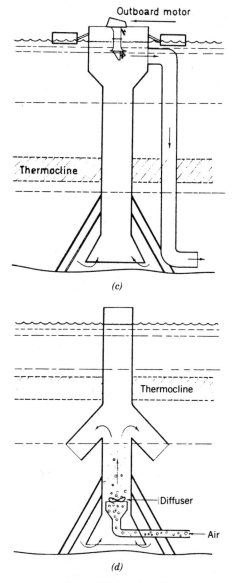

Figure 15.9 (*continued*)

Some of the methods mentioned in Chapter 16 for lake restoration will also have the effect of increasing oxygen and decreasing BOD concentration. But addition of nitrate for stabilization of the sediment should be included here, because the effect mainly is on the redox condition in the sediment. The nitrate will oxidize the organic matter (denitrification), giving a higher redox potential at the sediment surface. This method has some pronounced

Table 15.2 Features of Four Different Aeration Methods

Method	Is Thermocline Maintained?	Advantages	Disadvantages
Pressurized air in perforated tubes	No	Simple installation, moderate costs	Daily control required, clogging of perforated tubes, low efficiency, difficult to move
Use of diffusions (similar aeration to an activated sludge plan)	No	Good efficiency, movable, relatively simple installation	Daily control required, noisy
Aeration with turbines	No	Simple installation, good efficiency, movable	Some fish may be killed in the turbine, noisy
Mammut pumps or outboard motor	Yes	Good efficiency, movable	Costly, daily control required

advantages—quick, permanent, and effective results—but also some pronounced disadvantages—the risk for wrong dose is high, control of sediment after the addition is required, and it involves addition of a nutrient (nitrate), which may cause side effects with eutrophication. The method is often combined with the addition of iron (III) chloride, which increases the binding capacity of the sediments for phosphorus.

15.4   NEEDS FOR INCREASING THE REDOX POTENTIAL

In all the uses of ecotechnological methods, it is advantageous to set up a model to be able to give predictions on the effects of the applied methods. In chapter 16 on lake restoration a case study is described in which the effects of the various methods for lake restoration are compared, including aeration and addition of nitrate to the sediment. Here we only summarize the characteristic features of the cases, where it most probably will seem an advantage to use the methods presented in Section 15.3.

For streams it may be relevant to use aeration for a shorter period to oxidize soluble BOD_5, if conventional biological treatment of wastewater is not operating. Aeration should not replace the biological treatment, at least not permanently.

Streams with slow and/or little flow may show low oxygen concentrations in spite of sufficient biological treatment of wastewater discharged to the streams owing to a high accumulation of organic sludge. The accumulation

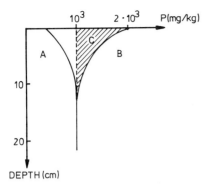

Figure 15.10 Possible profiles of phosphorus in lake sediments. The A profile will give only a small release of phosphorus, whereas profile B most probably will exchange phosphorus corresponding to area C.

of sludge is often caused by a previously insufficient wastewater treatment. Because the water flow is low, it does not contain sufficient oxygen to oxidize the organic sludge. Aeration may here solve the problem caused by previous omissions of wastewater treatment.

Many BOD–DO stream models do not include the BOD originated from bottom sediment, but it is obviously necessary to include BOD in sediments as a state variable, if a stream model should be able to give any predictions on the effects of aeration. Such models show clearly and not surprisingly that the lower the water flow, the more significant the organic matter in the sediment and the more pronounced the effect of the aeration.

Aeration or addition of nitrate to lakes is closely related to internal loading of phosphorus. If the sediments have a high content of phosphorus that may be released as the phosphorus input to the lake is reduced, internal loading may be significant. If, in addition, the retention time is long, it will be advantageous to reduce the internal loading by aeration, thereby achieving a more rapid decrease of the phosphorus concentration in the lake water.

Figure 15.10 shows two different phosphorus profiles for lake sediments. As demonstrated, the profile may give an indication of the expected internal loading.

Lakes covered by ice during the winter have a particular problem. The microbiological activity (and oxygen demand) is of course low during the winter, but because the lake water is separated from the atmosphere by ice even the very low decomposition rate may cause oxygen depletion. Figure 15.11 illustrates the concentration profiles for oxygen, methane, carbon dioxide, and hydrogen sulfide for a lake in southern Sweden after a severe winter (1970). The formation of hydrogen sulfide shows that the redox potential is at a very low and very critical level.

Ammonia (NH_3) is toxic to fish (but ammonium, NH_4^+, the ionized form, is harmless). It is therefore important for the redox potential to be sufficiently high to allow nitrification to occur. pH is high in eutrophic lakes during summer, which implies that the ratio NH_3/NH_4^+ increases. An ammonia concentration of 0.025 mg/L is considered an upper limit, for most fish ex-

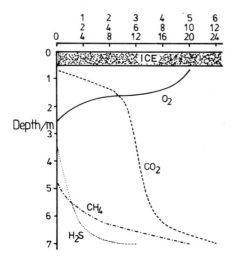

Figure 15.11 Profiles of oxygen, methane, hydrogen sulfide, and carbon dioxide in an ice-covered lake in Skaane, Sweden, April 1970 during a severe winter.

crete ammonia through the gills at this level. As seen in Table 15.3, this limit does not allow a high total ammonium + ammonia concentration at high pH values such as 8.5–9.5, which often may be found in shallow eutrophic lakes in the summer. Aeration may be urgently needed in such cases to accelerate nitrification.

In general it can be concluded that the following features may increase the advantage of aeration or nitrate addition:

1. high phosphorus concentration in sediment,
2. relatively large amount of phosphorus in the sediment in exchangeable form (compare with Figure 15.10),
3. long retention time,
4. formation of thermocline (anaerobic hypolimmion),
5. ice cover during winter, and
6. high ammonium concentration in lake water.

Table 15.3 Concentration of Ammonia (NH_3 + NH_4^+) Containing an Un-ionized Ammonia Concentration of 0.025 mg NH_3/L

Temp. (°C)	pH					
	7.0	7.5	8.0	8.5	9.0	9.5
5	19.6	6.3	2	0.65	0.22	0.088
10	12.4	4.3	1.37	0.45	0.16	0.068
15	9.4	5.9	0.93	0.31	0.12	0.054
20	6.3	2	0.65	0.22	0.088	0.045
25	4.4	1.43	0.47	0.17	0.069	0.039
30	3.1	1	0.33	0.12	0.056	0.035

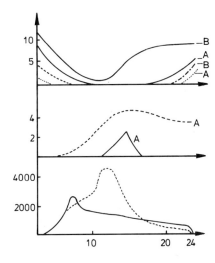

Figure 15.12 Results of aeration on Boltz Lake, Kentucky. Solid lines, test lake; dotted lines, control lake. (A) 15 m depth; (B) 10 m depth. Upper plot; mg O_2/L; middle plot; S^{2-}, mg/L; lower plot, algae counts per milliliter.

15.5   CASE STUDIES

Investigations into the application of aeration have been carried out on Boltz Lake (Kentucky) by Symons (1969). He concludes that artificial destratification improves several important water quality parameters: increased concentration of dissolved oxygen, decreased concentration of sulfide, and reduced algae growth, especially of the blue-green species. Parallel to this study, investigations were carried out on a control, Bullock Pen Lake, which has the same morphometry and water quality as Boltz Lake. The results are summarized in Figure 15.12.

The observations by Ambuhl (1967 and 1969) and Thomas (1970) on the restoration of Pfaffikersee in Switzerland are less positive. The concentration of dissolved oxygen improved during the first period of the experiment, but in September–October the oxygen condition deteriorated, as it did in the control lake, Greifensee. The higher temperature in the hypolimnion caused increased consumption of oxygen, owing to an accelerated microbiological decomposition rate of detritus. The aeration did not cause any reduction in the algae growth, probably because of the mixing with hypolimnetic water.

Other case studies (see, for example, Bernhardt, 1967 and Hooper et al., 1953) show the same tendency. Destratification might, in some cases, even have a negative effect on the water quality, owing to a higher temperature in the hypolimnion.

Lake Brunnsviken, Sweden, has been restored by means of hypolimnetic aeration (Figure 15.13). In 1970 wastewater was diverted from Lake Brunsviken, but the water quality improved only slowly. Phosphorus concentrations decreased and the oxygen concentrations in the epilimnion increased, but the transparency did not improve and the hypolimnetic water still con-

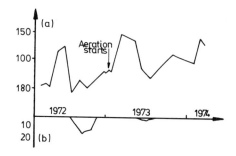

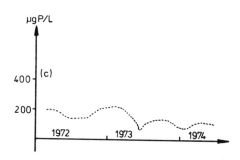

Figure 15.13 Restoration of Lake Bruuns-viken, Sweden. (*a*) Dissolved oxygen as percent of saturation; (*b*) hydrogen sulfide, mg/L; (c) phosphorus concentration, μg/L. Aeration started in the spring of 1973.

tained hydrogen sulfide. It was therefore decided to attempt hypolimnetic aeration.

The results were positive, as shown in Figure 15.13. Hydrogen sulfide production was reduced significantly, the oxygen condition improved, and the level of phosphorus concentration continued to decrease. However, primary production was only minimally reduced, probably because the nutrient concentrations were still above the saturation point.

15.6 CONCLUSIONS

A eutrophication model was not used in this case study. However, some model exercises since have shown that the minor phosphorus input during a period of heavy rain due to discharge of overflow water from the watershed is not insignificant for eutrophication. This point is often overlooked, but the modeling concept will always take the total nutrient balance into consideration. Other case studies on shallow lakes confirm these observations (Bengtsson et al., 1972).

From these case studies it must be concluded that destratification has certain disadvantages caused by the elevation of the hypolimnetic temperature. But positive effects on hydrogen sulfide, phosphorus, and oxygen are observed, although the effect on primary production seems minimal.

REFERENCES

Ambuhl, H. 1967. Discussion of Impoundment destratification by mechanical pumping. *J. San. Eng. Div., Proc. Am. Soc. Civ. Eng. 93*(SA4):141–143.

Ambuhl, H. 1969. Die neueste Entwicklung des Vierwaldstattersees. *Inst. Verein. theor. angew. Limnologie 17*:219–230.

Bengtsson, L., H. Berggreen, O. Meyer, and B. Verner. 1972. Restaurering av sjoar med kulturbetinget hypolimiskt syrgasdeficit. Project Report published by University of Lund (Limnological Department), Atlas Copco AB (Central Physical Lab.), Sweden.

Bernhardt, H. 1967. Aeration of Wahnbach Reservoir without changing the temperature profile. *J. Am. Water Work Association 63*:943–963.

Chen, K. Y., and J. C. Morris. 1979. Oxidation of aqueous sulfide by O_2. *Adv. Water Pollution Res.* p. III-32.

Hooper, F. F., R. C. Ball, and H. A. Tanner. 1953. An experiment in the artificial circulation of a small Michigan lake. *Trans. Am. Fish. Soc. 82*:221.

Hynes, H. B. N. 1971. *Ecology of Running Waters.* Liverpool University Press, Liverpool, England.

Poole, N. J., D. J. Wildish, and D. D. Kristmanson. 1978. The effects of the pulp and paper industry on the aquatic environment. *Crit. Rev. Environ. Control 8*:153.

Stumm, W. and J. J. Morgan. 1970. *Aquatic Chemistry* Wiley-Interscience, New York, 540 pp.

Symons, J. M. 1969. *Water Quality Behavior in Reservoirs* A compilation of published research papers, U.S.A. Public Health Service Publication No. 1930, Washington, DC. 616 pp.

Thomas, E. A. 1970. *Sjorestaureringer i Schweiz. VVV, 3*:17–41.

Welch, E. B. 1980. *Ecological Effects of Waste Water.* Cambridge University Press, Cambridge, 337 pp.

16

ECOTECHNOLOGICAL APPROACHES TO THE RESTORATION OF LAKES

Sven Erik Jørgensen

Department of Environmental Chemistry,
Institute of Chemistry AD,
The Royal Danish School of Pharmacy,
Copenhagen, Denmark

and

Lief Albert Jørgensen

The Engineering College of Copenhagen,
Copenhagen, Denmark

16.1 INTRODUCTION

Lakes may suffer from various pollution problems, including eutrophication, acidification, oxygen depletion, and toxic substance pollution.

Eutrophication. If the concentrations of nitrogen or phosphorus are high, they will cause a high concentration of phytoplankton, which in turn will lead to a low transparency, anaerobic conditions in sediment or certain zones such as in the hypolimnion during the summer stagnation period, high pH, fish kills due to high pH and ammonia or due to oxygen depletion, and other problems. A high input of phosphorus and/or nitrogen is caused by discharge of wastewater, where these components have not been removed, and/or by agricultural runoff.

Acidification. This problem is caused mainly by acid rain but may also be due to natural means such as a high concentration of humic acid. Low pH involves several serious problems such as low fertility of fish and zooplankton, low concentration of phytoplankton as carbon becomes the limiting factor, low ecological diversity in general, and high solubility of heavy metals. These problems are most pronounced in areas where there is a high industrial activity and the natural water has a very small buffer capacity— in Norway, Sweden, the northeastern United States, and Canada.

Oxygen Depletion. This may be a consequence of discharge of biodegradable organic matter into streams, but is most often an indirect effect of eutrophication. Organic matter produced by photosynthesis also requires oxygen for decomposition processes. As phytoplankton settles, a high concentration of dead phytoplankton is found in the hypolimnion and close to the sediment surface, and oxygen depletion may occur in this parts of the lake. The photosynthesis produces oxygen equal to the amount of oxygen required for the decomposition processes but this oxygen is produced during the bloom period and at the water surface and is therefore not available in the autumn or in the bottom water, where the decomposition takes place. The redox potential may be so low at the sediment surface that a formation of hydrogen sulfide takes place, causing in many cases a very effective fish kill.

Toxic Substance Pollution. Discharges of wastewater and agricultural runoff, which contain toxic substances either as heavy metals or toxic organic compounds (e.g., pesticides from agricultural runoff), will, of course, cause toxic pollution of lake waters. The consequences are not different from similar situations in other receiving waters: reduced ecological diversity, fish kills, inedible fish (The World Health Organization has published maximum concentrations for several toxic substances in human food), and water unusable as potable water.

These four problems of lake pollution can be solved by a wide spectrum of methods. The environmental technological methods include treatment of wastewater for removal of phosphorus (by chemical precipitation, biological treatment, or ion exchange), nitrogen (by nitrification and denitrification, ammonia stripping, and ion exchange), and toxic substances. These last substances are removed most advantageously from industrial wastewater before it is discharged to public sewers, because the concentration is highest here, which gives the lowest treatment costs. Agricultural runoff is much more complicated to treat because it is non-point pollution.

This chapter focuses on ecological engineering methods that can be used to solve these four problems. These methods must work hand in hand with environmental technology to achieve the best possible management plan. A review of the ecological engineering methods is given in the next section. These methods either deal with the non-point sources of pollution or attempt to assist the lake to return faster to its natural balance. Later in this chapter a case study of lake restoration is presented. The selection of restoration methods is based on a eutrophication model, which also has been used to determine the effects of the various alternative methods.

The application of ecotechnology for the restoration of lakes is very illustrative, because these methods have been widely applied to lakes. It can be clearly demonstrated that to achieve good results in lake management one must combine several methods—environmental technological as well as ecotechnological methods—simultaneously. Pollution of lakes is a very complex problem and it can rarely be solved by the use of one (simple) method. It is not surprising that a good solution to a complex problem requires a complex solution and that models have been widely used in the selection of lake management strategies to get the optimal overview of the complex problem involved.

16.2 ECOLOGICAL ENGINEERING METHODS

Table 16.1 gives an overview of the ecotechnological methods that are used in lake restoration. The names of the methods are indicated in the table, along with the problems they are used to solve and whether the methods change the forcing functions of the lake or attempt to reach the natural balance faster than otherwise would have been the case. In other words, the methods will *directly* change the mass balances of the lake or the methods will change the structure (the state variables) of the system. Indirectly, all methods will change the mass balance and the structure more or less. The methods that change the state variables of the systems but do it by by use of a new forcing function (e.g., aeration) are indicated in Table 16.1 by $F(+S)$.

Another overview of ecotechnological methods applied in lake restoration can be obtained by the use of models. From eutrophication models we know that algae growth (A) is a function of light (L), soluble inorganic nitrogen (N_S), soluble inorganic phosphorus (P_S), soluble inorganic carbon (C_S), the settling rate (S), the retention time (R), and the grazing rate (G):

$$A = f(L, N_S, P_S, C_S, S, R, G) \tag{16.1}$$

Table 16.1 Ecotechnological Methods Applied in Lake Restoration

Method	Problem to Solve	Direct Change in Forcing Function (F) or State Variables (S)
Aeration	Oxygen depletion in hypolimnion. Release of P and Fe from sediment (eutrophication)	F(+S)
Siphon of hypolimnion water	Removal of P and oxygen-poor water (eutrophication)	F
Nitrate to sediment	Reduce release of P from sediment (eutrophication)	F(+S)
Aeration with circulation	Eutrophication: depression of algae growth	S
Removal of upper sediment layer	Reduce release of P from sediment or remove toxic substances (eutrophication and toxic substance pollution)	S
Removal of P of algae from water	Eutrophication (reduce P/algae in water)	S
Wetland	Removal of N and P from non-point sources (eutrophication)	F
Decreased retention time	Reduce P and toxic substance concentration (eutrophication and toxic substances pollution)	F
Precipitation of P in lake	Eutrophication (reduce P in water)	S
Precipitation of P in inflowing water	Eutrophication (reduce P input)	F
Addition of $Ca(OH)_2$	Change of pH	F(+S)
Coverage of sediment	Eutrophication	F(+S)
Preimpoundment	Eutrophication (reduce P input)	F
Plastic beads or sheet dyes or soot	reduce high penetration (eutrophication)	F(+S)
pH-modification shock	Eutrophication (reduce inorganic carbon in water)	S

Restoration methods may also be classified according to which factor they change either *in* the lake or by decreasing input or increasing output, that is, by changing forcing functions (Table 16.2). The principle of this classification may also be applied to toxic substance and acidification problems.

Artificial reaeration of lake water has already been mentioned in Chapter 15 but is included in this survey because it is widely used in lake restoration

Table 16.2 Classification of Ecotechnological Methods for Lake Restoration, According to Factor Controlled

Factor Controlled	In Lake	By Changing Input/Output
Light	Plastic beads, sheets, dyes or soot, circulation	Shading the shores
Soluble inorganic nutrient	Removal of sediment algae or P.	Pre-impoundment, wetlands.
	Precipitation of P in water, NO_3^- to sediment	Precipitation of P in inflow, siphon of hypolimnion water
Soluble inorganic carbon	Change of pH	
Retention time		Reduction of retention time
Grazing rate	Biomanipulation	

to obtain better oxygen conditions. The same is the case with the other methods, which improve the redox conditions—for example, addition of nitrate to the sediment (Ripl, 1976).

Anaerobic sediments will generally release more phosphorus than aerobic sediments, for the latter contain iron in oxidation state three, whereas the former contain iron in the oxidation state two. Iron(III) has a higher adsorption capacity to phosphorus than iron(II), and iron(III) phosphate is much more insoluble than iron(II) phosphate. These processes may cause a "runaway" effect because increased eutrophication will mean a higher production of phytoplankton, which will settle and thereby give a high input of organic matter to the sediment. It therefore becomes anaerobic, which will mean enlarged release of phosphorus, which will cause more eutrophication, and so on. If the sediment becomes sufficiently anaerobic (low redox potential), sulfide formation may occur and, because iron(II) sulfide is highly insoluble, this will promote further the release of phosphorus. There is a closed relation between the redox conditions and the nutrients cycling in lakes.

Aeration may be used to depress eutrophication. It is achieved if the aerator circulates the water as shown in Figure 16.1. The water of the bottom layer is lifted by aeration and is replaced by the upper layer of water, which may be supersaturated with oxygen produced by the photosynthesis. The algae will be brought down to the dark hypolimnion, where light conditions are unfavorable for algae growth. Figure 16.2 shows the profile of phytoplankton and oxygen concentration before and after aeration with water circulation (Kojima 1987). Table 16.3 shows the effects on the eutrophication measured in reservoirs with different depths. If the depth is more than 5–10 m, the effect is very pronounced.

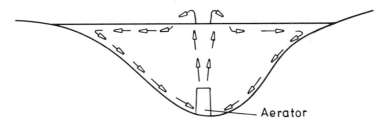

Figure 16.1 The circulation of a lake by use of an aerator.

Siphoning of hypolimnetic water serves two purposes—to improve the oxygen condition and to change the concentration of nutrients in the lake water. This is the case when hypolimnetic water is siphoned downstream, because hypolimnetic water generally has lower oxygen and higher nutrients concentrations.

Removal of the upper layer of the sediments may be used when the upper layer has higher nutrients concentrations than the deeper layers, which is most often the case. Figure 16.3 shows the phosphorus profile in a typical lake sediment. The phosphorus of the upper layer is either nonexchangeable because it is bound in insoluble chemical compounds or it is exchangeable because it is organic bound and after decomposition of the organic matter it is released. Consequently, it will be an advantage to remove the upper layer of the sediment and thereby reduce the release of phosphorus from

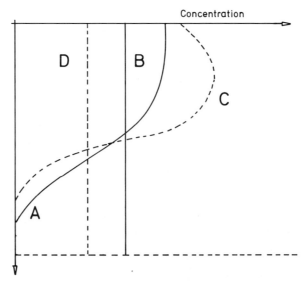

Figure 16.2 The oxygen profiles before (A) and after (B) aeration. The phytoplankton profiles are also shown before (C) and after (D) aeration.

Table 16.3 Effects of Aeration on Eutrophication as Function of Lake Depth

Depth (m)	% Reduction in Eutrophication
2–3	0–5
5–7	10–20
10	45–50
>10	>50

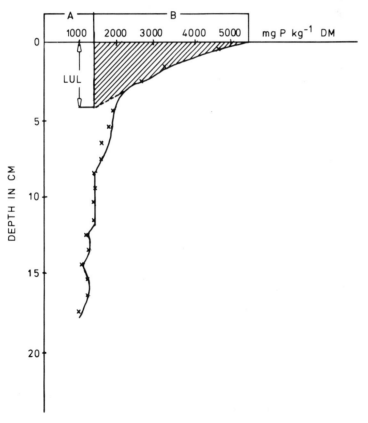

Figure 16.3 Phosphorus profile of sediment core from Lake Esrom, Denmark. The hatched area represents exchangeable phosphorus. LUL is the unstabilized area.

the sediment. This has been used only in one case—Lake Trummen in Sweden—and it was only partially a success (see discussion below). If the upper layer of the sediment contains toxic substances the method must be considered rather attractive, but because the method is expensive, it seems applicable only for small and very valuable lakes.

Reduction of the retention time may also have a positive effect on the reduction of eutrophication, as seen from Table 16.1, and it also implies a smaller concentration of toxic substances, if that is considered a pollution problem of the lake. The method is realized by the discharge of unpolluted water to the lake. For reservoirs it is possible to control the retention time by the dam. The use of dams enables changes not only in the retention time, but also in the water depth and water volume. In general, a longer retention time means increased eutrophication and a larger volume means a decreased eutrophication. It is therefore necessary to use models to make predictions of the overall effects of these changes in the hydrologic balance of a lake or reservoir. A further complication is related to the soil exposed to the lake water at increased depth. It will be in an oxidized state and therefore in general have a great capacity for nonexchangeable phosphorus. If this new sediment surface is significant in area, an increased water depth may cause an essential change in the phosphorus balance of the entire lake.

Removal of phosphorus directly from the lake water is also possible. The water is either pumped through an activated aluminum oxide column, which removes the phosphorus very effectively, or is filtered for removal of phytoplankton. The treated lake water is returned to the lake with a reduced concentration of soluble phosphorus or of phytoplankton, thereby reducing the total concentration of nitrogen and phosphorus as a result.

Precipitation directly in the water by use of aluminum sulfate or iron(III) chloride is also applicable. The method has, however, given less promising results in shallow lakes, because wind may stir up the flocs formed by the precipitation. Precipitation of phosphorus in inflowing water has also been used (Bernhardt, 1979, 1981 a,b).

The use of wetlands is often a workable method, which can be applied for removal of non-point pollutants. If the sources of lake pollution are diffuse, the problem is in general more difficult to solve. Wetlands often have anaerobic sediment and therefore a high denitrification rate. In addition, the wetland soil has a high capacity for uptake of phosphorus, pesticides, and other pollutants. Finally, the plants of the wetlands take up phosphorus and nitrogen and even other pollutants, and by harvest of the plants the pollutants are removed. Chapters 8 and 9 focus in more details on the application of wetlands as ecotechnological methods for nutrient control.

The use of pre-impoundments is based on a similar idea. Lakes and reservoirs have a phosphorus retention capacity, which implies that the phosphorus concentration of the water leaving the lake is lower than in the water entering the water body (Reckhow, 1979, Reckhow and Chapra, 1983). Phosphorus retention increases with decreasing hydraulic retention time and mean depth and the phosphorus retention is also less for a lake or reservoir with an anoxic hypolimnion. Consequently, construction of a man-made lake (called a pre-impoundment) in front of another lake may have a highly positive effect on the eutrophication. Models have been developed to evaluate the phosphorus retention of pre-impoundments (Uhlmann and Benndorf 1976, 1980) and independent data support the model well (Wilhelmus et al., 1978).

Reduction of light penetration by change of the extinction coefficient has also been used to depress phytoplankton growth (Jørgensen, 1980; Los et al., 1982). The presence of humic substances in lake water may have a pronounced effect on the eutrophication owing to a significant change in the extinction coefficient of the water. Humic substances are natural, colored organic matter, but also human-induced changes may alter the extinction coefficient, particularly for pollution from pulp and paper mills. Hartman and Kudrilicka (1980) have studied the use of yellow tetrazin and found that 1 mg/L was needed to increase the extinction coefficient 0.08. Although yellow tetrazin may be harmless, it is not considered a sound ecological method to add artificial compounds to lake water. Other possibilities are the shading of shores by trees or floating vegetation. The use of plastic non-transparent sheets, plastic beds, or soot has been tried rather successfully for small reservoirs and tanks.

Biomanipulation has been studied quite intensively (Harbacek et al., 1978). The idea is to increase grazing on phytoplankton, but it is still not clear whether the method can be used by itself or can be used only to bring the ecosystem faster to a new steady state after the forcing functions have been changed.

16.2.1 Choice of Methods

It is hardly possible to give any general directions on which combination of environmental technology and ecotechnology to use to solve a given pollution problem of a lake. It will require a development of a model in each case to be able to take into account the ecosystem, the problem, the environment, and the pollution sources. It is possible, however, to give some first directions based upon mass balance considerations.

Figure 16.4 illustrates the possibilities ecotechnology can offer by use of the mass balances. Figure 16.4a shows a typical situation for phosphorus and nitrogen balance of many shallow lakes in the industrialized countries. It is clear from this mass balance that the highest reduction in the phosphorus input is obtained by treatment of the wastewater. Figure 16.4b shows a mass balance shortly after the discharge of wastewater to the lake has ceased. It is shown here that the internal load is pronounced at first after the wastewater is cut off owing to the accumulation of phosphorus in the sediment. Figure 16.4c shows the mass balance after the lake has reached a new balance. In addition to the wastewater cutoff, a wetland is created in front of the main

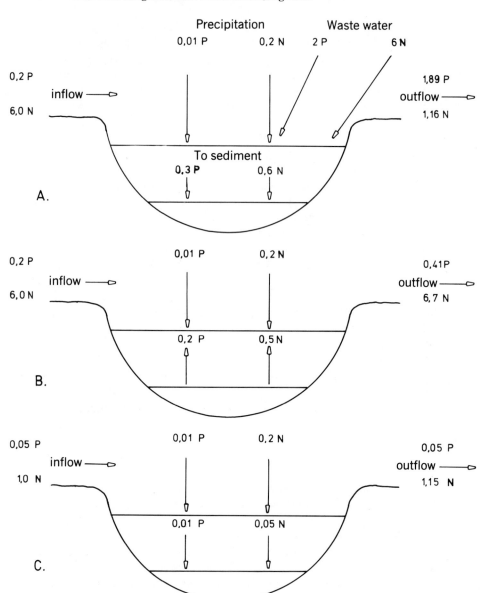

Figure 16.4 Nutrient budgets for a typical shallow lake: (A) with wastewater inflow; (B) after wastewater discharge has ceased; (C) some years later after a new steady state has been reached and a wetland has been constructed to cope with runoff. Lake (20 ha, 5 m deep) has a retention time of 1 yr and a catchment area of 10 km². All figures are ×1000 kg/yr.

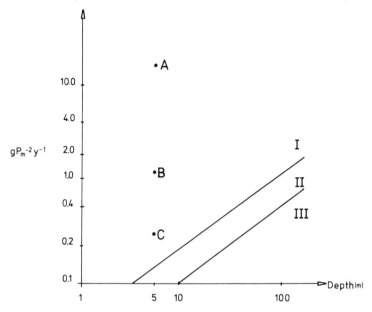

Figure 16.5 Vollenweider plots with data points corresponding to Figure 16.4. I is eutrophic lakes, II is mesotrophic lakes, and III is oligotrophic lakes.

tributary for removal of nitrogen and phosphorus. It is presumed that the wetland is harvested when the phosphorus and nitrogen concentrations of the vegetation are at their maxima. Comparison of the three mass balances shows that the very pronounced reduction in eutrophication is obtained only by application of the described combination of environmental technology and ecotechnology. The three levels are indicated in Figure 16.5 on a Vollenweider nutrient limitation plot; it is clear from this graph that the treatment or cutoff of the wastewater is insufficient and that it is required to consider additional methods. If the retention time of the lake is relatively long and it is considered important to get the lake into the new balance as quickly as possible, further use of ecological engineering should be considered. As seen in Figure 16.4b, the internal load coming from the sediment is significant, and a removal of the upper layer of sediment, discharge of drinking water to reduce the retention time, or coverage of the sediment could be considered. If the lake has a thermocline, hypolimnetic water could possibly also be used as output from the lake to get a faster removal of the phosphorus and nitrogen from the lake.

16.3   A MODEL USED FOR SELECTION OF ECOLOGICAL ENGINEERING METHODS

A very well examined model has been developed to document and predict lake restoration for Lake Glumsø, Denmark, a shallow lake with an average depth of only 1.8 m. The modeling activity was initiated in 1972 and in 1976 a prognosis based upon the model results was published. The prognosis considered a 90 or 98% removal of the phosphorus from the wastewater, which was only mechanically–biologically treated.

During the period 1976–1981 the model was improved further. The parameter estimation was improved by use of what is called an intensive measuring period—it means that all relevant state variables were determined every second day for a period of 3 weeks, when the spring bloom or summer bloom took place.

In 1981 the wastewater was cut off and discharged downstream of the lake. Two phosphorus reduction achieved was less than expected—from about 6.2 to about 1.6 g P/m^{2-} yr, but it was possible to compare the measurements with the prognosis published previously for a 90% reduction of the phosphorus input to the lake. It was found that the prognosis gave approximately the right values for the maximum phytoplankton concentration, the maximum and average production, the tranparency, and the concentration of the nutrients. With a correction of the prognosis in accordance with the real reduction of phosphorus—not 90% but about 75%—the prognosis was even improved slightly. A further improvement was obtained by changing the meterological forcing function from the average values that one is always forced to use for prognosis to the actual observed values.

A conceptual diagram of the model is shown Figure 16.6. The phosphorus cycle is shown, but similar diagrams are valid for the nitrogen and carbon cycle. More detailed information on the equations, the ecological considerations, the calibration, and validation results can be found in the following references: Jørgensen (1976), Jørgensen et al. (1978), Jørgensen et al. (1981), and Jørgensen et al. (1986a).

From this summary of the modeling development and the results obtained by use of the model, it can be concluded that the model can be used as a management tool and to set up a prognosis for changes in the eutrophication of a lake when changes in the forcing function or ecological structure are introduced. The model has therefore been used to see which further improvements we can achieve by use of various ecotechnological methods. In this context, it should be mentioned that the model has been used on several other eutrophication studies with good results, but in each case study it was necessary to introduce some modifications which consider the site-specific properties of the ecosystem (see Jørgensen et al., 1986b).

Many eutrophication models have been published since the early 1970s, but the model used here for examination of the applicability of ecotechnological methods has some characteristic features by which it differs from most other eutrophication models. These model properties are general for *all* the versions of the model that have been used on the various ecosystems as referred to above. The characteristic features of the model (see Figure 16.6) are as follows:

1. The sediment water exchange described is rather detailed. The input

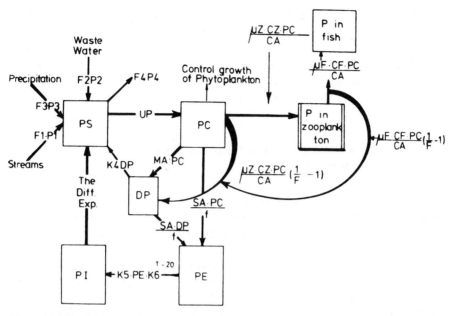

Figure 16.6 Phosphorus model for Lake Glumsø, Denmark. Terms for the equations are included in the diagram.

to the sediment from settling of detritus and phytoplankton consists of two fractions: exchangeable and nonexchangeable nutrients, that is, nutrients that after mineralization are able to return to the water phase, and nutrients that are bound to sediments as insoluble chemical compounds and do not return to the water phase. The exchangeable nutrients are returned to the water phase by a two-step process: mineralization and diffusion from interstitial water to the lake water. The first process is described as a first-order reaction and the rate of the second process is governed by the concentration gradient.

2. The growth of phytoplankton is described as a two-step process. The nutrients are first taken up by the phytoplankton and then the growth takes place as a first-order reaction, but limited by several factors. The first process is governed by the concentrations of nutrients in the lake water and in the phytoplankton cells, whereas the second process is determined by the temperature, the solar radiation, and the intercellular concentrations of nutrients as limiting factors.

3. It must also be considered a characteristic feature that the model is supported by intensive measurements, as mentioned above. This is not only the case for the Lake Glumsø study, but this approach has been used by most of the studies that have used this model.

4. The model has been constructed in the computer language CSMP. It

is a very flexible computer language, which facilitated the introduction of modifications in the model when it was used in another study.

The food web structure, the other process equations, and selection of forcing functions are not significantly different from what can be found in many other eutrophications models.

16.4 EXAMINATION OF THE APPLICABILITY OF ECOTECHNOLOGICAL METHODS

Models can be used to examine and compare alternative ecotechnological methods as in the case of Lake Glumsø. Before we go into the details on this examination, the characteristic features of this ecosystem should be given.

Until 1981, Lake Glumsø was a hypereutrophic lake with an annual production of 1050 g C/m^2 and a transparency of only 18 cm during the spring and summer blooms. After the reduction of the phosphorus input a significant improvement took place, as illustrated in Figures 16.7 and in Table 16.4, where a comparison between the prognosis and the observations also is included.

The reduction in the phosphorus input was not as significant as was expected and, because the nitrogen concentration in the tributary has increased

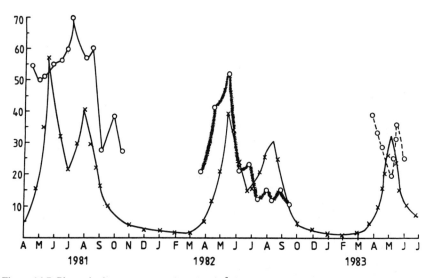

Figure 16.7 Phytoplankton concentrations (mg/m^3) in Lake Glumsø after phosphorus input was reduced. Shown are measured values (o) and model output (x).

Table 16.4 Comparison of Model Prognosis and Measured Data for Lake Glumsø, Denmark

	Prognosis approximately (Case A, 92% P reduction)	Measurement approximately (88% reduction)
Minimum transparency first year	20 cm	20 cm
Minimum transparency second year	30 cm	25 cm
Minimum transparency third year	45 cm	50 cm
g C/24 h–m^2 maximum first year	9.5 ± 0.8	5.5 ± 0.5
g C/24 h–m^2 maximum second year		
(spring)	6.0 ± 0.5	11 ± 1.1
(summer)	4.5 ± 0.4	3.5 ± 0.4
(autumn)	2.0 ± 0.2	1.5 ± 0.2
g C/24 h–m^2 maximum third year		
(spring)	5.0 ± 0.4	6.2 ± 0.6
Chlorophyll (spring) maximum mg/m^3 first year	750 ± 112	800 ± 80
Chlorophyll (spring) maximum mg/m^3 second year	520 ± 78	550 ± 55
Chlorophyll (spring) maximum mg/m^3 third year	320 ± 48	380 ± 38

considerably, it was appropriate to examine the possibilities of controlling the non-point sources of nutrients and to use ecotechnological methods in general.

Some of the methods have been examined and compared by use of the model described above. The results are shown in Table 16.5, including the simulated method, the changes introduced in the model to account for the use of the method, and the results obtained by use of the method. As seen from the results, coverage or removal of the sediment reduces the annual production. However, the maximum phytoplankton is not changed, although the phytoplankton concentration is reduced substantially in some periods of the year.

Simulated precipitation of phosphorus directly in the lake gives only a minor reduction in the phytoplankton concentration the first and second year

Table 16.5 Comparison of Lake Restoration Methods for Additional Restoration of Lake Glumsø, Denmark

Method	Model Changes	Results
Coverage of sediment	No P and N release from "old" sediment	Primary production decrease 30–40% on annual basis, but no change in maximum phytoplankton concentration
Removal of sediment	Exchangeable P and N set to zero at day zero	
Precipitation of P in lake water	Soluble P is removed to sediment at day zero	Less than 10% reduction in primary production 1st year. No changes after 2nd year
Use of wetland for nutrient removal from non-point sources	Input of P and N from tributary reduced 50% and 90% respectively	A continuous reduction of primary production to a level of 20% at year four and following years
Reduction of retention time	Inflow of water, but not of P and N, is increased	Less than 20% reduction in primary production

after the precipitation took place, whereas there is no reduction the third and following years. This seems not to be a very useful method to apply, which other studies also have revealed (Smidth, 1973).

The reduction of the retention time by pumping groundwater with only minor concentrations of nitrogen and phosphorus through the lake gives a certain effect, but it is probably too small to justify the costs.

By far the best method seems to be the installation of an artificial wetland for removal of the nutrients before the inflowing water of the tributary enters the lake. The reductions in the phytoplankton and nutrients concentrations are really significant. This is illustrated in Figure 16.8, where the phytoplankton concentration and the concentrations of nutrients are shown for a 5-year period. During the third or fourth year, the phytoplankton concentration has reached a new steady state, five times lower than before.

16.5   CONCLUSIONS AND FURTHER RESEARCH NEEDS

The case study presented here illustrates two important points:

1. Modeling is a very useful tool in the selection of a proper ecotechnological methods for the restoration of lakes; and
2. It is possible to obtain significant results by application of ecotechnological methods, if the prognosis resulting from the model is approximately correct.

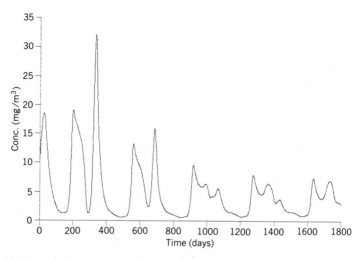

Figure 16.8 Phytoplankton concentrations (mg/m³) in Lake Glumsø after nutrients from non-point sources are controlled by a wetland with a 40% reduction of nitrogen and a 50% reduction of phosphorus. Time scale is in days.

Lake restoration methods are illustrative examples of the interaction between modeling and ecotechnology and of the basic ideas behind the application of ecotechnological methods. Our present experience in lake restoration by use of ecotechnology was reviewed in Section 16.3 and although it is rather comprehensive, more experience is urgently needed in the future to be able to make more general conclusions that are valid for a large spectrum of lakes. We particularly need more experience in the use of prognosis based on models to be able to evaluate the accuracy of such predictions in an ecological engineering context.

REFERENCES

Bernhardt, H. Ed. 1979. Von der Theorie zur Praxis-Entwicklung der Massnahmen gegen die Eutrophierung. *Zeitsch. Wasser-Abwasser Forsch. 12*(2–4):124 pp.

Bernhardt, H. 1981a. Gewässerschutz und Landwirtshaftliche Produktivität im Konflikt—die Nährstof-feliminierungsanlage das Wahnbachtalsparrenverbandes als Lösung.-*Wasser Abfallwirtschaft Nordrhein-Westfalen* 80/81:58–69.

Bernhardt, H. 1981b. Recent developments in the field of eutrophication prevention. *Zeitsch. Wasser-Abwasser Forsch. 14*:14–26.

Habacek, J., B. Desortova, and J. Popovsky. 1978. The influence of fishstock on the phosphorus–chlorophyll ratio. *Verh. Intern. Verein. Limnol. 20*:1624–1628.

Hartman, P. and J. Kudrlecka. 1980. Prevention of gill-necrosea of fish by controlling photosynthetic assimilation of pond phytocoenosis. *Bul. Vurh Vodneny:* 11–15.

Jørgensen, S. E. 1976. A eutrophication model for a lake. *Ecol. Modelling 2:*147–165.

Jørgensen, S. E. 1980. Lake Management. Pergamon, Oxford.

Jørgensen, S. E., H. F. Mejer, and M. Friis. 1978. Examination of a lake model. *Ecol. Modelling 4:*253–279.

Jørgensen, S. E., L. A. Jørgensen, L. Kamp-Nielsen, and H. F. Mejer. 1981. Parameter estimation in eutrophication modelling. *Ecol. Modelling 13:*111–129.

Jørgensen, S. E., L. Kamp-Nielsen, T. Christensen, J. Windolf-Nielsen, and B. Westergaard. 1986a. Validation of a prognosis based upon a eutrophication model. *Ecol. Modelling 32:*165–182.

Jørgensen, S. E., L. Kamp-Nielsen, and L. A. Jørgensen. 1986b. Examination of the generality of eutrophication models. *Ecol. Modeling 32:*251–266.

Kojima, T. 1987. Aeration of Lake Water. A lecture held in Sao Carlos, Brazil April 1–4, 1987, at the Symposium "Management of Lakes and Reservoirs."

Los, F. J., N. M. Derooij, J. G. C. Smits, and J. H. Bigelow. 1982. Policy analysis of water management for the Netherlands. Vol. IV. Design of Eutrophication Control Strategies. Rand Note M-1500/6 Meth., Rand Corporation, Santa Monica, CA.

Reckhow, K. H. 1979. Empirical lake models for phosphorus development applications, limitation and uncertainty. In D. Scavia and A. Robertson Eds., *Perspectives on Lake Ecosystem Modelling*, Ann Arbor Sci., Ann Arbor, MI, pp. 193–221.

Reckhow, K. H. and S. C. Chapra. 1983. Confirmation of water quality simulation models. *Ecol. Modelling 20:*113–133.

Ripl, N. 1976. Biochemical oxidation of polluted lake sediments with nitrate—a new lake restoration method. *Ambio 5:*132–135.

Smidth, F. L. 1973. An internal report on "Precipitation of phosphorus in Lyngby Lake." Copenhagen, Denmark.

Uhlmann, D. and J. Benndorf. 1976. Prognose der Wassergüte Hydrischer Ökosysteme. *Die Technik 31*(8):486–492.

Uhlmann, D. and J. Benndorf. 1980. The use of primary reservoirs to control eutrophication caused by nutrient inflows from non-point sources. In N. Duncan and I. Rzoska, Eds., *Land Use Impacts on Lake and Reservoir Ecosystems*, Facultas-Verlag, Vienna, pp. 152–188.

Wilhelmus, B., H. Bernhardt, and D. Neumann. 1978. Vergleichende Untersuchungen über die Phosphoreliminierung von Versperren-Vermindurung der Algenentwicklung in Speichberbecken und Talsperren. *DVGW, Schriftenreihe Wasser. 16:*140–176.

17

INTEGRATED FISH CULTURE MANAGEMENT IN CHINA

Yan Jingsong

Nanjing Institute of Geography and Limnology
Academia Sinica, Nanjing, China

and

Yao Honglu

Jiangsu Provincial Freshwater Fisheries Research Institute
Nanjing, China

17.1   INTRODUCTION

Integrated fish culture management in ponds is ecological engineering because it uses a series of ecotechniques in a semiartificial ecosystem for integrative utilization of natural light, heat, dissolved oxygen, food organisms, and minerals, in order to greatly increase fish production. These ecotechniques include many measures for environmental, structural, and biological adjustment in the fishpond ecosystem, such as improving the construction of the fishpond; regulating water quality; providing feedstuffs and fertilizer; choosing a mix of species with different living habits; adjusting the composition and proportion of fishes; and constructing food webs. We apply the main principles of biotic structure, together with efficiency in utilizing natural resources and economic benefit as found in natural ecosystems, to attain the purpose of multilayer and gradational utilization of limited materials including the waste produced.

17.1.1   History of Fish Culture in China

China was the first country in the world to use fish culture. In the Yin Dynasty, about 3100 years ago, fishpond culturing began. In 460 BC, Fani wrote a famous book, *Treatise on Pisciculture*, which is the earliest monograph on fish culture in the world. It summed up the experience to date on rearing common carp. During the Tang Dynasty, 618–907 AD, people were prohibited from catching and eating common carp, because in Chinese the common carp is called "li," which was the same as the pronunciation of the surname of the emperor. As a result, the culturing of common carp came to a stop, and polycultures of grass carp, silver carp, bighead carp, and black carp gained development. Through a great deal of practice of pond culture for more than 1000 years since the seventh century, Chinese fish culturists have accumulated a wealth of experience on integrated pond fish culture. On the basis of this experience, the key measures and techniques on integrated pond fish culture are summarized by eight words (in a systematic and complete way), namely, "water," "fingerling," "food," "density," "polyculture," "prevention," "rotation," and "management" (Committee of Summarizing Fresh Water Fish Culture Experience of China, 1961). The production of pond fish culture increased in China with the spread and practice of these advanced techniques from one region to another. Now the net yield of pond fish culture is 7.5–15 metric tons/ha-yr or more in most counties and provinces.

Most of these techniques, accumulated from long practice, are based on observation and experience. In order to raise them to a theoretical level demonstrating the principles and laws and quantitative relationships among rearing fishes and environmental elements and agents in the semiartificial ecosystem, these techniques are being studied and analyzed from ecological, especially from ecosystematic, points of view and methods. It is believed that the results of such studies will help us not only to understand the principles deeply and thus be able to explain them, but also, particularly in applying the knowledge from such studies actively, to improve, develop, and spread these techniques for ecological, economic, and social efficiency.

17.1.2   Integrated Fish Culture as Ecological Engineering

In the broad sense, integrated fish culture is ecological engineering because such semiartificial co-ecosystems (or combined production systems of aquaculture and animal husbandry, agriculture, and/or some other industry) make a nearly endless beneficial cycle of materials for the full multilayer and gradational utilization of various resources, including wastes within this co-ecosystem. The production by this ecotechnology includes high output, low consumption, good quality, and high benefits; the ecological, economic, and social efficiencies are often obvious. According to nature and the main structure of ecological engineering, the ecotechnologies fall roughly into four categories: (1) utilization and purification of wastewater by fish culture; (2) cycling and regeneration of substances, such as the mulberry grove–fishpond model (Figure 17.1); (3) multilayer and gradational utilization of crop fodder, poultry, livestock, and mushrooms, earthworms, maggots, fishes, and their excretions in a co-ecosystem; and (4) multifunctional agro–aquaculture–industrial (combined) production systems.

For example, in production systems combining the agro-food processing industry, animal husbandry, and aquaculture, the principles of endless beneficial cycling of materials and regeneration of organisms in the ecosystem

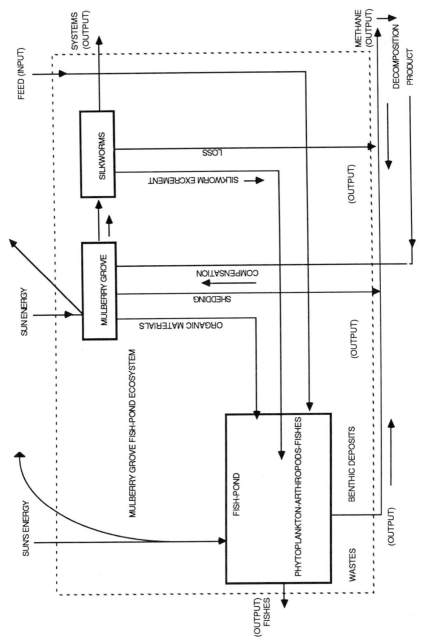

Figure 17.1 Mulberry grove–fishpond model (Ma, 1985b).

may be applied to processing factories, with agricultural products as raw materials. Wastewater and residue from food processing factories, such as meat, beans, starch, and sugar beet, may be utilized as feeds for fish, livestock, and domestic fowl. The excrement of the livestock and poultry may be used as an important part of the culture media for mushrooms, earthworms, and/or maggots or as fertilizer for farmland and fish ponds. The earthworms and maggots are a good feed for black carp, common carp, hens, and ducks, and residues of the culture medium for cultivating mushroom, earthworms, and maggots may be thrown into fishponds or farmland for fertilizer; the bottom materials of fishponds may be used as manure for farmland. In this manner, the combined production system of a region not only may supply the market with many kinds of products such as food fish, agriculture products, livestock, domestic fowl, mushrooms, bean curd, wine, sugar, sheet jelly, and noodles, but also may keep the ecobalance of the region relatively stable, with reduction of wastes, prevention of pollution, and improvement of rural environments.

17.2 BASIC ECOLOGICAL PRINCIPLES

The basic ecological principles on which integrated fish culture ecotechnology depends include the following:

1. The principle of adaptation of organisms to each other and to their environments is fundamental to the design and practice of integrated fish culture ecotechnology. In particular, the law of limit of tolerance, Liebig's law of the minimum, and the principle of symbiosis are the main theoretical bases for environmental and structural regulation in integrated fish culture ponds.
2. The foundations of a suitable fish density rotation are that individual and population growths are regulated by both density-dependent and density-independent factors particular to the conditions of different species of cultures fish.
3. Intra- and inter-specific interactions, such as competition, association, and mutualism should be considered in polyculture ecotechnology.
4. The transformation, decomposition, and regeneration of substances based on the multilayer trophic structures play leading roles in the dynamic processes of ecosystems (Ma, 1985a).
5. Normal metabolic processes in a homeostatic ecosystem, such as the combination of materials and energy transfer, should be in accordance with a definite hierarchy of structure and in a given quantitative proportion among materials or substances (Ma, 1982).
6. In a homeostatic ecosystem, organisms not only have established many mutually-conditioned relationships but also have become specialized

in their living habits so as to (1) occupy different ecological niches in the ecosystem, and (2) make gradational utilization, according to their own particular needs, of the substances offered by the natural world.

It is precisely for these reasons that earth's finite space can support so many kinds of organisms and that they can be maintained in a relative steady state. Analyzed from an economic standpoint, biotic communities are efficient in utilizing natural resources. Consequently it is desirable in the construction of integrated fish culture to apply the above principles of biotic structure with high economic benefit as found in natural ecosystems. This will make full use of limited resources and speed up the development of the human economy while protecting our ecological environment (Ma, 1982).

17.3 CHARACTERISTICS AND ESSENTIAL FACTORS OF INTEGRATED FISH CULTURE

The main characteristic of integrated fish culture is rearing structures that are adjusted to environmental conditions in the fishpond. The net productivity of integrated fish culture is almost 10 times that of extensive culture as practiced in China. The products of an integrated fishpond include various species of food fishes that are low in fat content and high in protein.

There are eight essential factors of integrated fish culture, as mentioned in the introduction. In general, characteristics and essential factors in integrated fish culture include (a) the rearing structure in the food chain or web to utilize fully various kinds of food including waste; (b) the space structure for full vertical utilization of the fishpond; (c) the time structure for maintenance of high productivity of cultivated fishes in the pond during most of a year; (d) the control of organisms to maintain a rational proportion of those organisms occupying various ecological niches, including different trophic levels and different layers of pond water; and (e) preserving appropriate living conditions in the pond environment.

17.4 BIOLOGICAL CHARACTERISTICS OF MAJOR REARING FISHES IN CHINA

It is very important in developing an integrated fish culture to understand the biological characteristics of major rearing fishes, because it is the base of selecting species for stocking structure for polyculture, reasonable stocking and rearing density with proportion, application of feeds and fertilizer, planning and adjusting rearing, space and time structure. The main rearing fishes in China include silver carp (*Hypophthalmichthys molitrix*), bighead carp (*Aristichthys nobilis*), grass carp (*Ctenopharyngodon idella*), black carp (*Mylopharyngodon piceus*), common carp (*Cyprinus carpio*), crucian carp

(*Carassius auratus*), silver crucian carp (*Carassius auratus gibelio*), Wuchang fish (*Megalobrama amblycephala*), tilapia (*Sarotherodon nilotica* ♂ × *Sarotherodon mossambica* ♀). Their chief biological characteristics, after Wu et al. (1963) and Lei et al. (1981), are summarized in Table 17.1.

17.5   FOOD CHAINS WITHIN INTEGRATED FISH CULTURE PONDS

In rearing ponds, the chief primary producers are always algae, especially phytoplankton, whereas aquatic macrophytes are usually rare or absent. The standing crop, species, and production vary with conditions of throwing in baits, applying fertilizer, species, and density of rearing fish. The standing crop and production of phytoplankton in integrated cultured fishponds can reach 40–100 mg/L and 100–150 metric tons/ha-yr fresh weight, respectively. The secondary producers are commonly zooplankton, zoobenthos, and fishes. Some are primary consumers, including: most species of Protozoa, Rotifera, Cladocera, freshwater Lamellibranchia, and silver carp, which feed chiefly on phytoplankton; freshwater Gastropoda, some species of chironomid larvae and aquatic earthworms, which take mainly sessile algae and sediments from detritus and plankton; and, grass carp, Chinese bream, and Wuchang fish, which feed mainly on aquatic and land macrophytes. Others are secondary consumers, such as most species of copepods; a few species of Rotifera; bighead carp, which prey chiefly upon zooplankton; all species of Tanypodiinae larvae; and black carp, which take mainly zoobenthos. The omnivores, which may be regarded as a mixture of the trophic levels of the first and secondary consumers, include common carp and crucian carp, which feed on zoobenthos and also on phytoplankton, zooplankton, detritus, and a few higher plants, and silver crucian and Nile tilapia, which feed chiefly on phytoplankton but sometimes on suspended detritus and zooplankton as well. The decomposers are mostly bacteria, but some species of saprotrophic invertebrates may be regarded as partial decomposers, such as most aquatic earthworms, especially *Tubifex* and *Limnodrillus*, and some species of Chirominae and Orthocladiinae larvae, which mainly utilize detritus and play the role of decomposers as well. The species, density, and production of secondary producers and decomposers vary with the conditions, such as throwing in bait, applying manure, and predation intensity of relevant predators.

One of the main differences between integrated culture and extensive culture is that the yield of aquatic products from the former depends not only on a species of fish and throwing in bait, as does the latter, but especially on cultivated fishes and on full utilization of limited materials and energy offered by the natural world. It should be an important ecotechnique to adjust the structure of reared fish and the food chain or food web to make the relationship more diverse and complex in the semiartificial ecosystem, namely, within the rearing pond. This should be based on stimulating biotic

Table 17.1 Biological Characteristics of Major Freshwater Cultivated Fish in China

Characteristic	Silver Carp	Bighead Carp	Grass Carp	Black Carp	Common Carp	Wuchang Fish
MORPHOLOGY						
Body shape	Compressed	Similar to silver carp but with big head	Almost cylindrical, with flat head	Like grass carp, with pointed head	Compressed dorsal project in arch shape	High, compressed, and lozenge shaped, small and short
Scale	Small	Small	Big and circular	Big and circular	Round, big, and thick	Round
Gill membrane	Unconnected to isthmus	Unconnected to isthmus	Connected to isthmus	Connected to isthmus	Connected to isthmus	Connected to isthmus
Gill rakers	Dense	Dense and separated, delicate and sabre shape	Small and short, scattered	Short	Scattered	
AGE AND GROWTH						
Average body weight (kg)						
Age 1 yr	0.42–0.49	0.2–0.27	0.5–0.8	0.46	0.065–0.15	0.03–0.05
Age 2 yr	1.23–2.03	2.0–3.0	2.0–3.3	2.93	0.2–1.25	0.1–0.4
Age 3 yr	2.20–3.50	5.0–6.0	5.0–6.0	7.63	0.8–2.2	
Age 4 yr	4.20–5.30	10.1–11.0	6.5–7.5	12.78	3.4–4.5	
Age 5 yr	5.50–7.60	11.8–13.5	8.0–8.5	16.65	3.9–7.0	
Age 6 yr	8.50–10.8	16.6–18.0		20.23	3.3–8.75	

	2nd year in both body length and weight	2nd–3rd year in body length and after 3rd year in body weight	1st–2nd year in body length and 2nd–3rd year in body weight	1st–2nd year in body length and 3rd–4th year in body weight		1st–2nd year in both body length and weight
Period of fastest growth						
Largest body weight (kg)	20	40	35	50	13	
Intestinal length (times that of body length)	6–10	about 5	2.3–3.3	1.2–2.0	1.5–2.0	about 2.7
FEEDING AND HABITAT						
Feeding habits	Filter	Filter	Herbivorous	Carnivorous	Omnivorous	Herbivorous
Main food (fry)	Zooplankton	Zooplankton	Zooplankton	Zooplankton	Zooplankton	Zooplankton
Main food (adult)	Phytoplankton with zooplankton	Zooplankton with phytoplankton	Aquatic vegetation	Benthos, especially snails and Corbicula	Benthos, such as snails, young clams, and zooplankton, with detritus of aquatic plants and seeds	Aquatic grass with some algae detritus
Water layers for habitat	Upper	Middle and upper	Middle and upper	Bottom	Bottom	Middle and lower
WATER TEMPERATURE (°C)						
Range of adaptation	0.5–38	0.5–40	0.5–40	0.5–30	0.5–30	0.5–30

383

Table 17.1 Biological Characteristics of Major Freshwater Cultivated Fish in China (Continued)

Characteristic	Silver Carp	Bighead Carp	Grass Carp	Black Carp	Common Carp	Wuchang Fish
Optimal for food intake and growth	25–32	25.0–32	25.0–32			
At diminished appetite	<15	<13	<15	<15	<13	<15
At appetite loss	<7	5	7	7	5	7
Water quality for optimal habitat	Fertile	Fertile	Clear water	Fertile	All kinds of fresh water	Clear water
Highest tolerated BOD	30	30	15	15	30	30
DISSOLVED OXYGEN (MG/L)						
At great feeding intensity	>4–5	>4–5	>4–5	>4–5	>3–4	>4–5
At appetite loss	1	1	1	1	0.5	1
REPRODUCTION						
Maturity age of fish years (body weight)	4 (5 kg)	5 (12.5 kg)	4–5 (3 kg)	5–6 (9.5 kg)	3 (2–2.5 kg)	3 (0.4 kg)
Spawning time	Late April to July	Late April to late July	Late April to July	May to July	April to May	Late April to May
Spawning water temperature (°C)	18–30	20–27	22–28	22–28	17	20–28
Natural spawning bed	Running water—river	Running water—river	Running water—river	Running water—river	Still water	Standing water

community structure, complete with its symbiotic functions, including the multilayer and gradational utilization of materials and energy, according to actual needs as found in natural homestatic ecosystems.

For example, mixing grass carp with silver carp, bighead carp, and common carp is beneficial for all species. Grass carp has a great appetite but, for lack of fibrous enzymes in its intestine, can digest only broken plant cells ground by its pharyngeal teeth. So its excrement contains plenty of undigested plant detritus, which directly supply food for Wuchang fish, common carp, Nile tilapia, crucian carp, and some benthic organisms and helps to fertilize the pond and nourish plankton, resulting in increasing indirectly the food organisms for silver carp and bighead carp. If grass carp is not raised together with planktivorous filter-feeding fishes and omnivorous fishes, the output is reduced and the water becomes too fertile for grass carp to live in. Black carp should be added to the pond for integrated culture; otherwise mollusks would multiply and consume great quantities of plankton, thus hindering the growth of silver carp and bighead carp. By raising 75–150 individuals of black carp per hectare the reproduction of snails and clams can be controlled.

Chinese bream and Wuchang fish mostly resemble the grass carp in feeding habit. They chiefly feed on macrophytes, filamentous algae, and larvae of midges, but their feeding capacity is smaller than that of grass carp. However, there is more advantage than disadvantage to mixing Wuchang fish with silver carp, bighead carp, common carp, and grass carp, because Wuchang fish may occupy that trophic niche in the food web within an integrated fish culture pond instead of grass fish. The disease resistance is stronger, and commercial value is higher in Wuchang fish than in grass fish, and Wuchang fish are more marketable in China. Moreover, most Wuchang fish feed upon the excrement of grass carp and the remainder of food devoured by the grass carp, although Wuchang fish are also herbivores. Mixing Wuchang fish and grass carp in an integrated fish culture pond can improve the multilayer and gradational utilization. Common carp mainly feeds on detritus and benthos. There is no competition for food among common carp, silver carp, bighead carp, and grass carp. Common carp likes to turn up mud in seeking food, thereby accelerating the decomposition of the fertilizer accumulated at the pond bottom. This fertilizes the pond and nourishes plankton directly and increases food organisms, bringing about rapid growth of silver carp and bighead carp and other species. The relation between silver carp and bighead carp is that silver carp feeds chiefly on phytoplankton and sometimes on zooplankton and detritus; bighead carp takes mainly zooplankton and also some phytoplankton. In order to ensure the normal growth of these two species of fishes in integrated culture, different methods are adopted. The ideal way is to control the stocking density of bighead carp, and to keep the ratio of the number and weight of bighead carp to silver carp to be about 1:3. This density has no ill effect on either species. However, by catching and stocking in rotation for adjusting the rearing structure and density, fewer silver carp should be raised from June to September when bighead carp grow most rapidly, whereas more silver carp should be stocked from October to April when bighead carp grows comparatively slowly. By doing so, the output of both species can be high.

17.6 ECOTECHNIQUES OF INTEGRATED FISH CULTURE

Integrated fish culture in ponds is actually a type of ecological engineering for aquaculture, namely, making a semiartificial pond ecosystem that is not only used as a cultivated fish pond, but also as a pond of cultivation of food organisms—the productive base of food for fishes. The pond is also a workshop of oxygenation by photosynthesis, and a biological treatment pond for sewage. To develop such a type of ecological engineering, according to studies on pond ecology and the experiences from practicing pond fish culture, comprehensive ecotechniques can be summarized. These ecotechniques include (1) choosing different species of rearing fishes and size of fingerlings in polycultures; (2) arranging reasonable stocking and proportional rearing densities; (3) catching and stocking rotation; (4) applying feeds and fertilizers; (5) maintaining adequate water supply and good water quality protection; (6) rearing management; and (7) preventing diseases and eliminating enemies. These ecotechniques are summarized below.

17.6.1 Polyculture

Polyculture is one of the important ecotechniques for integrated fish culture. The principle purpose of polyculture is to bring the productive potential of the water body into full play to ensure a good harvest. Thus the prerequisite of polyculture is as much vertical utilization of various strata of water in the fishpond as possible. The structure of the food chain or web should be adjusted and rearranged for the multilayer, gradational and full utilization of natural foods for aquaculture. In a community of mixed species of fish stocked in the same pond, the choice of fish species should be made according to their biological characteristics, especially their residential habitat and feeding habits. A mixed community of fishes in the same pond will require species that dwell in the upper, medium, and lower layers of the pond, as well as species with different feeding habits, that is, plankton feeders, herbivorous fishes, benthic mollusk feeders, benthos feeders, and omnivorous fishes.

In general, one to three species of fishes are reared as the dominant species of fish stock in a polyculture, and the others are secondary or companion species. At present in China, generally five to eight or sometimes more than 10 species are stocked in adult fish culture ponds. Which species should be reared as the main or secondary ones and how many individuals are suitable depend on the availability and quality of local feed for the various feeding habits of the rearing fish, the biological characteristics of the fishes, especially their feeding habits and inhabitation; the economic value (commercial) of the different sorts of cultivated fishes; the conditions of the pond, such as its fertility, which has great effect on the composition of plankton and the production of natural food organisms; the source of fingerlings; and geographical location.

For example, silver carp and bighead carp may be reared as the major species in fishponds in which the water is comparatively fertile and rich in plankton and where sources of abundant organic fertilizers are within and/or near the aquaculture farm. In this type of rearing structure, with plankton feeder fishes as the main species, the secondary species may be herbivorous fishes such as grass carp, Chinese bream, and Wuchang fish, and omnivores and benthic feeders such as common carp, crucian carp, and a few black carp, which mainly feed on snails and/or clams (Table 17.2).

Another type of rearing structure involves raising grass carp and Wuchang fish as the main species where aquatic and/or land grasses are abundant and the water is infertile. In this type of polyculture, a given number of silver carp, bighead carp, common carp, crucian carp, and a few black carp are mixed as companion species, as shown in Table 17.3. The main species here are herbivorous, with plankton feeder fishes as secondary species.

In a third type of polyculture black carp is reared as the main species in regions of abundant mollusks, especially snails and clams, and infertile water. The companion species are silver carp, bighead carp, Chinese breams, Wuchang fish, and common carp, as shown in Table 17.4.

17.6.2   Reasonable Stocking and Rearing Density With Proportion

One of the principles of ecosystem ecology is that to maintain a homeostasis with high efficiency in the multilayer and gradational use of limited materials and energy in an ecosystem, it is necessary not only to have harmonious coordination of structure with function in the ecosystem, but also a suitable quantitative proportion among these agents. This is because homeostasis is conditioned by a definite order of time and space, by a given hierarchy of structure, and also by fixed ratios among agents of material and energy transfer in normal metabolic processes within an ecosystem.

Based on this ecological principle, another important ecotechnique for constructing this type of ecological engineering for pond aquaculture is to plan rational stocking and rearing densities of proper proportion among the fishes cultivated. Density means both the number and standing crop (the amount of body weight) of fishes within a given area or volume of water.

In a rearing pond, density has much effect on the output of cultured fishes, because the production of fish depends on the product obtained by multiplying average individual growth rates of fishes by their density. But from the point of view of critical standing crop (CSC) and carrying capacity, density also has a great influence on the growth of individuals and populations. Rationally increasing the density increases the yield of fish. How-

Table 17.2 Stocking Species, Density, and Harvest in a Model of High-Yield (15 metric ton/ha-yr) Integrated Fish Cultures with Silver Carp and Bighead Carp as the Principal Fish Reared[a]

Species	Stocking				Harvest				Net Growth Rate (times)
	Size (g/ind.)	Number (ind./ha)	Density (kg/ha)	Survival Rate (%)	Size (g/ind.)	Gross Production (kg/ha)	Net Production (kg/ha)	% of Total Production	
Silver carp	250–350	3,375	975	96	600	1,950	975		
	100–150	3,375	450	96	600	1,950	1,500		
	30	3,750	112.5	92	600	1,875	1,762.5		
	1	4,500	4.5	82	165	600	595.5		
Total of silver carp		15,000	1542			6,375	4,833	31.2	3.1
Bighead carp	250–350	1,200	375	98	700	825	450		
	100–150	1,200	150	96	650	730	600		
	1	1,500	1.5	86	175	225	223.5		
Total of bighead carp		3,900	526.5			1,800	1,273.5	8.2	2.4
Tilapia	1.5	75,000	112.5		100–150	6,450	6,337.5	40.9	56.3
Silver crucian carp	20	6,000	120		150	525	405	2.6	3.4
Crucian carp	25–50	3,000	112.5		100–150	525	412.5	2.7	3.7
Common carp	200–300	1,200	300	85	750–1000	825	525	3.4	1.8
Wuchang fish	60	7,500	450	85	150–250	1,200	750	4.8	1.7
Grass carp	400–500	1,200	540	85		1,500	960	6.2	1.8
Total		112,800	3703.5			19,200	15,406.5	100	4.2

[a] Ind. = individual.

Table 17.3 Stocking Species, Density, and Harvest in a Model of High-Yield (15 metric ton/ha-yr) Integrated Fish Cultures with Grass Carp and Wuchang Fish as the Principal Fish Reared[a]

| Species | Stocking | | | | Harvest | | | | Net Growth Rate (times) |
	Size (g/ind.)	Number (ind./ha)	Density (kg/ha)	Survival Rate (%)	Size (g/ind.)	Gross Production (kg/ha)	Net Production (kg/ha)	% of Total Production	
Grass carp	400	1,500	600	90	1750	2,362.5	1,762.5	14.2	2.9
	50	3,000	150	80	400	960	810	6.5	5.4
	50–400	4,500	337.5	90	375	1,518.8	1,181.3	9.5	3.5
Wuchang fish	10	3,000	30	80	150	360	330	2.7	11.0
	200	4,500	900	95	600	2,565	1,665	13.5	1.9
Silver carp	50	1,500	75	95	600	855	780	6.3	10.4
	2	4,500	9		200	882	873	7.1	97.0
	175	1,500	262.5	95	650	926.3	663.8	5.4	2.5
Bighead carp	2	1,500	3	95	175	257.3	254.3	2.1	84.8
Silver crucian carp	25	6,000	150	95	150	540	390	3.1	2.6
Crucian carp	25–50	3,000	112.5		150	450	337.5	2.7	3.0
Common carp	50	2,250	112.5	95	400	855	742.5	6.0	6.6
Black carp	500	150	75		2500	375	300	2.4	4.0
Tilapia	2	15,000	30		150	2,250	2,200	17.9	74.0
Other fish						75	75	0.6	
Total		51,900	2,847			15,231.9	12,384.9	100.0	4.4

[a] Ind. = individual.

389

Table 17.4 Stocking Species, Density, and Harvest in a Model of High-Yield (15 metric ton/ha-yr) Integrated Fish Cultures with Black Carp and Wuchang Fish as the Principal Fish Reared[a]

Species	Stocking			Harvest					Net Growth Rate (times)
	Size (g/ind.)	Number (ind./ha)	Density (kg/ha)	Survival Rate (%)	Size (g/ind.)	Gross Production (kg/ha)	Net Production (kg/ha)	% of Total Production	
Black carp	750	1,800	1350	98	2500	4,410	3,060	22.0	2.3
Common carp	150	4,200	630	85	750	2,677.5	2,047.5	14.7	3.3
Wuchang fish	15	6,000	90	99	425	2,524.5	2,434.5	17.5	27.1
	40	6,000	240	93	290	1,620	1,380	9.9	5.8
Grass carp	400	450	180	80	1500	540	360	2.6	2.0
	60	750	45	87	450	292.5	247.5	1.8	5.5
Silver carp	50	4,800	240	96	650	2,987.3	2,747.3	19.8	11.4
Bighead carp	50	1,200	60	100	675	807	747	5.4	12.5
Silver crucian carp	15	6,000	90	64	135	522	432	3.1	4.8
	1	10,500	10.5	47	25	124.5	114	0.8	10.9
Crucian carp	20	1,500	30	92	195	262.5	232.5	1.7	7.8
Other fish						84.8	84.8	0.6	
Total		43,200	2965.5			16,852.6	13,887.1	100.0	4.7

[a] Ind. = individual.

ever, growth is slow when the density is too high. There is a definite density at which the growth of certain species of fish is hampered; this is determined by their feeding and living habits and limits of tolerance. There are many factors that are affected by the infinite increase of density of fish, the main ones being dissolved oxygen content, feeding, and space for vital activities. Changes in these factors affect the growth, production, and even life of cultured fishes. For example, exceeding a reasonable density of fishes in a rearing pond can result in a dissolved oxygen deficit. If the dissolved oxygen content becomes less than the limit of tolerance for oxygen of certain species of fish, it may check the growth, feeding, production, and yield; raise the food coefficient; and even cause fish kills.

The most economical stocking density is not necessarily that which results in the highest average growth rate of fishes, but rather that which results in the highest yield per area. This can be explained by the following example from our survey in China. The average growth rate of fish in a pond stocked with 15,000 individuals of fish fry (mainly silver carp) per hectare is 322 g per individual, and the net yield is 4833 kg/ha. In another pond with an increased density of 112,800 individuals per hectare (with the proportion of various species of stocking fish and stocking size similar) the net yield is 15,500 kg/ha, but the average growth rate is only 137 g per individual. In such a case, it is more economical to stock 112,800 fish in spite of the reduced growth rate of fishes.

However, the density cannot be infinitely increased. In another example, when the stocking density is raised to 200,000 individuals per hectare (the proportion of stocking of fish species and stocking size is also similar to former), the net yield is only 15,000 kg/ha, and the average growth rate of an individual reduces to 75.0 g/individual. The net yield is not more than that of the second example but the costs for stocking fish, oxygenation, and supplying feeds all increase. The commercial value of reared fishes is reduced because the size of the individual fish is less than marketable size.

Density means both the number and the total body weight of fish per unit area or volume. Rational stocking and rearing density are most important for high output, and their ecological bases are mainly the critical standing crop (CSC) and carrying capacity. As indicated by Hepher and Pruginim (1981): "Since below CSC the growth rate of fish of a given weight reaches a maximum, the yield is proportional to density. The higher the density the higher the yield. When CSC is reached the growth rate decreases and the log yield-log density function deviates from a linear proportion. As long as growth rate decreases at a smaller rate than that increase in density, yield increases. Above a certain standing crop the average growth rate decreases sharply and reaches zero at carrying capacity." In a rearing pond, both the critical standing crop and carrying capacity are closely related to factors such as food and dissolved oxygen. Hepher (1975) estimated the critical standing crop and carrying capacity for common carp at different feeding

Table 17.5 Effects of Feeding Level on Critical Standing Crop (CSC) and Carrying Capacity for Common Carp (Hepher, 1975)

Feeding Level	CSC (kg/ha)	Carrying Capacity (kg/ha)
Nonfeeding and no fertilization	65	130
Fertilization but no feeding	120	480
Fertilization and feeding with cereal grain	550	2000
Fertilization and feeding protein-rich pellets	2400	—

levels from the Fish and Aquaculture Research Station, Dor, Israel, as shown in Table 17.5.

Very high carrying capacities have been reported by Van der Lingen (1959) for tilipia (*Sarotherodon mossambicus*) in Rhodesia: 896 kg/ha in unfertilized ponds, 2128 kg/ha in fertilized ponds, and 6160 kg/ha in fertilized and fed ponds.

Our many years of study of integrated fish culture ponds have shown that during the period July to September, when water temperatures range from 28 to 32°C, dissolved oxygen and water quality are always the limiting factors. Critical standing crop is 6000 kg/ha and the carrying capacity is 9000 kg/ha in ponds used mainly for cultivating silver carp, bighead carp, and common carp. Critical standing crop is 11,250 kg/ha and carrying capacity is 15,000 kg/ha in those ponds used mainly for rearing black carp and common carp.

The difference in densities of fishes in rearing fishponds from those in natural ponds is that the former can be adjusted by humans. The fish density in a fish pond, first of all, is controlled by the stocking density of fingerlings and the fish spawns in number, weight, and size at the beginning of every year. It depends on what the main species are and which rearing schedules are selected, on the critical standing crop and carrying capacity of the total and of each species of rearing fish, and on the conditions of the fishpond. Stocking densities of various rearing schedules are shown in Table 17.2, 17.3, and 17.4 and are described in the case study below.

The proportion of various species of cultivated fishes is determined mainly according to quantitative relationships among the food cycle. For example, grass carp is such a tremendous eater that one individual 1 kg in body weight takes about 1 kg of grass per day. It produces a great quantity of excrement, about 60–70% of weight of the food ingested. According to our studies and practical experience, with such a large amount of excrement of grass carp plus remainders from the food devoured by the grass carp, for every ton of net weight produced, there may be enough food for Wuchang fish to increase by 0.2 ton and for common carp and crucian carp to increase by 0.5 ton.

Furthermore, excrement and remainders from the secondary fish and other species of rearing fishes in the same pond can fertilizer the pond and indirectly promote increased production of plankton. This plankton production may be enough food for silver carp, bighead carp, tilapia, and silver crucian carp to increase by 1.5 tons in net weight. Based upon similar quantitative relationships among food cycles, the proportion among various species of rearing fishes in polyculture may be adjusted in different rearing schedules. Proportions of various kinds of fishes are given in Tables 17.2–17.4 and illustrated in the case study below.

17.6.3   Catching and Stocking in Rotation

The standing crop of fishes in a cultivated fishpond increases with time and the growth of individual fishes during the rearing period. In order to avoid exceeding or attaining the critical standing crop and carrying capacity of fishes in the fishpond, an ecotechnique has been developed for step by step fish production, catching and stocking in rotation to maintain rational densities of fishes throughout the rearing period. The method involves stocking enough fingerlings of different sizes and species at one time, and as the fish grow and the pond becomes crowded, harvesting the big ones for supplying the market and adding small ones. This method decreases the standing crop raised by the growth of individuals of fish and ensures enough space for normal growth and a high growth rate. By doing so, a proper density of fish can be maintained and the water is fully utilized.

The frequencies of catching and stocking are determined according to different species, stocking size, marketable size, feeding quantity and quantity, and so on. Data are introduced in the case study below.

17.6.4   Application of Feeds and Fertilizer

High yield in an integrated fish pond is based not only on the full utilization of natural food organisms, but also on the quantity and quality of feed. The kinds of feeds vary with the different type or schedule of cultivated structure. Generally, there are four kinds of feeds, namely, green fodder, animal feeds, fine feeds, and others.

17.6.4.1   Green Fodder

The term *green fodder* refers to all kinds of nontoxic green plants that are edible for herbivorous fishes such as grass carp, Chinese bream, and Wuchang fish. It is necessary in integrated fish culture to throw in great quantities of green fodder including various kinds of tender xerophilous and aquatic plants. The kinds of plants used as feed for fish vary with local conditions according to the availability and source of the local plants in the region. Land grass used as feeds for fish are commonly harvested from the land nearby, from the fish farm or from the banks around the fish pond, where *Lolium pereme, Sorghum sudanense*, and some species belonging to Gramineae, Leguminosae, and Commelinaceae are cultured. Discarded leaves and stems of many kinds of vegetables may also serve as green fodder. Aquatic vegetation is gathered from lakes, rivers, streams, and other ponds in the local region. In China these include mainly submerged plants such as *Vallisneria spiralis*, curly pondweeds (*Potamogeton crispus*), pondweeds (*P. malanius*), *P. maackianus, Hydrilla verticillata, Najas minor*, and so on; some floating vegetation such as duckweed (*Lemna minor, L. perpusilla, Wolffia arrhiza, Spirodela polyhiza*), *Azolla filiculoides, A. imbricata*, and especially the water hyacinth (*Eichhornia crassipes*, Zhang et al., 1986); and a few tender emergent plants such as reed (*Phragmites*) and wild rice (*Zizania caduciflora*).

17.6.4.2   Animal Feed

Snails, Corbicula, and clams belonging to Mollusca are the main feed for black carp and common carp. The flesh of the snail in proportion to its whole body is 19.2–27.4%, and the food coefficient for black carp to snail is about 45; the flesh of Corbicula in proportion to its whole body is about 13% and the food coefficient for black carp to corbicula is about 60. They are always caught from lakes, rivers, and streams in the nearby region. The main species in China are *Bellamys purificata, B. aerugimosa, B. angularis, B. guadrata, B. dispiralis, Corbicula flluminea*, and *Sphaerium lacustre*, In some regions of China, earthworms and maggots have also served as animal bait for carnivorous and omnivorous fishes. These living baits are cultured with garbage and waste from certain processing food factories and sideline production. Such procedures result not only in increased living animal bait with high contents of protein but also in the efficiency of waste reduction and the prevention of pollution. Fresh silkworm pupae have been used as animal bait for a long time in Jiangsu, Zhejiang, Guangdong, and Guangxi provinces of China. The protein and fat contents of fresh silkworm pupae are 17.1 and 9.7%, respectively. They are a superior food for fish.

17.6.4.3   Fine Feeds

Fine feeds include the grain of barley, peanut cakes, rape cakes, rice bran, wheat bran, the grain of corn sauce dregs, bean dregs, and wine lees. Since the 1970s, granulated feeds have been widely utilized in many region in China.

17.6.4.4   Other

All the wastewater and residue from food processing factories may be utilized as feeds or fertilizer for fish. Organic fertilizers include all kinds of green manure, human excrement and urine, animal manure, and domestic sewage. These possess comprehensive nutritive salts, such as nitrogen phosphorus, potassium, and calcium. Because organic manure is long lasting, quite cheap, and abundant, organic fertilizer is chiefly applied to fishponds in China. Inorganic fertilizers have a quick effect and offer mono-nutrition. They include nitrogen fertilizers such as ammonium sulfate, urea, and ammonia water; phosphate fertilizers such as calcium superphosphate and potash fertilizer; and potassium fertilizers such as potassium sulfate or potassium chloride. The application ratio is commonly $4(N):4(P):2(K)$, but the proportion may change for different ponds according to actual conditions of various nutrients, the ratio of N to P needed for phytoplankton, namely, 7–14:1, and Liebig's law of the minimum. Now a bottle for slow-release inorganic fertilizer is being used, which can make the fertilization last longer than by sprinkling.

17.6.5 Adequate Water Supply and Good Water Quality

17.6.5.1 Construction of Fishpond

Within a certain range, output increases with increased water depth in a fish pond. In shallow ponds, space for activity of the fish is small, and dissolved oxygen content and water temperature are easily affected by the external environment. But requiring deep water does not mean that deeper is better. The depth should be limited. Experience proves that the proper depth is 2.5–3 m.

The pond area should be fairly large. A common Chinese saying is, "A large water area brings big fish." A large water area not only provides a wide space for the vital activities of fish, but also, owing to the greater effect of wind, helps increase oxygenation and favors turbulence, steady flow, and mixing between the upper and lower layers of pond water, resulting in the improvement of water quality of different water layers. Based on experience, an area of 0.4–1.5 ha is suitable for an adult fishpond.

A round or rectangular shape (3×3.3) of the pond and a flat bottom with a slightly oblique slant toward the water exit is convenient for management, operation, and harvesting. A gentle slope of 60° favors growth and reproduction of plankton. It is desirable that the length of pond should be along an east–west line, so as to benefit from longer hours of sunshine for the promotion of photosynthesis of photoplankton and increased dissolved oxygen content.

Ponds should, if possible, be sited near a source of water of good quality, free from drought and floods, and separate from irrigation and drainage canals.

17.6.5.2 Temperature control

In integrated fish culture ponds, water quality, including physical, chemical, and biological characteristics, has much effect on the growth and output of

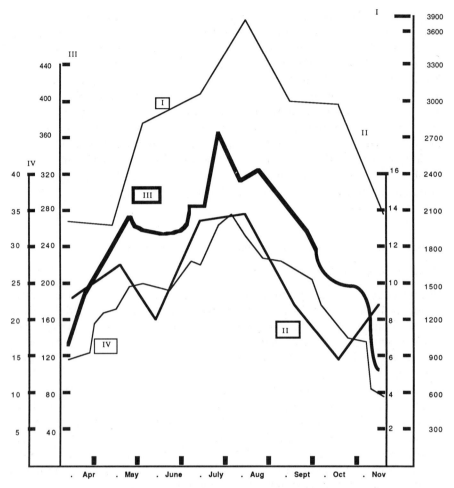

Figure 17.2 The relationships among monthly primary productivity (I, kg O$_2$/ha-month), solar energy (II, 10^4 kcal/cm^2-month), accumulated water temperature (III, °C), and monthly average water temperature (IV, °C) in a high-yield (15 metric ton/ha-yr) integrated fish culture pond in Suzhou, China, from April to November, 1984.

fish. As we know, the metabolic intensity, growth, and production of photoplankton vary with seasonal changes in radiant energy and water temperature (Figure 17.2). The optimal temperature for fast growth of most Chinese cultured fishes is about 22–32°C. Generally, these fishes scarcely feed below 10°C and stop feeding entirely below 5°C. In order to prolong hours of sunshine in a fishpond to help raise the water temperature, promoting intensity of fish feeding and increasing production of phytoplankton, the length of the pond should be east–west as mentioned above.

Table 17.6 Annual Production of Phytoplankton in Three Types of High-Yielding (15 metric ton/ha) Fish Culture Ponds

Principal Species	Gross Production of Oxygen (metric ton O_2/ha-yr)	Net Production of Phytoplankton (metric tons fresh weight/ha-yr)
Black carp, common carp	26.1	127
Grass carp, black carp	27.9	136
Silver carp, bighead carp	31.7	155

17.6.5.3   Control and Improvement of Water Quality

Dissolved oxygen, nutrients, oxygen demand by organic matter, acidity, and alkalinity are the chief ecological factors that control and affect the metabolism, growth, and production of cultured fishes and food organisms.

Dissolved oxygen content and its change result in physical, chemical, and biological processes. Dissolved oxygen also is a comprehensive indicator of the water quality, properties and function of fish production, and management in a fish culture pond. The dissolved oxygen content in a pond is determined by the oxygen budget. The sources of oxygen in a pond are mainly three, including reaeration from air, photosynthesis by aquatic plants, especially phytoplankton, and sometimes the inflow of water. According to our survey, among these sources of oxygen in a cultured fishpond, photosynthesis by phytoplankton is dominant. The annual net production of oxygen produced by phytoplankton in three type of fish culture ponds is about 26–32 metric tons/ha (Table 17.6). According to a survey of a pond with an area of 0.6 ha and a depth of 2.7 m, the daily production of oxygen by phytosynthesis of phytoplankton was 86.0% of the total sources of oxygen (Table 17.7).

There are five pathways of oxygen loss in a fishpond: water respiration, including the respiration of all aquatic organisms except fish; chemical changes such as nitrification and oxidation of some materials; uptake by bottom muds, including the respiration of benthos, and bacteria in the mud; the release and diffusion from water surface to air owing to oversaturation of oxygen in the upper layers of a water body; and sometimes output of oxygen through the outflow of pond water. Among these pathways water respiration is always the dominant for oxygen consumption, as shown in Table 17.7, where water respiration occupies 72% of the total amount of output and consumption of oxygen in a pond.

The seasonal changes of solar energy and water temperature have much effect on the primary production and photosynthesis of algae and other aquatic plants. Because of these, there are obvious seasonal changes in dissolved oxygen content in the fish culture ponds (Figure 17.4). The results

Table 17.7 Daily Budget of Dissolved Oxygen in a High-Output Fish Culture Pond Surveyed on September 15th 1984 in Suzhou, China[a]

	Input		Output and Consumption of Dissolved Oxygen		
Source	g O_2/ m²-day	%	Sink	g O_2/ m²-day	%
Photosynthesis	14.33	86.0	Water respiration	12.0	72.0
Diffusion from air	2.34	14.0	Fish respiration	3.67	22.0
			Bottom mud	0.48	2.9
			Release from water to air	0.52	3.1
Total	16.67	100.0		16.67	100.0

[a] This was a fine day with wind force 3–4 and a temperature of about 25°C.

of our survey of the seasonal change of dissolved oxygen in the high-yield (15 metric ton/ha-yr) fish culture ponds in Suzhou, China are given in Table 17.8.

Figures 17.3–17.6 show the results of our surveys, including the vertical distribution of primary productivity, the day–night changes of dissolved oxygen in the upper and lower (depth of 2.7 m) layers of water, the vertical distribution of oxygen saturation with time during day and night, and the synthesis of horizontal and vertical distribution of dissolved oxygen in a high-yield (15 metric tons ha/yr) integrated fish pond.

It is important for maintenance and control of good water quality to understand these changes. For example, in order to prevent dissolved oxygen

Table 17.8 Seasonal Change of Dissolved Oxygen (Monthly Average mg O_2/L) in High-Yield (15 metric ton/ha-yr) Integrated Fish Culture Ponds in Suzhou, China, 1984[a]

Principal Species	Feb.	Mar.	Apr.	May	June	July	Aug.	Sept.	Oct.	Nov.
Black carp, common carp	—	—	9.2	7.0	5.8	4.6	5.3	3.6	4.0	6.5
Grass carp, black carp	—	—	7.0	6.5	4.8	5.6	4.3	4.6	4.2	5.6
Silver carp, bighead carp	9.3	10.1	9.3	7.9	5.6	5.3	5.0	5.1	5.9	8.1

[a] All sampling was done at nightfall.

light   underwater

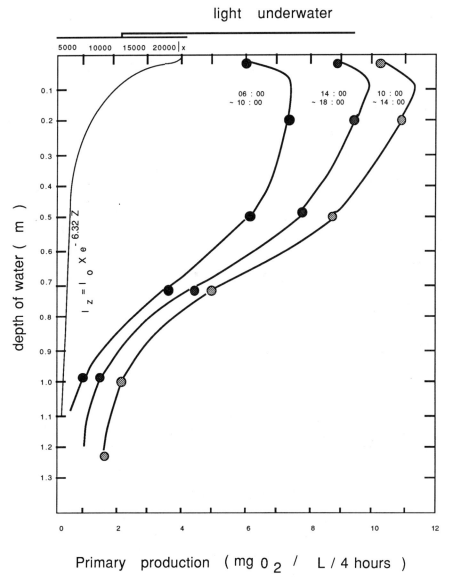

Primary   production $(\text{mg } O_2 / \text{ L } / \text{ 4 hours })$

Figure 17.3 Vertical distribution of primary production (mg O_2/L in 4 hours) in a pond with black carp and grass carp as the main species in Suzhou, China, September 15–16, 1984.

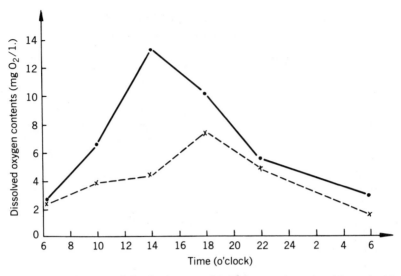

Figure 17.4 Diurnal change of dissolved oxygen (Mg/L) in upper (—·—) and lower (----) layers of water in a high-yield (15 metric ton/ha-yr) integrated fish culture pond in Suzhou, China, September 15–16, 1984.

deficits in the pond, the dissolved oxygen content at night or for the following morning can be forecast using the following equation (Yao, 1988):

$$DO_t = DO_f + DO_{df} - DO_r - DO_m - DO_w \qquad (17\text{-}1)$$

where DO_t is the predicted dissolved oxygen content for the next morning;

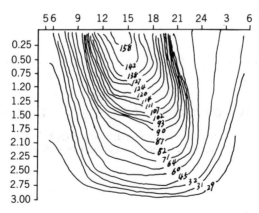

Figure 17.5 Vertical distribution with day–night change of dissolved oxygen saturation in a high-yield (15 metric ton/ha-yr) integrated fish culture pond in Suzhou, China, September 15–16, 1984. Graph indicates hour of day (horizontal) versus water depth in meters (vertical) (after Yao, 1988).

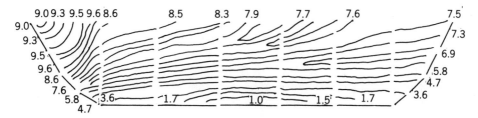

Figure 17.6 Synthesis of horizontal and vertical distribution of dissolved oxygen (mg/L) in a high-yield (15 metric ton/ha-yr) integrated fish culture pond in Suzhou, China, nightfall, September 15–16, 1984 (after Yao, 1988).

DO_f is the dissolved oxygen content measured at dusk; DO_{df} is the diffusion of dissolved oxygen, positive value when diffusing from air to water, and negative value when diffusing from water to air; DO_r is the oxygen consumption by fish respiration; DO_m is the oxygen consumption by bottom mud; and DO_w is the dissolved oxygen consumption by water respiration. When DO_t is above 3 mgO_2/L, it shows safe conditions for cultured fishes in the pond, while a DO_t decrease to 1.5 mgO_2/L warns of the danger of surfacing of cultured fishes. When DO_t is below 1.0 mg O_2/L, it is a warning that many cultured fishes in the pond may die of suffocation. Some measures of reaeration should be taken or not taken, depending on the varying of the DO_t value.

17.6.6   Rearing Management

17.6.6.1   Pond Inspection

An inspection is made of the ponds every morning to observe the manner of the surfacing of fish. If the fish are still at the surface 1–3 hours after sunrise, it means that the pond water is deficient in dissolved oxygen owing to overfertilization or food remnants. Feeding and manuring should be stopped or reduced. If surfacing continues, fresh water should be introduced into the pond. Generally, surfacing of fishes occurs in dissolved oxygen contents below 1 mg O_2/L.

17.6.6.2   Application of Feeds and Fertilizer

The principle for the application of feeds and fertilizer is "more frequent but in less quantity." It should be flexibly adopted according to the weather, the feeding activities of fish, water quality, and change of seasons. The main purpose of the inspections at noon and dusk is to observe the feeding activities of fish, so as to decide on the application amount of feeds and fertilizer.

17.6.6.3   Daily Management

Drainage and irrigation of pond water should be well managed, so as to maintain a proper water level and keep the water fresh and clean. Measures should be taken to prevent the fish from escaping. The weeds around the ponds and the remnants of grass in the ponds should be removed on a regular basis, so as to ensure a good environment for the fish.

17.6.7   Prevention of Diseases, Toxication, and Deterioration of Water Quality

17.6.7.1   Prevention of Diseases

The policies that should be adopted are ''prevention is more important than cure'' and ''all around prevention and active cure.'' Thus prevention seems to be the key link in fish disease control. General preventive measures have mainly two aspects, (1) fortifying the resistance of fish and (2) abolishing pathogens and controlling their spread. The resistance of fish plays an important role in the prevention of disease. The following are some additional points in fish culture: selection of health and strong fry; proper density and rational polyculture, careful management, uniform rations and fresh food, good water quality, and prevention of the fish body from injury. To abolish pathogens and control their spreading, the following measures can be taken: thorough pond clearing and disinfection, prohibition of diseased fish from outside, adoption of a strict quarantine, medicinal prevention by application of medicated feeds, and disinfection of fry and fingerlings and pond water.

17.6.7.2   Prevention of Toxication

Instruments for spraying pesticides should not be washed in the fish pond. When a pesticide is applied to nearby fields, water containing the pesticide should be prevented from flowing into the pond. It is absolutely forbidden to introduce into the ponds poisonous industrial wastewater from such factories as paper mills, tanneries, steel plants, and pesticide plants.

17.6.7.3   Prevention of Deterioration

Deterioration of water occurs mostly in sandy ponds on cloudy and rainy days in late spring and early summer. It can be recognized when the water in the ponds is very clear and the bottom is visible; the dissolved oxygen content therein drops below 1 mg O_2/L; the fish wander about on the water surface all day long, and water fleas are plentiful along the edges of the pond, but phytoplankton are very few. The reason is that the density or the size of bighead carp stocked is so small that the multiplication of water fleas cannot be restricted. The water fleas eat up most of the phytoplankton and the equilibrium between organisms and the physicochemical environment is destroyed.

One of the important methods of preventing this condition is the removal of excessive humus during pond clearing. Pond clearing is an important ecotechnique for this type of ecological engineering. It has many advantages, such as adding to fertility; eliminating fish enemies and other aquatic organisms; reducing disease; permitting a greater stocking density; maintaining water levels; and providing fertilizers to farmland. After the pond is drained, the topsoil of the sunlit pondbottom becomes loose, improving the aeration and accelerating the changing of organic matter into nutritive salts. In China, a common chemical used to clear ponds is quicklime. When quicklime (CaO) absorbs water, it transforms easily into $Ca(OH)_2$, which increases the pH of the water to 11–12, releasing heat so as to kill fish enemies, depleting predators and food competitors, clearing away certain aquatic organisms such as *Spirogyra zygnema, Mougeotia*, water plants, and tadpoles, which are detrimental to fish, and helping to kill or remove most of the latent bacterial pathogens, parasites and their eggs, and harmful aquatic insects. The bottom deposits are removed from the ponds to the farmland, and as the ooze is cleared off, the volume of water in the pond may increase, as does the stocking density. The humus collected from the pond bottom is a very good manure for crops.

17.7   CASE STUDY

The information and data described below were gathered during studies by us in Nanjing, Suzhou, and Yangzhou, China, except that the mulberry grove–fishpond model is from Zhong (1982) and Ma (1985b).

17.7.1   A Model of Integrated Fish for High Yield

The main differences among various models of integrated fish culture in ponds are the principal reared fish and the stocking density. Three models of integrated fish cultures in ponds are described as a part of this study. They are high-yield (15 metric tons or more of fishes per hectare per year) integrated fish cultures in ponds used primarily for reared fish with (1) herbivorous fish such as grass carp and Wuchang fish; (2) planktivorous fish such as silver carp and bighead carp; and (3) benthos-eating fish such black carp and common carp. Their stocking species, density, and harvest are listed in Tables 17.2 to 17.4, respectively. According to these models, the first stocking times were all in the period from the end of February to March.

In the first model, the biomass of cultivated fishes increases from 2.85 metric ton/ha in March to 4.5 metric ton/ha in May and 6.0 metric ton/ha in June. The first catches were made at the end of June, some yielding silver carp and bighead carp, which had attained 0.6 kg/individual in accordance with the marketable size. They were taken from the pond to July to reduce the standing crop to 3.5 metric ton/ha. Some fry of silver carp with body weights of about 2 g/individual were secondarily stocked to a density of 4500 individuals/ha in August. A second catch was made from the end of August to September. The main objective was to reduce again the standing crop to about 4.0 metric tons through harvesting of silver carp, bighead carp, grass carp, Wuchang fish, and tilapia, which had attained marketable size. The standing crop of cultivated fishes in the pond can grow rapidly and continuously from 4.0 metric ton/ha in August to about 7–8 metric ton/ha in September. The last harvest is at the end of December or early January. The annual net and gross production of the pond are 12.38 and 15.23 metric ton/ha-yr, respectively, and the average annual weight increases 4.4 times.

It is necessary, in the second model of integrated fish culture, to drain and clear the pond and to fertilize the water in winter, to prevent the surfacing of fish and to improve the water quality in spring and summer, and to fertilize more frequently but with less quantity in summer and autumn. The monthly net production of cultivated fishes from June to October is 1.5–3.0 metric ton/ha. Catching in rotation in this period should be done 15–18 times to adjust and reduce the standing crop to 5.25–7.50 metric tons/ha, below the critical standing crop. The harvest every month from June to October about equals the range of monthly net production, about 1.5–3.0 metric tons/ha.

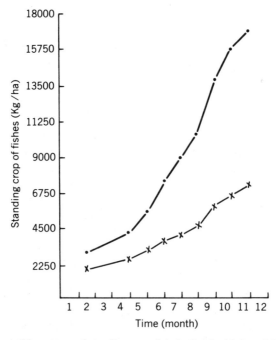

Figure 17.7 The monthly pattern of standing crop fish (kg/ha) in third model of the pond. The upper line is the total standing crop; the lower line represents standing crop of black carp.

Table 17.9 Transformations and Transfer Efficiencies of Energy Subsidies into Aquaculture Products for Different Models of Fish Culture Ponds in China

Kind of Energy	Silver Carp and Bighead Carp as Main Species (Nanjing)	Black Carp and Wuchang Fish as Main Species (Yangzhou)	Black Carp and Common Carp as Main Species (Suzhou)	Salmon as Sole Species
BIOENERGY (GJ/ha.yr)				
Fertilizer	118.47	94.78	—	—
Green fodder	32.58	117.99	192.60	
Fine feeds	89.41	84.75	175.72	1,291.45
Snails	—	—	179.86	—
Fry and fingerlings	12.24	6.55	10.54	44.97
Total and bioenergy	252.70	304.07	558.72	1,336.42
PRODUCTIVE ENERGY (GJ)				
Prevention of disease	0.3	0.3	0.3	3.24
Labor	4.73	4.73	6.09	15.19
Electric power and transport	11.00	9.92	13.54	23.55
Maintenance	—	—	—	16.2
Total of productive energy	16.03	14.95	19.93	58.1
Total energy subsidy	268.73	319.02	578.68	1,394.60
Total of output of aquaculture products (GJ/ha·yr)	75.53	49.78	81.19	161.90
Ratio of total output of products to total input of energy subsidy (%)	28.11	15.60	14.03	11.61
Ratio of total output of products to total input of bioenergy subsidy (%)	36.60	16.34	14.53	12.11
Production cost (GJ/metric ton)	14.00	29.73	34.55	55.78

The monthly change of standing crop of fish in the third pond model is shown in Figure 17.7. During the high-temperature (28–32°C) period from August to September the standing crop of fish in the pond attained nearly 11.25 metric ton/ha, the critical standing crop.

17.7.1.1 Energy Subsidies

In order to obtain high yields of fish in integrated fish culture ponds, it is important to add an energy subsidy into the semiartificial ecosystem. The transformation and transfer efficiency of the energy subsidy into aquaculture varies with the rearing fish architecture. This is illustrated in Table 17.9. The cycle of nitrogen in the integrated fishpond is shown in Figure 17.8. The ratio of productive energy of fishes to total input of subsidy energy in three models of integrated fish culture pond ranges from 14.03 to 28.11%.

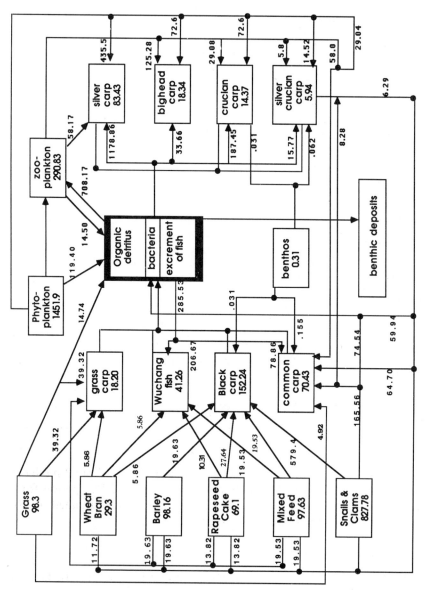

Figure 17.8 The nitrogen cycle in an integrated fish culture pond in Suzhou, China. Flows are in kg N/ha-yr and storages are in kg/ha.

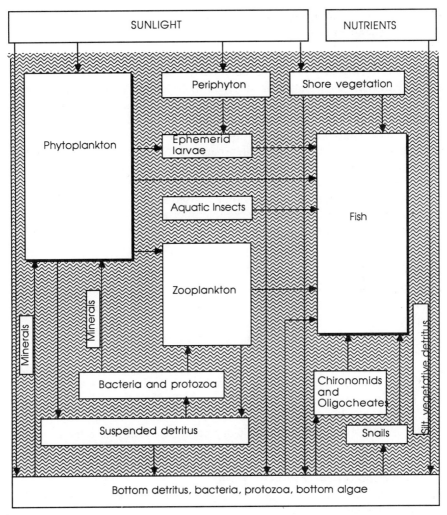

Figure 17.9 Theoretical pathways of nitrogen, phosphorus, carbon, and energy conversion in a fish culture pond.

The production energy cost of fishes, which is the consumption of subsidy energy for producing a ton fresh weight of fishes, is 14.4–34.55 GJ, only one-fourth to two-thirds that of culture ponds in Britain. The utilizable ratio of subsidy protein to product of fish protein in the model of polyculture with black carp and common carp as main species is 3.044. The high ecological efficiency and economic benefit of integrated fish culture ponds is obvious because of the shorter culture cycle and multilayer and gradational utilization of substances and energy shown in Figure 17.9.

17.7.2   The Cycle and Regeneration of Substances: The Mulberry Grove–Fishpond Model

The mulberry grove–fishpond system is an effective, multiobjective measure in agriculture (Figure 17.1). It has been popularized and become an effective measure of increasing output in the main fish-producing areas in the Pearl River Delta of China. The white mulberry (*Morus alba*) tree produces organic substances (mulberry leaves, etc.) through photosynthesis. These are used to feed silkworms that, in turn, produce their silk and chrysalides (Zhong, 1982; Ma, 1985b). The fallen parts of the mulberry tree, as well as the excrement of the silkworms, are applied to the fishpond, where they are converted into fish biomass through the process of another food chain. The excrement of the fish, as well as other unused organic matter and bottom mud, after being broken down by benthic microorganisms, are returned to the mulberry grove as fertilizer.

REFERENCES

Committee of Summarizing Experience on Fresh Water Fish Culture. 1961. A corpus of experience for fresh water fish culture in China. Sci. Publ. House. Beijing. 612 pp.

Hepher, B. 1975. Supplementary feeding in fish culture. *Proc. Int. Congr. Nutr.* 9(3):183–198.

Hepher, B. and Y. Pruginim. 1981. *Commercial Fish Farming* Avelley, New York, 253 pp.

Lei, H. S., R. L. Jiang, and D. C. Wang. 1981. *Pond Culture.* Sci. and Tech. Publ. House of Shanghai, Shanghai. China. 282 pp.

Ma, S. J. 1982. Holistic and economic approaches on ecological balance. *Beijing Agric. Sci.* (4):1–2.

Ma, S. J. 1985. An introduction to ecological engineering application of ecosystem principles. *Kexue Tongbao* 3(4):534–538.

Ma, S. J. 1985b. Ecological engineering: Application of ecosystem principles. *Environ. Conservation* 12(1):331–335.

Pitcher, T. J. and P. J. B. Hart. 1982. *Fisheries Ecology.* Croom Helm, London, 414 pp.

Van der Lingen, M. I. 1959. Some preliminary remarks on stocking rate and production of tilapia species at the Fisheries Research Centre. Proc. 1st fish Day in S. Rhodesia, Aug. 1957. Gov. Printer, Salisbury, pp. 54–62.

Wu, X. W., G. L. Yang, P. Q. Yao, and H. J. Huang. 1963. Fauna of economic animals in China: fresh water fishes. Sci. Publ. House, Beijing. 360 pp.

Yao, Honglu. 1988. Fluctuation of dissolved oxygen in integrated fish culture ponds. *Acta Hydrobiol. Sinica* 12(3):199–211.

Zhang, Yushu, Day, Q. Y., Li, X. Q., and Zhang, X. G. 1986. Studies on the water hyacinth used as fodders for cultured fish. *Jiangsu Ecology* 2:129–136.

Zhong, G. F. 1982. Some problems about the Mulberry-Dike-Fish-Pond ecosystem in the Zhujiang Delta. *J. Ecol. (China)* 1:1–3.

18

ECOLOGICAL
CONSEQUENCE
ASSESSMENT: PREDICTING
EFFECTS OF HAZARDOUS
SUBSTANCES UPON
AQUATIC ECOSYSTEMS
USING ECOLOGICAL
ENGINEERING

John Cairns, Jr., and David Orvos

Department of Biology and University Center for Environmental and Hazardous Materials Studies
Virginia Polytechnic Institute and State University,
Blacksburg, Virginia

18.1   INTRODUCTION

Graduate students are incredulous when they hear that there was a time following the industrial revolution when biological evidence was not considered mandatory for predicting the effects of hazardous substances in aquatic ecosystems. This was particularly evident in the period before World War II, when measurements made were practically entirely chemical and physical. A lengthy discussion of the evolution of ecological hazard assessment methodologies can be found in a more extensive paper (Cairns, in press), so it will not be repeated here. However, it was eventually realized that only living material can be used to measure toxicity, which is the inherent property of a substance that will produce harmful effects to an organism (or community or ecosystem) after exposure of a particular duration at a specific concentration. This information, coupled with the environmental concentration of the chemical or substance (which is that actual or predicted concentration resulting from all point and non-point sources as modified by the biological, chemical, and physical processes acting on the chemical or its by-products in the environment), can be used to determine the hazard or risk. Risk is the probability of harm from an actual or predicted concentration of a chemical in the environment. Safe concentrations are those for which the risk is acceptable to society. As a consequence, the assessment of hazard requires both a scientific judgment based on evidence and a value judgment of society and/or its representatives.

Sloof (1983) stated, "Around the turn of the century, aquatic toxicology was born as an illegitimate child of classical (mammalian) toxicology." It is certainly true that the tentative first steps in developing an aquatic toxicological methodology involved testing one species at a time, usually fish, in jars. There was some justification for the choice of fish: (1) of all aquatic organisms, fish are the best known; (2) fish are the end users in the aquatic food chain; and (3) fish are best known to legislators, law enforcement people, and so on.

Unfortunately, although these arguments appeared persuasive initially, it soon became apparent that just because a group of organisms was well known did not make it any better or any worse a predictor of the response of a wide variety of other organisms. Additionally, although it is true that in natural systems the population of fish would almost certainly be affected by changes in populations of organisms in the food chain leading to them, it might take far too long for this ultimate effect to be noticed. Also, this reasoning applied to biological assessments of natural systems, not laboratory single-species tests involving fish only. Finally, some of the least known organisms, such as the algae and macroinvertebrates, were providing valuable toxicological information that could not be readily obtained from fish. As a consequence, tests of other organisms became increasingly common, although only in relatively recent times has there been any indication that they might displace fish. Even when non-fish species were used, they were still tested individually, and therefore the interactions among species were rarely part of laboratory toxicity testing. Most important, although such terms as *biologically safe concentration, no-observable-effects level*, and the like were used, validations of the predictions made with these laboratory tests in natural systems were exceedingly rare.

In the article on the evolution of ecological hazard assessment, Cairns (in press) notes the factors leading to the separation of laboratory toxicity testing and field investigations. This is probably responsible for the rather curious evolutionary process of laboratory tests becoming more sophisticated and the methodology more complex and, most important, more isolated from mainstream ecology. At the same time, theoretical ecologists shunned involvement with the field of toxicology, and, as a consequence, the two fields had virtually nothing to do with each other. In fact, they did not view themselves as related in any substantive way. This is abundantly clear in Brungs' review (1986) of the book *Multispecies Toxicity Testing*. Brungs states, "Herein lies the basic difference that resulted in little consensus for the conclusions of this workshop. The fields of toxicology and ecology are rarely associated, either academically or professionally (there are efforts to create an ecotoxicological science but its impact has yet to be demonstrated)." Despite the estrangement noted by Brungs, the ultimate goal of applied or regulatory toxicity tests is to ensure the well-being of natural systems. Unless the two fields have more interaction than they have had in the past, it is unlikely that any meaningful validation of the predictions of safety or harm based on laboratory tests will be effectively validated in natural systems. The only way that this can happen is for each group (i.e., toxicologists and ecologists) to learn more about the other and to collaborate in the validation or confirmation of the ecological predictions based on single species or even multispecies laboratory toxicity tests. *Only then will the risk assessment or hazard evaluation component of ecological engineering be adequate!*

A major event leading to questioning existing practices in toxicity testing was the "Pellston series" of books on the process of hazard evaluation. This series (Cairns et al., 1978; Dickson et al., 1979, 1982; Maki et al., 1980) was the first to focus on the critical relationship between the environmental fate of chemicals and their biological effects. Because most readers will have difficulty in identifying Pellston even if they are U.S. citizens, it is worth noting that Pellston is a small town near the northern tip of the lower peninsula of Michigan and is near the University of Michigan Biological Station where the series of workshops on hazard evaluation was launched. Developing predictive models for the environmental fate of chemicals focused attention on both the environmental realism of the laboratory systems used to make these predictions and the need to validate the predictions in natural systems. Furthermore, if the two types of information (i.e., environmental fate and biological response) were to be closely coupled in the decision-making process, more attention had to be given both to the predicted events in natural systems and to the validation of laboratory-based predictions. The first book called for a sequence in determining the biological response, most notably stated by Kimerle et al. (1978), which had the following sequence: (1) screening or range-finding tests, (2) predictive tests, (3) validating or confirming tests, and (4) monitoring. The identification of the sequence necessary for an orderly and systematic hazard evaluation process is as important a contribution from the Pellston series as was the coupling of environmental fate of chemicals to the biological response to these chemicals.

The Pellston series raised some serious questions about the adequacy of single-species toxicity tests for making predictions of ecosystem response. The executive summary of the National Research Council (1981) report, *Testing for Effects of Chemicals on Ecosystems*, states "Single-species tests, if appropriately conducted, have a place in evaluating a number of phenomena affecting an ecosystem. However, they would be of greatest value if used in combination with tests that can provide data or population interactions and ecosystem processes." Cairns (1983a) asks, "Are single species toxicity tests alone adequate for estimating environmental hazard?" and states (Cairns, 1983b) that "The belief that multispecies, community, and ecosystem toxicity tests are second-order tests that can only be carried out after single-species tests is an assumption that deserves serious attention. There is no compelling evidence that single-species tests can be used to reliably predict responses at more complex levels of organization." More recently, the National Research Council (1986) report, *Ecological Knowledge and Environmental Problem Solving*, affirms the need for considering the attributes of the systems themselves as well as the components (i.e., single species) of the systems. In contrast, Tebo (1983) states, "The debate over ecological relevance of laboratory evaluation reminds me of the caption on an 'ivory tower' cartoon by Mount which states that ivory tower retreat occurs 'where usefulness is a handicap and practicality is a terminal disease.' The converse of the argument about the ecological relevence of laboratory data is that we provide the individual discharger a target level of toxicity reduction based on ecological parameters.'' In our opinion, the condition of the receiving system is a sine qua non of environmental management and pollution regulation. However, Tebo (1983) states, "Unfortunately, in my opinion, receiving water studies of populations and communities have very limited real-world utility in evaluating and regulating the impact of individual point source discharges." This is clear documentation from a member of a regulatory agency of the degree of estrangement of toxicity testing and ecology.

An oversimplification of the polarized positions might be that laboratory toxicologists should either limit their predictions of ecological safety well below the scope the public has been led to believe is now in effect or acquire the scientific justification for making the predictions. On the other hand, ecologists will have to abandon their "holier than thou" attitude toward the practical and applied if they want sound ecological concepts to be incorporated into toxicology. Only when these two events occur will the word *ecotoxicology* be justified.

18.2 RANGE-FINDING TOXICITY TESTS

As mentioned in the introduction, the first book of the Pellston series (Cairns et al., 1978) recommends the following sequence for the biological effects component of the hazard evaluation process: (1) screening or range-finding toxicity tests, (2) predictive tests, (3) validating or confirming tests, and (4) monitoring. The hazard evaluation process is not complete until the predictions have been validated.

Although range-finding tests are relatively crude, inexpensive tests not greatly different from the basic methodology developed by Hart et al. (1945) over 40 years ago, it will come as a shock to many people that it is still necessary to defend the use of biological testing in the control of toxic water pollutants. A recent paper by Wall and Hanmer (1987) states: "Traditionally, EPA has pursued a chemical-specific approach to regulate discharges of toxic pollutants. This approach requires industrial dischargers to analyze their wastewaters for a number of widely-used toxic compounds. Current effluent testing requirements may extend to 126 'priority list' toxic compounds and other 'non-conventional' toxicants such as chlorine and phenols. Release of unacceptable quantities of these compounds must be controlled in accordance with technology-based requirements such as 'best available technology' and state water quality standards.''* Astonishingly, Wall and Hanmer (1987) found it necessary to state in the opening sentence of their article: "The U.S. Environmental Protection Agency (EPA) has found, through laboratory and field research, that biological testing is a scientifically

---

* Quoted with permission of the *Journal of the Water Pollution Control Federation*.

valid approach to control toxics in wastewater discharges."* It is worth noting that Thomas M. Wall is an environmental scientist with the Permits Division, Office of Water Enforcement and Permits, EPA. Rebecca W. Hanmer is Deputy Assistant Administrator for Water, EPA. It is remarkable that these two highly placed officials found it necessary to state that biological testing is a scientifically valid approach in an article in the *Journal of the Water Pollution Control Federation*, a publication of one of the largest organizations of this type in the world. If EPA had attempted to validate the chemical-specific approach in a scientifically justifiable way when the organization was first formed and compared the results in an objective way to the predictions of environmental safety including the biota resulting from biological testing, the EPA might have realized what the scientific community realized more than 40 years ago. This is not to say that some scientists in the EPA are not aware of the benefits of biological testing, but rather that the official policy of the EPA is just beginning to recognize what has been regarded as a truism for many years by the scientific community.

The biological effects portion of hazard evaluation protocols usually begins with acute single-species laboratory toxicity tests using lethality as an end point. Sometimes, tests of a limited duration (i.e., 1 or 2 weeks) are called chronic tests because the organisms used (e.g., *Ceriodaphnia dubia* and *Selenastrum capricornutum*) have a short life cycle. Using the same definition, one might conceivably call a toxicity test of less than 1 day a chronic test if the organism divides in a matter of a few hours and thereby completes the life cycle of that individual. Many protists or bacteria would easily fit this definition. Ecologically, however, chronic exposure would mean taking a community of organisms or an ecosystem or a surrogate thereof through a normal seasonal cycle of events to which it must adjust or adapt on a regular basis. Therefore, when EPA uses the term "chronic testing," as it was used in the fifth paragraph of the Wall and Hanmer (1987) article, one should be very wary of the term and not assume that because an organism with a very short life cycle of a month or less has tolerated the exposure that a community or ecosystem will tolerate the same concentration for a much longer period of time.

Similarly, Wall and Hanmer (1987) did not define their particular use of the word "validated" in their article. Does "validate" mean that others could carry out the toxicity tests they describe? Or does it mean that they have confirmed or validated that the tests they describe can reliably and accurately predict responses of much more complex systems such as communities and ecosystems?

Although Cairns and Dickson (1978) produced one of the first pairs of field and laboratory biological response protocols (the basis for the paper was a report, "Protocols for Evaluating the Effects of Munition Wastes on Aquatic Life," prepared in 1973 for the U.S. Army Medical Research and

---

* Quoted with permission of the *Journal of the Water Pollution Control Federation*.

Development Command), it later occurred to Cairns (1980–1981) that the sequential arrangement of tests that were used from the simple to the more complex possibly reflects, in a broad, general way, the historic development of the field. As a consequence, tests with which there is a long familiarity are placed early in the sequence and the more recent and more sophisticated tests that are still in the experimental stage of development are placed last. This is an extremely important point because the purpose of the screening or range-finding tests is either to make a judgment that no more evidence is needed because the biological effects concentration is above the environmental concentration of the chemical, providing an enormous degree of safety, or to determine what additional tests are necessary to predict accurately the hazards that accompany the intrusion of this chemical into complex natural systems. If one makes the assumption that it is difficult to extrapolate responses from one level of biological organization to another (e.g., single species to ecosystem), then one can make a case for simultaneous toxicity testing at different levels of biological organization (Cairns, 1983b) at the range-finding stage of hazard evaluation. That is, if the purpose of toxicity testing is to produce an ecological consequence assessment resulting from the intrusion of a particular chemical into natural ecosystems, then the types of tests described by Wall and Hanmer (1987) may not always be appropriate. Such tests may not even be cost effective because the cost goes beyond the price of the test and must include the cost of a bad management decision (Cairns, 1986a). As Mount (1983) notes, the cost of biological testing is small compared to the cost of building and operating waste-treatment facilities. Mount states that $50,000 may appear to be an astronomical sum of money to ecotoxicologists, but it shrinks to a paltry amount when compared to the cost of an advanced water treatment facility for a large chemical production plant. Not only that, Mount also notes that the value of the impact area and the consequences of an impact must always enter into the cost accounting.

The *Challenger* space shuttle disaster in the United States should have taught us that the reasoning should be spelled out carefully preceding decisions for whch the outcome is uncertain and the price for failure is high. The investigation into the *Challenger* incident has shown the difficulty in reconstructing the reasoning process that led to a particular decision. The same is true of the selection of organisms, test methodology, and end points for range-finding tests. An explicit statement of the decision being made and the way in which the range-finding information will be used to make this decision must be given. Wall and Hanmer (1987) discuss in great detail setting discharge limits, establishing control priorities, and identifying environmental problems. However, they stop short of saying precisely what ecosystem qualities are protected by the measures they espouse. This particular article is singled out because: (1) it was the feature article of the largest organization in North America devoted to water pollution control and therefore presumably represents the latest "state of the art" policy and procedures for water pollution control; (2) the authors are high-ranking employees of the EPA who are charged with responsibility for safeguarding natural systems; and (3) it is representative of much of the very recent literature produced by regulators in the United States. In this article the regulatory/administrative aspects of the decision-making process are fairly explicitly stated; however, the key issue, namely, the degree to which the environment or U.S. ecosystems are protected, is only implied. Water quality, which is given significant attention in the article, is extremely important, but it is not identical to or a substitute for ecosystem quality. The EPA presumably has a broader mission than water quality; otherwise it would not have replaced the Federal Water Quality Administration that preceded it.

18.3 PREDICTIVE TESTS

Although it may be platitudinous to say so in view of current regulatory practices in the United States, it seems essential to reiterate that, as in the screening or range-finding tests, the nature of the decision being made should influence the type of toxicity test chosen, the duration of the test, the type of biological system used, and the degree of environmental realism. If the tests are designed to produce information to protect the structure and function of an ecosystem, there is no substantial body of scientific information in the peer-reviewed literature that demonstrates a high probability of success from simple laboratory toxicity tests with low environmental realism, despite the implications in the Wall and Hanmer (1987) article that this is possible.

Even if the decision being made justifies the use of single-species tests only, one might question the use of a photosynthetic alga (*Selenastrum*) in a headwater stream where most of the energy utilized by the aquatic organisms is derived from land-based detritus (e.g., leaves and other organic matter). Of course, one might say that the leaves are the result of photosynthesis, but they came from photosynthetic terrestrial organisms, not aquatic ones.

The nature of the predictions being made upon which the decision of safety will be based should be explicitly stated. For example, is there a commercially valuable (e.g., oyster) or recreationally valuable (e.g., rainbow trout) key organism that must be protected? In that case, one should either use the organism itself in the test or have evidence that the response pattern of the surrogate organism is sufficiently well correlated with that of the organism at risk to justify making an extrapolation.

In another instance, one might wish to protect the entire ecosystem. In this case, one should then identify key structural and functional characteristics (Barrett and Rosenberg, 1981; Odum, 1985) of this particular ecosystem that ecologists feel are important and suitable as end points in the test. These tests might then be carried out in surrogates for natural ecosystems [e.g., the mesocosms described by Odum (1984), in field enclosures as described by Kuiper (1982) and the National Research Council (1981), or at experimental sites (Levin, 1982)].

It is worth noting that microcosms and mesocosms are not miniature ecosystems but rather test systems with sufficient environmental realism to simulate key cause–effect pathways of more complex systems. Some illustrative examples are furnished by the National Research Council (1981, 1986). A number of examples of test methodology at higher levels of biological organization may be found in Cairns (1985), Cairns and Pratt (1988), and Hammonds (1981).

Some good examples of single-species toxicity testing methodology may be found in Rand and Petrocelli (1985), in *Standard Methods for the Examination of Water and Wastewater* (1985), or in the American Society for Testing and Materials Annual Publications on Standards for Water. Few of these methodologies will provide a detailed description of how the information can be used to make predictions of ecological risk or on how these predictions can be validated. For example, Mayer and Ellersieck (1986) found that no one species, family, or class was the most sensitive to all chemicals all the time. They also concluded that no single approach correctly predicts acute toxicity under all situations. However, the book does not have a substantive discussion of effects at higher levels of biological organization than single species. One of the primary purposes of examining these issues at length is to acquaint readers with major weaknesses in the field of hazard assessment, with the expectation that once identified they will soon be corrected.

The development of predictive tests has been driven more by regulatory convenience than by sound ecological principles. Thus the EPA has placed reliance on relatively few species that were tested individually without persuasive evidence that would stand peer review in a major scholarly journal that ecosystem protection results inevitably or even frequently as a result of decisions based on single-species range-finding or predictive tests. Neither is there persuasive evidence that such tests are invariably inaccurate or wrong. However, in science, lack of evidence and confirming evidence are entirely different things. Presumably, the methodology, reasoning, and decision making should be based on well-established scientific principles. Until this happens, ecological consequence assessment will be more a matter of conforming to regulations than sound protection. As a result, we may well lose some of the ecosystem services described by Ehrlich and Mooney (1983).

One might argue, as have many (Kimerle et al., 1986), that the laboratory acute and chronic tests are reasonable models of toxicity in the receiving water under similar exposure conditions. However, one might immediately point out that there is no direct scientific evidence for ecosystem protection that would pass peer review as validating the predictions made on the basis of single-species toxicity tests. Without an explicit statement of precisely what is being protected and scientifically justifiable validation of predictions of safety, we lack direct evidence of the efficacy of both range-finding and predictive toxicity testing. Although the news almost daily has some evidence of the unfavorable impact of chemicals upon human health and the environment, one could still say that the majority of tests are working quite well. The reason for this may not be so much the tests themselves but rather the use of an application factor in conjunction with them. The application factor reduces the concentration actually introduced into the environment well below the levels obtained in the toxicity tests themselves. Application factors have been discussed extensively by Buikema et al. (1982). Application factors are designed to compensate for a variety of deficiencies in the toxicity tests themselves. An illustrative list of deficiencies follows (Cairns, 1984)*:

1. Sensitivity in the test species not measured in the test population.
2. Changes in water quality (e.g., hardness, pH, dissolved oxygen concentration) that would mediate toxicity.
3. Responses of other life history stages of the test species not included in the tests.
4. Response of other species not included in the tests.
5. Response of other levels of biological organization not included in the tests. For example, a single-species toxicity test application factor should include responses at community and ecosystem levels, if such protection is implied.
6. Interactions with other chemicals that might make the response additive or synergistic.
7. Effects of condition factors (organismal) not included in the tests. For example, if practically no control mortality is mandatory, the animals are likely to be in better condition than they are in many natural systems. However, poor conditions might produce more susceptibility. Control organisms are often parasite free or may have had parasitism reduced by prophylactic treatment that would make the organisms more resistant to a toxicant than they might otherwise be.
8. A longer period of exposure than was possible in the test. Few tests are carried out for the lifetime of vertebrates, or even for many macroinvertebrates. Tests carried out for several generations are exceedingly rare. However, if exposure might be for a long period of time, extrapolation to the no-response threshold following long exposure must be made.
9. Deficiencies caused by problems of scale. The test container may alter the response because it is too small or in some way different from a natural system.

* Quoted with permission of The Academy of Natural Sciences of Philadelphia.

10. Lack of environmental realism in the test container or device.
11. Margin of safety the public or their representatives consider desirable.

This list provides some insight into the weaknesses of the range-finding and predictive toxicity tests now being used, but application factors may frequently, if they allow for all deficiencies simultaneously in an additive manner, be markedly overprotective because not all may be important in a particular test or for a particular chemical. Alternatively, if the application factor allows for only a fraction of the deficiencies or underestimates the effects of one or more of the components on the list, the result may be significantly underprotective. In any case, it is quite clear that a single number designed to compensate for deficiencies in so many dissimilar areas will almost certainly be imprecise. However, until a substantive attempt is made to validate the predictive toxicity tests in natural systems using a variety of parameters endorsed by ecologists, there will be no error control or feedback loop that will help correct deficiencies in the predictive toxicity tests. It is this error control loop that should be driving the development of predictive toxicity tests rather than regulatory convenience.

Three of the chief criticisms leveled against multispecies toxicity tests are that (1) they are difficult to replicate, (2) quality assurance programs are not in place, and (3) they are enormously more costly than single species. Giesy and Allred (1985) have shown that replication is not nearly the problem it was once thought to be and may not be appreciably more difficult than replication of single-species tests if one considers all the variability of natural populations of even a single species. Taub (1985) had the beginning of a quality assurance program in place in 1985, and improvements and further development have continued since that time. It also appears that quality assurance for multispecies toxicity will not be the problem it was once thought to be. Finally, there are some multispecies tests that are no more expensive than a sophisticated single-species test (Cairns et al., 1986), and if one considers validation as part of the cost of the test itself, multispecies tests may be no more expensive than single-species tests as they are presently carried out. Even community toxicity tests (Cairns, 1986) or functional tests (Cairns and Pratt, 1988) are available in the peer-reviewed scientific literature, and presumably after a reasonable developmental period, at least some of them will appear in modified form as formally endorsed standard methods. Some tests with a greater degree of environmental realism than is common are already in use by industry (Woltering, 1985).

18.4 VALIDATION OF PREDICTIONS

In the fifth paragraph of their article, Wall and Hanmer (1987) indicate that their biological tests to measure the toxicity of complex effluents have been developed, validated, and published. In the sentence containing the word "validated," there is no reference given that will enable the reader to determine precisely how this word was used.

There are at least three phases to the validation process: phase 1, scientific validation; phase 2, general applicability validation; phase 3, site-specific validation.

18.4.1   Phase 1: Scientific Validation

As stated previously, the overall hazard evaluation protocol consists of four parts: (1) screening or range-finding toxicity tests, (2) predictive toxicity tests, (3) validating or confirming tests, and (4) monitoring. The overall purpose of the validation process is to determine if the predictions of biological effect made in the laboratory toxicity tests can be validated in natural systems or surrogates thereof. In order for the validation process to be scientifically justifiable, the predictions made must be explicitly stated, the laboratory evidence for the predictions provided, and the evidence for validation in natural systems, including methodology, must be explicitly stated. At present, there is no standard method for validation of predictions based on laboratory tests in natural systems. There is repeated reference to natural systems because the general public and legislators expect the EPA and its counterparts elsewhere in the world to protect natural systems. If something else is being protected, this should be explicitly stated and the responsibility for natural systems assigned to some other organization other than EPA. If EPA is indeed protecting natural systems, then validation should occur in a natural system or a surrogate.

The word "validate" is defined in *Webster's International Dictionary,* third edition, as "to confirm the reliability of a method." The Wall and Hanmer statement that the biological tests proposed to avert water pollution were validated and published was clearly intended to enhance their credibility. Because no references were appended to that crucial sentence, a curious omission for a feature publication in a professional peer-reviewed journal, validation seems to be a crucial issue, though undocumented. The references included in the literature cited for the Wall and Hanmer article (1987) show no evidence of predictions of community or ecosystem response or validation at higher levels of biological organization than single species, even if the predictions were not explicitly stated. If the regulations of the EPA are to have scientific credibility, then this process must be explicitly stated and published in a peer-reviewed ecological journal.

A few illustrative predictions and validations follow. It seems reasonable to start with a prediction that should be important to regulatory agencies, ecologists, and knowledgeable citizens. The prediction is that the important ecosystem level processes of energy flow and nutrient spiraling are unaffected by concentrations of the test chemical or mixture of chemicals that result in no adverse biological effects using test methodologies and safety factors espoused in the Wall and Hanmer (1987) article. Methodology for energy transfer is available in peer-reviewed literature (Odum, 1982) as it is for nutrient spiraling (Webster et al., 1975). If there is no deviation in these parameters from the nominative state (as defined by Odum et al., 1979) in either natural systems or surrogates thereof (Odum, 1984), then the prediction can be considered valid. At the lowest level of biological organization, one might predict that a recreationally important species (e.g., the large mouth bass or channel catfish) suffered no deleterious effects (e.g., lethality, reproductive success, growth rate) in natural systems at the same concentrations of the chemical or mixture of chemicals deemed to have no adverse biological effects by the biological tests proposed by Wall and Hanmer (1987).

Biological tests espoused by the EPA for widespread general use, such as those in the Wall and Hanmer (1987) article, should have not only validation in the way just described but additional validation to show that they are applicable in the wide range of ecosystems found in the United States. Even in a state small in area such as Rhode Island or Delaware, the aquatic ecosystems are not homogeneous.

18.4.2 Phase 2: General Applicability

Some of the tests, such as those just discussed, are intended to be used in a wide variety of ecosystems in quite different geographic areas and climatic zones, as well as in a variety of water qualities inhabited by different species. Once the predictions have been checked in one area to show that there is, in fact, some merit to them, it would be well to determine whether they are equally meritorious in ecologically dissimilar systems. Some modification of the methodology might be essential because, for example, energy flow in a first-order headwater stream would almost certainly be quite different from that of an estuary. Similarly, nutrient cycling might follow different paths in different ecosystems. Thus in designing these tests, sound ecological principles must be involved.

The National Research Council volumes (1981, 1986) provide some guidelines for the types of aquatic ecosystem attributes that might be useful and have a number of literature citations that provide additional evidence. Bormann (1985) provides some good examples of forested ecosystem end points. Again, it is important to note that there is no need to test every ecosystem, but rather a random selection of ecologically different ecosystems to determine the efficacy of the test in predicting responses in each of these ecosystems.

18.4.3 Phase 3: Site-specific Validation

As Mount (1983) indicates, the cost of biological testing is small compared to the cost of constructing and operating a waste treatment plant for a major chemical industry. Similarly, the cost of biological testing is considerably less than some of the legal settlements of recent years. Therefore, site-specific validation makes good sense because management decisions based on inappropriate information are likely to be very expensive, either because the system is overprotective (resulting in needless waste treatment expenditures) or underprotective (resulting in fish kills and other unfortunate events).

It is important to note that this is a validation calling for considerable judgment on the part of a trained ecologist (Barrett, 1985). That is, what are the important characteristics or qualities reviewed from a holistic management perspective (Risser, 1985) of the local ecosystem that deserve protection, and are the predictions of the single-species tests sufficiently accurate to make management decisions?

This validation need not be carried out each time the laboratory tests are performed but only often enough to validate that the predictions made from them apply to this particular ecosystem. If they err consistently in a particular direction (either high or low), one might still make the decision to use this information but could develop a "K factor" to compensate for the error. Alternatively, one might select a test giving more accurate predictions that is more easily and readily validated in that particular system. It is probably better to do the latter than the former because, if there is no correspondence between the laboratory and field results, it is unlikely that the error will be constant. At the very least, some evidence should be provided to show that the assumption of constant error has a basis in the evidence available.

18.5   MONITORING

Even if the predictive tests turn out to be entirely correct and the validations of them are scientifically persuasive, there are still a number of situations where continuous biological monitoring is justified. A primary consideration is that change is characteristic of all natural systems, and many change quite rapidly. Some changes are quite predictable; others are likely to be unexpected (Loucks, 1985). Additionally, there are episodic events so unusual or infrequent that they are unlikely to be incorporated in a meaningful way into either the predictive testing or the validation process (Holling, 1985). These include drought, floods, aberrant temperatures, atmospheric deposition, fire, and the like (Vogl, 1980).

The tests provided by Wall and Hanmer (1987) focus on specific point source discharges, and there is clear-cut evidence that non-point source discharges may be responsible for at least half the pollutants in some aquatic ecosystems. In addition to the non-point source discharges, the NPDES permitting system does not adequately account for the aggregate ecological effect of non-point source discharges.

Biological monitoring is designed to provide an information feedback loop from both natural ecosystems and waste lines entering them, to indicate that something has gone wrong. The management decisions will be more effective if the information is generated rapidly (Cairns et al., 1970a). More information on the field of biological monitoring is given in a series of six articles in *Water Research* (Buikema et al., 1982; Cairns and van der Schalie, 1980; Cairns, 1981a; Cherry and Cairns, 1982; Herricks and Cairns, 1982; Matthews et al., 1982). A substantial body of the older literature is covered in this series. More recent literature can be found by using the keywords in any of the customary literature search programs.

Hellawell (1978) defines *biological monitoring* as "surveillance undertaken to ensure that previously formulated standards are being met." In short, there is an intent to invoke management intervention when the monitoring system signals that there is a marked deviation from the expected. In the United States and even in parts of Europe, the term *biological monitoring* is used to cover not only Hellawell's definition but also two other definitions used by Hellawell (1978). These are *surveillance*, "a continued programme of studies systematically undertaken to provide a series of observations in time," and *survey*, "an exercise in which a set of standardised observations (or replicate samples) is taken from a station (or stations) within a short period of time to furnish qualitative or quantitive descriptive data."* The more precise definitions of Hellawell (1978) should be more generally used.

If corrective action is to be taken, time of information generation is the main determinant in selecting the end point to be used. This should be selected to provide the least number of either false positives or false negatives. That is, the system should not signal malfunction when it, in fact, is functioning normally, and it should not signal that it is functioning normally when it is not. Therefore, the final selection will be a compromise between the speed of information generation and the tolerable number of false positives or negatives.

Biological monitoring should be used: (1) whenever the environmental concentration of a chemical is very close to the biologically adverse response threshold of a chemical, (2) when changes in environmental quality will markedly affect the biological response even if the concentration of the chemical remains constant, (3) when interactions with pollutants from upstream or elsewhere in the system are likely to interact synergistically with the waste being discharged, (4) when the ecosystem involved is very fragile with a long recovery time if placed into disequilibrium, (5) when the quality or quantity or both of the waste vary markedly, or (6) when there is any reason to distrust either the accuracy of the predictions based on laboratory tests or their validations in the field.

If biological monitoring is used as part of a quality control system to fit

---

* Definitions reprinted with kind permission of the Water Research Centre, United Kingdom, and Dr. J. M. Hellawell, Severn Trent Water, United Kingdom.

the definition of Hellawell (1978), then time of information generation becomes a major determinant. An earlier article (Cairns et al., 1970a) discusses this at some length for both industrial and regulatory decision making.

The case for managing an entire aquatic ecosystem, as opposed to fragments thereof, has also been discussed (Cairns, 1975). It is important to note that, even on a single monitoring site for a single waste discharge, the signal that a deleterious material has entered the waste line provides three management options: (1) shunt the waste to a holding pond, (2) recycle the waste for additional treatment, and (3) shut down the plant or scale down operations until the signal disappears or the problem has been corrected. Managing on a system-wide basis adds two additional management options: (4) flow augmentation—release of water from an upstream reservoir (the temporary solution to pollution is dilution); and (5) because the wastewater is the aggregate of all waste discharges affecting the biota in a receiving system under many circumstances, it might be possible for all industries on the system to reduce temporarily their discharges to offset temporary malfunction of one discharger. Then either reciprocity could be called into play, with the expectation that each industry on the system would someday need such help, or certain charges could be levied, representing the additional cost of withholding wastes or treating them beyond regulatory requirements.

Although direct interfacing of biological information systems with computers has not proven popular despite the increasingly reduced costs of small computers and the ease of doing so, the methods have been around for some time for both early warning in-plant monitoring systems using fish behavior (Lubinski et al., 1977), respiratory rhythm (Cairns et al., 1970b), or even for counting complex communities (Cairns et al., 1972). A more recent paper (Doane, 1984) has shown that respiratory rhythm is not only a good parameter for direct computer interfacing, but is markedly more sensitive than a variety of other parameters that n.ight be used to assess pollutional stress on fish. The use of minicomputers will probably not become common until the major regulatory focus is on the condition of the receiving system [as evidenced by direct measurements on it instead of extrapolation to the system from three selected species that may or may not be key species in the system (Wall and Hanmer, 1987)]. There are a number of specially designed statistical methods for using minicomputers (Sydor et al., 1982; Thompson et al., 1982). Therefore, the hardware, software, general methodology, and so on are in place for systems for integrated resource management, but the regulatory authorities are still focused on the single-species, single-discharge approach rather than using the system itself as the main source of evidence for environmental quality control.

In an era where systems management is a sine qua non in every industrial society on earth, it is curious that the archaic fragmented approach of quality control is still in practice for the environment. Probably the reason for this is that the heads of most regulatory agencies are lawyers and sanitary engineers rather than scientists accustomed to ecosystem studies. Also, it is quite evident from an examination of the *Journal of the Water Pollution Control Federation* that the primary focus of that journal is on waste treatment techniques. When there is an article on ecosystem protection, it is closely tied to single-species laboratory toxicity tests rather than based on receiving system standards.

18.6   ASSIMILATIVE CAPACITY

The pivotal issue in ecological consequence assessment is whether ecosystems have an assimilative capacity for certain types of anthropogenic wastes. Included in the definition of anthropogenic wastes are the non-point source discharges due to societal activities, acid rain, municipal and industrial point source discharges, storm drainage from urbanized areas, mobile point sources such as automobiles, and the like. *Assimilative capacity* is defined as that range of concentrations of a substance or mixtures of substances that will cause no deleterious effects upon the receiving ecosystem (i.e., the ecosystem into which the materials intrude). Odum et al. (1979) have produced a very interesting paper that discusses the fact that a substance might act as a subsidy to an ecosystem under certain circumstances but under others might become a stress. Readers interested in this subject would do well to read that particular paper. Those interested in an exchange in the professional journals on this subject should examine the article by Campbell (1981) in which he criticizes two articles on assimilative capacity (Cairns, 1977a,b) and one of Westman's (1972) on another subject. The response to the Campbell article (1981) has also been published (Cairns, 1981b; Westman, 1981). It is, nevertheless, astonishing that there is so little discussion in the open literature on this important question, because it should be central to the entire regulatory strategy if that strategy is indeed designed to protect the environment or ecosystems.

Because the determination of assimilative capacity is an ecosystem level phenomenon and because it is pivotal to the development of an integrated resources management policy, it is curious how little discussion it receives. If one takes the position as Campbell (1981) did, for at least one substance, that all introductions into natural systems cause change and all change not of natural origin is deleterious, we still cannot avoid determining the degree of change. We cannot stop our industrial technology upon which the earth's very sizable population depends without causing ecosystem disruptions that will probably exceed those that occur when the system is under control. If one takes the position of Odum et al. (1979) that there are a series of gradients, some acting as subsidies, others as stresses, or the position of May (1977) that there are multiple thresholds, we must then determine which of these is a good indicator of ecosystem health or condition and keep discharges or environmental concentrations of chemicals at levels below those that perturb or displace ecosystem structure and function.

It may well be that the question of assimilative capacity has been so studiously ignored because it focuses attention on the ecosystem rather than on waste treatment plant operation and single-species toxicity tests. This is quite apparent in the feature article of Wall and Hanmer (1987), where the word "ecosystem" does not even appear, although the article implies that ecosystems are being protected.

18.7 LEGISLATION COVERING HAZARDOUS SUBSTANCES

The U.S. Toxic Substances Control Act (TSCA) provides that no person may manufacture a new chemical substance or process a chemical substance for a new use without obtaining clearance from EPA. One of the main purposes of TSCA was to establish a procedure for estimating the hazard to human health and the environment before widespread use of a new chemical occurs. Although the legislation was passed many years ago, it is either not being implemented effectively or else the word "environment" is being interpreted as something other than the ecosystems that most of us associate with the environment.

Essentially what legislation attempts to do in the United States and elsewhere is to exert control on presumably deleterious materials entering the environment with the expectation that, if this is done effectively, the environment will be in satisfactory condition. There is virtually no direct ecologically sound evidence to support this assumption. Furthermore, there is no national inventory of ecosystem condition that will enable us to know the rate at which ecosystems are deteriorating.

The work of John Harte (personal communication) of the University of California at Berkeley and his colleagues and students carried out at the Nature Conservancy's Mexican Cut Preserve, which is supervised by Rocky Mountain Biological Laboratory (RMBL), shows that acid rain is having a deleterious effect upon some of the aquatic ecosystems in the preserve. It is worth noting that RMBL itself is at 9500 feet near Crested Butte, Colorado, and access to the Mexican Cut Preserve from RMBL was blocked for all of the summer of 1986 by an avalanche in Schofield Pass. Therefore, for practically all the summer of 1986, access was by skis, snowshoes, or later on foot. It could hardly be called a highly industrialized and highly traveled area.

One might presume that such a remote area with such difficult access would be in good condition. In fact, casual observations even by trained ecologists could easily lead to this conclusion. However, detailed studies by Harte and his colleagues (1985) have shown a decline in the reproductive success of a resident amphibian population that is sensitive to episodic acidification that occurs at snow melt. The only thing unique about this situation is that some rather subtle, but now well-documented, ecological effects have occurred that would not have been apparent unless there had been careful studies carried out by trained professionals.

If such things can happen on a Nature Conservancy tract a long distance from industry that would escape detection unless skilled professionals examine the system over a period of time, it seems highly probable that systems closer to various industrial sources, for example, might be undergoing equal or greater stress that is undetected owing to lack of thorough observation.

It is worth emphasizing that Mexican Cut has been visited sporadically by some of the country's leading ecologists over a long period of time. Eventually, they would undoubtedly have made the connection between cause and effect that Harte has made, but not with casual observations.

Legislation based on ecosystem condition is needed in order for the EPA to manage the environment so that its quality is maintained within certain parameters. Fragmenting the approach or using indirect evidence simply will not work, given the multiple sources of pollutants. Furthermore, the burden of proof should be on the dischargers to show that they are not damaging the receiving system. They should be permitted to do this in the most ecologically and scientifically sound way. They should not be forced to use species not indigenous to the system or to use standard tests recommended by EPA that are mechanically followed and have little or no ecological significance. All of this will be more costly than present testing. Continuing on our present course will be environmentally costly and the damage will be difficult, perhaps impossible, to repair.

18.8 INTEGRATED RESOURCE MANAGEMENT

Once the methodology is in place, we can begin practicing integrated resource management, which is one of the goals of ecological engineering. It seems to us that the next 10 years will bring a concerted effort to use the results of hazard evaluation and aquatic toxicology, as well as other types of environmental and ecological information, in the process of integrated resource management. It is important to note that the purpose of hazard evaluation is to prevent damage—the purpose of integrated resource management is to get optimal performance from the system. This requires different information and a different management strategy.

Although much is written about the management of large natural systems such as drainage basins, little effective action has been taken. Institutional arrangements fragment the responsibility so that federal and state regulatory agencies and other organizations are often at odds and sometimes in direct conflict in their attempts to optimize that portion of resource management assigned to them. Although some quite predictable infighting over "turf" will probably occur, there has been a heartening indication of a willingness by the heads of governmental agencies to address this problem and a concomitant indication of an awareness of the need.

Integrated resource management will require that ASTM and other organizations developing methodology now give attention to how their standard methods can be used for systems level management activities. At the same time, there will be a need to determine how single-species laboratory toxicity tests and other types of conventional or customary information can be used effectively in systems level management. Institutional arrangements will be considerably more difficult to modify because responsibility once fragmented is exceedingly difficult to reassemble. In addition, much new information will be required, both scientific (e.g., predicting the environmental outcome of particular courses of action) and institutional (e.g., setting up a systems level quality control organization involving industry, government, and academic institutions). None of this will be successful unless the degree of mutual respect and trust among the three groups just mentioned is increased.

Integrated resource management will require that some of the enormous resources now being spent on litigation be redirected into problem solving. In addition, some of the prescriptive regulations, which now pit the regulators against the regulated in such ways that obvious solutions to simple problems cannot be implemented, must be changed. Because there is very little scientific and institutional information about integrated resource management, both of which will be essential for systems level quality control, and because funding to generate this information is not likely to come from traditional sources, some acknowledgment of the value of generating this information must be incorporated in the planning.

For example, if the three groups mentioned can get together on a systems level study to provide new information, rewards in the form of tax breaks, subsidies, variances in the statutory standards and the like should be awarded in proportion to the value of the information being generated. Although all of this sounds utopian, we have seen the major societal, industrial, and regulatory changes that were possible when the energy crisis occurred in the early 1970s, and it seems at least possible that similar redirection of the old way of doing things may be possible for integrated resource management as well.

It is essential that those engaged in environmental assessment and related activities become much more active in developing integrated resource management approaches. Unless this is done soon and skillfully, even the finest standard methods will not be used effectively. Without integrated resource management, information will be fragmented, often inappropriate for the decision being made, or in apparent conflict with other equally valid information. In short, those engaged in environmental surveillance, monitoring, and hazard prediction must pay more attention to the structure of the management systems into which their methods will flow. A method can be properly used and still generate inappropriate information if the context in which the information is generated is unsound.

18.9 CASE STUDY

18.9.1 Background

Genetic engineering of microorganisms is currently being used to develop organisms capable of performing specific commercially desired tasks, including those of potential use in ecological engineering. Genetically engineered micoorganisms (GEMs) will have such diverse ecotechnological applications as fixing nitrogen, ore leaching, and degrading hazardous wastes. Despite this potential usefulness, however, there is a lack of development of adequate risk assessment methodologies for those GEMs that are deliberately introduced into the environment (Gillett, 1986; Cairns and Pratt, 1986a,b).

Such environmental release of GEMs as originally banned under National Institutes of Health (NIH) guidelines (Karny, 1987) and even now have been limited to small-scale trials, owing to lack of knowledge about their impact on ecosystems and increased public awareness and concern over the release of novel organisms. The first approved release, of *Pseudomonas syringae* and *P. fluorescens*, occurred in the spring of 1987 and is expected to progress without incident (Crawford, 1987). It is generally agreed, however, that these releases will not reduce future legal action against additional experiments in the environment (Crawford, 1987). Established scientists have presented plausible and, often, conflicting arguments over the types of risks, if any, the environmental release of GEMs will create (Alexander, 1985).

Before examining a case history, environmental implications of such a GEM release must be considered so that a decision to release or not to release a GEM is based upon scientific ecotechnology. A number of factors must be considered before a GEM is released into the environment in large quantities or at a number of different locations. Before considering potential effects of a particular GEM upon the environment, survival of the GEM must be determined. If it cannot survive, it will probably do little permanent harm (Alexander, 1985). Microbes will not survive for many reasons, including excessive solar radiation, changes in pH, starvation, predators, and various abiotic factors (Alexander, 1985). A GEM, produced in a laboratory, encountering any one unfavorable condition may die; it is at a competitive disadvantage when compared to indigenous bacteria. Therefore, success of an organism under laboratory conditions does not ensure success in natural ecosystems. Conversely, the ability to contain and control species in the laboratory does not preclude their establishment, survival, and adverse effects outside the laboratory. Evaluation of survival, colonization potential, and persistence of a GEM is critical in initial testing.

Survival and/or colonization of microcosm or environmental habitats by GEMs can be ascertained by several detection methods, each having advantages and disadvantages. One of the most frequently used methods is the plate count. Although the method has a high potential and, too often, realization for error (Hamilton, 1979), it continues to be widely used because of relatively low cost and time investment, and because, by using various media, certain groups of bacteria can be selected for. Although no medium is truly selective all of the time (Mills and Bell, 1986), it does allow enumeration of many different groups with good replicability between samples. Use of various media with one or more antibiotics added also allows detection of relatively low numbers of released GEMs, although indigenous organisms that acquire the resistance gene through mutation or directly from the GEM may severely limit detection (D. R. Orvos, unpublished data). Development of detection methods that do not require antibiotic use is underway.

If the GEM does indeed survive, even if transiently, then possible effects upon the ecosystem must be considered. Although the potential number of ecological parameters that may be affected by the introduction of a novel species is great, several are important. The first possibility is that GEMs may cause alterations in community structure at some degree of complexity, especially at the microbial level. Effects of the GEM upon indigenous plants, fungi, and animals must be considered. Animals such as macroinvertebrates may also serve as vectors for the GEM. Abnormal alteration in species numbers or diversity may have adverse ecological consequences.

When viable, GEMs may compete with and displace native populations, thereby affecting the vital processing of materials in soils and sediments (Cairns and Pratt, 1986c). Complex ecological effects resulting from the displacement of native bacterial populations may occur, especially when the GEMs are derived from common free-living taxa (Liang et al., 1982). Even if structural effects are below detection, functional effects such as alteration of nutrient cycling or enzyme activities may be present.

Finally, potential genetic modification of the GEM after release as well as its ability to disseminate its own recombinant DNA must be ascertained. Bacteria, in particular, may be promiscuous and transfer of genetic information to other species is common, especially because the species concept, as applied to bacteria, is limited.

Data from all of these experiments must then be analyzed using some form of ecological engineering risk assessment. To date, no one satisfactory scheme exists. Careful development of relevant testing schemes that may be standardized must be done because it is not possible to do every test that might possibly be significant or to explore all possibilities of GEM dispersal. Ecotechnological assessment schemes may be derived from those used for chemicals (Fig. 18.1). Developing an assessment protocol for GEMS, however, is different from that developed for chemicals because, unlike toxic chemicals whose toxicities can be evaluated biologically and whose fates may be modeled based on physicochemical processes, the effects of GEMS may be more difficult to assess (Cairns and Pratt, 1986c).

Once data have been collected and subjected to ecotechnological assessment, provisions for field validation must be devised. Validation of a test

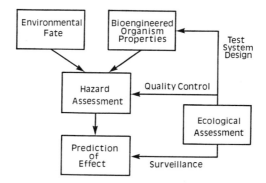

Figure 18.1 Simple hazard assessment scheme for environmental impact of genetically engineered microorganisms.

system is the "ability to predict the relationship between the response of the artificial laboratory system and the natural system" (Cairns, 1986b). Although still controversial in the ecotechnological assessment of chemicals, validation presents even more severe problems when applied to GEM studies in that the microbe is often not allowed, by law, to be released into the environment. Therefore, procedures may be developed that use surrogate strains, such as wild-type antibiotic-resistant spontaneous mutants, for the field validation experiments.

The *validation protocol* (a term used herein to mean an orderly and systematic gathering of certain types of evidence) should consist of three major components for each GEM: (1) colonization potential, (2) persistence in natural ecosystems, and (3) potential for causing ecological disequilibrium.

Routine ecological assessment methods will have to be selected for use by biotechnology industries to ensure the soundness of future economic development of biotechnology products and to provide necessary information to regulatory agencies. Current case-by-case criteria will soon become useless if the present expansion of biotechnology industries continues. Such testing will have to be scientifically justifiable and have general transferability to several potential receiving ecosystems. Testing schemes will have to be easily interpretable by industry and regulatory analysts, as well as producing few false negative or positive results. Microcosm-based tests examining critical functional attributes of ecosystem compartments will have general applicability to a variety of ecosystems.

18.9.2 Study Description

A genetically engineered form of *Erwinia carotovora* subspecies *carotovora* was used as a model for GEM risk assessment. *Erwinia carotovora* is an important plant pathogen that annually results in the loss of over $100 million from agricultural soft rot. *E. carotovora*, a member of the Enterobacteri-

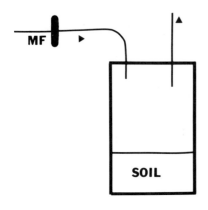

Figure 18.2 Terrestrial microcosm design. MF—Membrane filters for sterilizing air.

aceae, is often found in both terrestrial and aquatic environmental samples (McCarter et al., 1984; Jorge and Harrison, 1986), although it is normally associated with decaying plant tissue (Perombelon and Kelman, 1980). The regions of the genome producing the pectate lyase enzymes required for soft rot have been well documented (Allen et al., 1986). Therefore, an *Erwinia* with these enzyme-coding DNA sequences removed may have use as a biological control agent.

In this case history, we examine key characteristics of a strain of *Erwinia* that has been genetically engineered to remove some of these enzyme-coding sequences to determine its potential safety if it is ever actually released into the environment as a biological control agent. Two genetically engineered strains of *Erwinia* were used in this analysis—L-827, an ampicillin-resistant pathogenic strain, and L-833, a kanamycin-resistant, reduced-pathogenic strain. In addition, a rifampin-resistant spontaneous mutant, O-112, was used to monitor the activity of the wild-type *Erwinia*. This organism has growth and biochemical characteristics identical to that of wild-type *Erwinia* while allowing easy detection because of its antibiotic resistance. A selective medium for *Erwinia*, crystal violet–pectate medium (CVP), was also used to monitor populations.

18.9.3   Survival of Genetically Engineered *Erwinia*

Erwinia were released into terrestrial microcosms containing soil from a nearby agricultural area. Microcosm design is shown in Figure 18.2. *Erwinia* strain L-827 (ampicillin-resistant) survival for more than 60 days in soil is shown in Figure 18.3; its concentration declined three orders of magnitude during this time from the initial concentration of about 10^8 colony-forming units (CFU)/g dry soil. It is interesting to note that most of the decline occurred in the first 30 days. Survival of genetically engineered, kanamycin-resistant (L-833) *Erwinia* and the rifampin-resistant, spontaneous mutant (O-112) is shown in Figure 18.4. Both strains decline about two orders of

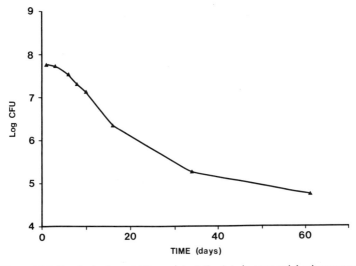

Figure 18.3 Survival of ampicillin-resistant *Erwinia* in terrestrial microcosms.

magnitude at the same rate over 15 days, indicating an inability to compete with indigenous organisms and/or abiotic conditions. The experiment was terminated, however, for reasons discussed below. Survival in aquatic microcosms is still being evaluated at this time, but preliminary results indicate similar declines with migration of the GEM into the sediment layer. Aquatic microcosms were collected from nearby lakes so as to contain sediment and placed into environmental chambers to duplicate ambient daylight and tem-

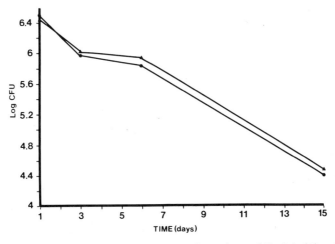

Figure 18.4 Survival of kanamycin-resistant genetically engineered *Erwinia* (▲) and rifampin-resistant mutants (●) in terrestrial microcosms.

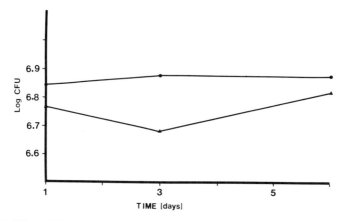

Figure 18.5 Effect of kanamycin-resistant GEM (▲) and rifampin-resistant mutant (●) on actinomycete populations.

perature. The design for aquatic microcosms was similar to that of terrestrial microcosms and allowed for both water and sediment to be included.

18.9.4   Structural Effects upon Indigenous Microbes

The effects of genetically engineered *Erwinia* and the mutant upon indigenous populations of Actinomycetes (Figure 18.5), total Gram-negative bacteria including *Erwinia* (Figure 18.6), and total populations of soil bacteria (Figure 18.7) revealed no statistically significant differences between the

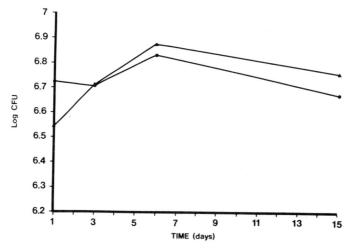

Figure 18.6 Effect of kanamycin-resistant GEM (▲) and rifampin-resistant mutant (●) on total populations of gram-negative bacteria.

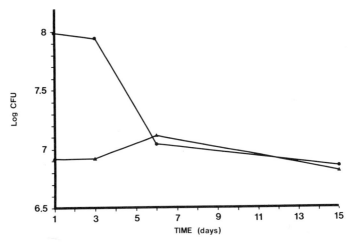

Figure 18.7 Effect of kanamycin-resistant GEM (▲) and rifampin-resistant mutant (●) on total populations of indigenous bacteria.

GEM and mutant. Some fluctuations in *Pseudomonas* sp. were observed (data not shown), but these studies are being repeated with increased replicates. Populations of other genera and functional groups (e.g., proteolytic bacteria) are being incorporated into present and future studies.

18.9.5 Functional Effects

Effects of GEM addition upon nutrient cycling, soil enzyme activities, redoxocline, production/respiration ratios, and substrate utilization are planned. Data are inconclusive at this time.

18.9.6 Potential Genetic Exchange

Indigenous bacteria were shown to demonstrate kanamycin resistance, the marker antibiotic of the GEM, 21 days after study initiation in two separate experiments. Initial attempts to demonstrate genetic exchange by DNA:DNA hydridization have not been successful; however, technical difficulties may have led to false negatives and these studies are continuing. It is also possible that the antibiotic resistance was the result of spontaneous mutation by the soil indigenous bacteria even though no external antibiotic stress was applied in the microcosms. Both conjugation and transduction of plasmid DNA have been demonstrated in *Pseudomonas* spp. in aquatic habitats (Genthner et al., 1988; Saye et al., 1987); we have not yet determined this for *Erwinia*.

18.9.7   Case Study Conclusions

Erwinia carotovora, genetically engineered to delete the genes for pathogenicity and insert genes for antibiotic resistance, appears to be at a competitive disadvantage in the environment. This has been shown for other GEMs, such as *Pseudomonas fluorescens* (Crawford, 1988). No functional effects studies have yet been completed but may yield important data. Because indigenous bacteria are expressing resistance to the marker antibiotic, DNA transfer is possible; this has not been demonstrated. Therefore, based upon preliminary data, it appears that *Erwinia* will have minimal environmental impact and may be useful in ecological engineering. However, more data, including field validation studies, are needed before actual release.

18.10   FUTURE APPLICATION OF ECOTECHNOLOGY TO GEM RELEASE

To prevent environmental damage from GEM release, future testing procedures must have a broad scope to ascertain realistically the probability of such damage before environmental or economical damage is incurred by either private or public factions. Ecological engineering studies, using microcosm, greenhouse, and field data, will generate basic ecological information concerning the fate of GEMs and their recombinant DNA in nature.

Ecotechnological assessment has at least two roles in the regulation of GEMs intended for environmental release (Cairns and Pratt, 1986c). First, ecosystem surrogates, such as microcosms and mesocosms, should be designed to evaluate adequately the potential for adverse effects prior to the actual release of the GEM. This design will vary depending on the potential release site and intended use of the GEM. Although the GEM may be designed for use in an agricultural area, testing should include material from areas of potential infiltration from an accidental release. The second role of assessment is ascertaining if an ecological effect has indeed occurred after the release of a GEM. For ecotechnologists to accomplish this, they must understand the molecular biology and microbial ecology of the GEM, the mechanisms of transport, the sites involved, and pathogenicity, as well as ecological principles similar to those encountered when ascertaining impact from a chemical release. Few applied ecologists presently have this training, and few institutions have the necessary resources to even offer it (Gillett, 1986). Ecotechnology must be able to adapt to the ever-changing potential problems posed by humans.

18.11   FUTURE RESEARCH NEEDS

As Macek (1982) notes, the amount of data generated by environmental toxicologists is overwhelming. He is also correct in stating that our knowledge has not greatly increased. This is unlikely to happen by developing more single-species tests, finding more sensitive species, or trying to develop application factors to cover the deficiencies of existing methodology. For the knowledge about ecotoxicology to increase, we must find out as quickly as possible how accurately responses in present laboratory toxicological test systems correspond to events in natural systems at not only the single-species level of biological organization but higher levels such as communities and ecosystems as well. If the field of ecological engineering is to be truly that, we must be able to predict the effect of toxic substances upon ecosystems using parameters, end points, or attributes at this level of biological organization.

It may well be that extrapolations from single-species response to ecosystem response are possible in practical terms, but there are persuasive reasons already in the literature to indicate they are not as reliable as the term ecological engineering implies. It is not yet certain that multispecies toxicity tests will provide a better base for making such extrapolations, but with more environmental realism they should. At the very least, we need to determine this. If the laboratory methods for ecotechnology at various levels of biological organization (e.g., single- or multispecies toxicity tests) are inaccurate, then developing monitoring procedures for natural systems becomes all the more crucial. Under these circumstances, it will be virtually mandatory for ecologists to develop the predictive capability espoused by the well-known Welsh ecologist John Harper (1982). The degree to which environmental engineering remains primarily a laboratory-based profession, becomes primarily a field-based profession, or ideally becomes a highly integrated combination of the two will depend on how satisfactorily predictions made with laboratory-generated evidence can be validated in natural systems.

REFERENCES

Alexander, M. 1985. Ecological consequences: reducing the uncertainties. *Issues Sci. Tech.* Spring:57–68.

Allen, C., V. K. Stromberg, F. D. Smith, G. H. Lacy, and M. S. Mount. 1986. Complementation of an *Erwinia carotovora* subsp *carotovora* protease mutant with a protease-encoding cosmid. *Mol. Gen. Genet. 202:*276–279.

American Public Health Association, American Water Works Association, and Water Pollution Control Federation. 1985. *Standard Methods for the Examination of Water and Wastewater*, 16th ed. Washington, DC, 1134 pp.

Barrett, G. W. 1985. A problem-solving approach to resource management. *BioScience 35:*423–427.

Barrett, G. W., and R. Rosenberg, Eds. 1981. *Stress Effects on Natural Ecosystems.* Wiley, New York, 305 pp.

Bormann, F. H. 1985. Air pollution and forests: An ecosystem perspective. *Bio-Science 35*:434–441.

Brungs, W. A. 1986. Review of *Multispecies Toxicity Testing*. *BioScience 36*:677–678.

Buikema, A. L., Jr., B. R. Niederlehner, and J. Cairns, Jr. 1982. Biological monitoring, part IV: Toxicity testing. *Water Res. 16*:239–262.

Cairns, J., Jr. 1975. Critical species, including man, within the biosphere. *Naturwissenschaften 62*:193–199.

Cairns, J., Jr. 1977a. Aquatic ecosystem assimilative capacity. *Fisheries 2*:5–7.

Cairns, J., Jr. 1977b. Quantification of biological intregrity. Pages 171–187. In L. K. Ballentine and L. J. Guarria, Eds., *Integrity of Water*. U.S. Environmental Protection Agency Office of Water and Hazardous Materials, U.S. Government Printing Office, Washington, DC. Stock #055-001-01068-1.

Cairns, J., Jr. 1980–1981. Guest editorial: sequential versus simultaneous testing for evaluating the hazard of chemicals to aquatic life. *Mar. Environ. Res. 4*:165–166.

Cairns, J., Jr. 1981a. Biological monitoring, part VI: Future needs. *Water Res. 15*:941–952.

Cairns, J., Jr. 1981b. Discussion of: a critique of assimilative capacity. *J. Water Pollut. Control Fed. 53*:1653–1655.

Cairns, J., Jr. 1983a. Are single species toxicity tests alone adequate for estimating environmental hazard? *Hydrobiologia 100*:47–57.

Cairns, J., Jr. 1983b. The case for simultaneous toxicity testing at different levels of biological organization. In W. E. Bishop, R. D. Cardwell, and B. B. Heidolph, Eds., *Aquatic Toxicology and Hazard Assessment: Sixth Symposium, STP 802*. American Society for Testing and Materials, Philadelphia, PA, pp. 111–127.

Cairns, J., Jr. 1984. Factors moderating toxicity in surface waters. In J. Wilson, Ed., *The Fate of Toxics in Surface and Ground Waters, Proceedings of the Second National Water Conference*. Academy of Natural Sciences, Philadelphia, PA, pp. 49–64.

Cairns, J., Jr., Ed. 1985. *Multispecies Toxicity Testing*. Pergamon, New York, 253 pp.

Cairns, J., Jr. 1986a. *Community Toxicity Testing, STP 920*. American Society for Testing and Materials, Philadelphia, PA, 350 pp.

Cairns, J., Jr. 1986b. What is meant by validation of predictions based on laboratory toxicity tests? *Hydrobiologia 137*:271–278.

Cairns, J., Jr. In press. Where is the ecology in toxicology? *Curr. Pract. Environ. Sci. Eng.*

Cairns, J., Jr. and K. L. Dickson. 1978. Field and laboratory protocols for evaluating the effects of chemical substances on aquatic life. *J. Test. Eval. 6*:81–90.

Cairns, J., Jr. and J. R. Pratt. 1986a. Developing a sampling strategy. In B. G. Isom, Ed. *Rationale for Sampling and Interpretation of Ecological Data in the Assessment of Freshwater Ecosystems, STP 894*. American Society for Testing and Materials, Philadelphia, PA, pp. 168–186.

Cairns, J., Jr. and J. R. Pratt. 1986b. Factors affecting the acceptance and rejection of genetically altered microorganisms by established natural aquatic communities.

In T. M. Poston and R. Purdy, Eds., *Aquatic Toxicology and Environmental Fate, Ninth Symposium, STP 921*. American Society for Testing and Materials, Philadelphia, PA, pp. 207–221.

Cairns, J., Jr. and J. R. Pratt. 1986c. Ecological consequence assessment: Effects of bioengineering organisms. *Water Resources Bull. 22:*171–182.

Cairns, J., Jr., and J. R. Pratt, Eds. 1988. *Functional Testing for Hazard Evaluation. STP 988*. American Society for Testing and Materials, Philadelphia, PA.

Cairns, J., Jr. and W. H. van der Schalie. 1980. Biological monitoring, part 1: early warning systems. *Water Res. 14:*1179–1196.

Cairns, J., Jr., K. L. Dickson, R. E. Sparks, and W. T. Waller. 1970a. A preliminary report on rapid biological information systems for water pollution control. *J. Water. Pollut. Control Fed. 42:*685–703.

Cairns, J., Jr., R. E. Sparks, and W. T. Waller. 1970b. Biological systems as pollution monitors. *Res. Develop. 21:*22–24.

Cairns, J., Jr., K. L. Dickson, G. R. Lanza, S. P. Almeida, and D. DelBalzo. 1972. Coherent optical spatial filtering of diatoms in water pollution monitoring. *Arch. Mikrobiol. 83:*141–146.

Cairns, J., Jr., K. L. Dickson, and A. W. Maki, Eds. 1978. *Estimating the Hazard of Chemical Substances to Aquatic Life, STP 657*. American Society for Testing and Materials, Phildelphia, PA, 278 pp.

Cairns, J., Jr., J. R. Pratt, B. R. Niederlehner, and P. V. McCormick. 1986. Simple, cost-effective multispecies toxicity tests using organisms with a cosmopolitan distribution. *Environ. Monit. Assess. 6:*207–220.

Campbell, I. C. 1981. Commentary: a critique of assimilative capacity. *J. Water Pollut. Control Fed. 53:*604–607.

Cherry, D. S. and J. Cairns, Jr. 1982. Biological monitoring, part V: Preference and avoidance studies. *Water Res. 16:*263–301.

Crawford, M. 1988. Monsanto marker shows promise in field test. *Science 239:*972.

Crawford, M. 1987. California field test goes forward. *Science 236:*511.

Dickson, K. L., J. Cairns, Jr., and A. W. Maki, Eds. 1979. *Analyzing the Hazard Evaluation Process*. American Fisheries Society, Washington, DC, 159 pp.

Dickson, K. L., A. W. Maki, and J. Cairns, Jr., Eds. 1982. *Modeling the Fate of Chemicals in the Aquatic Environment*. Ann Arbor Science, Ann Arbor, MI, 413 pp.

Doane, T. R., J. Cairns, Jr., and A. L. Buikema, Jr. 1984. Comparison of biomonitoring and techniques for evaluating effects of jet fuel on bluegill sunfish (*Lepomis macrochirus*). In D. Pascoe and R. W. Edwards Eds., *Freshwater Biological Monitoring*. Pergamon, Oxford, pp. 103–112.

Ehrlich, P. R. and H. A. Mooney. 1983. Extinction, substitution, and ecosystem services. *BioScience 33:*248–254.

Genthner, F. J., P. Chatterjee, T. Barkay, and A. W. Bourquin. 1988. Capacity of aquatic bacteria to act as recipients of plasmid DNA. *Appl. Environ. Microbiol. 54:*115–117.

Giesy, J. P., and P. M. Allred. 1985. Replicability of aquatic multispecies test systems. In J. Cairns, Jr., Ed., *Multispecies Toxicity Testing*. Pergamon, New York, pp. 187–247.

Gillett, J. W. 1986. Risk assessment methodologies for biotechnology impact assessment. *Environ. Management 10*(4):515–532.

Hamilton, R. D. 1979. The plate count in aquatic microbiology. In J. W. Costerton and R. R. Colwell, Eds., *Native Aquatic Bacteria: Enumeration, Activity, and Ecology, STP 695*. American Society for Testing and Materials, Philadelphia, PA, pp. 19–28.

Hammonds, A. S. 1981. *Methods for Ecological Toxicology: A Critical Review of Laboratory Multispecies Tests*. Ann Arbor Science, Ann Arbor, MI, 307 pp.

Harper, J. L. 1982. After description. In E. J. Newman, Ed., *The Plant Community as a Working Mechanism*. Blackwell, London, pp. 11–25.

Hart, W. B., P. Doudoroff, and J. Greenbank. 1945. The Evaluation of the Toxicity of Industrial Wastes, Chemicals and Other Substances to Fresh Water Fishes. Waste Control Laboratory, Atlantic Refining Company, Philadelphia, PA.

Harte, J., G. P. Lockett, R. A. Schneider, H. Michaels, and C. Blanchard. 1985. Acid precipitation and surface-water vulnerability on the western slope of the high Colorado Rockies. *Water Air Soil Pollut. 25:*313–320.

Hellawell, J. M. 1978. *Biological Surveillance of Rivers*. Water Research Centre, Stevanage, England. 332 pp.

Herricks, E. E., and J. Cairns, Jr. 1982. Biological monitoring, part III: receiving systems methodology based on community structure. *Water Res. 16:*141–153.

Holling, C. S. 1985. Resilience of ecosystems: local surprise and global change. In T. F. Malone and J. G. Roederer, Eds., *Global Change*. Cambridge University Press, Cambridge, UK., pp. 228–269.

Jorge, P. E., and M. D. Harrison, 1986. The association of *Erwinia carotovora* with surface water in northeastern Colorado. I. The presence and population of the bacterium in relation to location, season and water temperature. *Am. Potato J. 63:*517–531.

Karny, G. M. 1987. Regulation of the environmental applications of biotechnology. *Environmental Law:* spring. pages 1–3

Kimerle, R. A., W. E. Gledhill, and G. J. Levinskas. 1978. Environmental safety assessment of new materials. In J. Cairns, Jr., K. L. Dickson, and A. W. Maki, Eds., *Estimating the Hazard of Chemical Substances to Aquatic Life, STP 657*. American Society for Testing and Materials, Philadelphia, PA, pp. 132–146.

Kimerle, R. A., W. J. Adams, and D. R. Grothe. 1986. A tiered approach to aquatic safety assessment of effluents. In H. L. Bergman, R. A. Kimerle, and A. W. Maki, Eds., *Environmental Hazard Assessment of Effluents*. Pergamon, New York, pp. 247–264.

Kuiper, J. 1982. The Use of Enclosed Plankton Communities in Aquatic Ecotoxicology. Ph.D. dissertation. Agricultural University, Wageningen, The Netherlands. 256 pp.

Levin, S. A. 1982. *New Perspectives in Ecotoxicology*. Ecosystems Research Center, Cornell University, Ithaca, New York, 134 pp.

Liang, L. N., J. L. Sinclair, L. M. Mallory, and M. Alexander. 1982. Fate in model ecosystems of microbial species of potential use in genetic engineering. *Appl. Environ. Microbiol. 44:*708.

Loucks, O. L. 1985. Looking for surprise in managing stressed ecosystems. *BioScience 35:*428–432.

Lubinski, K. S., K. L. Dickson, and J. Cairns, Jr. 1977. Microprocessor-based interface converts video signals for object tracking. *Computer Design* Dec.: 81–87.

McCarter, N. J., G. D. Franc, M. D. Harrison, J. E. Michaud, C. E. Quinn, I. A. Sell, and D. C. Graham. 1984. Soft rot *Erwinia* bacteria in surface and underground waters in southern Scotland and in Colorado, United States. *J. Appl. Bacteriol. 57:*95–105.

Macek, K. J. 1982. Aquatic toxicology: anarchy or democracy? In J. G. Pearson, R. B. Foster, and W. E. Bishop, Eds., *Aquatic Toxicology and Hazard Assessment: Fifth Conference, STP 766.* American Society for Testing and Materials, Philadelphia, PA, pp. 3–8.

Maki, A. W., K. L. Dickson, and J. Cairns, Jr., Eds. 1980. *Biotransformation and Fate of Chemicals in the Aquatic Environment.* American Society for Microbiology, Washington, DC, 150 pp.

Matthews, R., A. L. Buikema, Jr., J. Cairns, Jr., and J. H. Rodgers, Jr. 1982. Biological monitoring, part II: receiving system functional methods, relationship and indices. *Water Res. 16:*129–139.

May, R. M. 1977. Thresholds and breakpoints in ecosystems with a multiplicity of stable states. *Nature 269:*471–477.

Mayer, F. L., Jr., and M. R. Ellersieck. 1986. Manual of Acute Toxicity: Interpretation and Data Base for 410 Chemicals and 66 Species of Freshwater Animals. Fish and Wildlife Service, United States Department of the Interior, Washington, DC, Resource Publication 160. 508 pp.

Mills, A. L., and P. E. Bell. 1986. Determination of individual organisms and their activities in situ. In R. L. Tate, Ed., *Microbial Autecology.* Wiley, New York, pp. 27–60.

Mount, D. I. 1983. Principles and concepts of effluent testing. In H. L. Bergman, R. A. Kimerle, and A. W. Maki, Eds., *Environmental Hazard Assessment of Effluents,* Pergamon, London, pp. 61–65.

National Research Council. 1981. *Testing for Effects of Chemicals on Ecosystems.* National Academy Press, Washington, DC, 103 pp.

National Research Council. 1986. *Ecological Knowledge and Environmental Problem-Solving.* National Academy Press, Washington, DC, 388 pp.

Odum, E. P. 1984. The mesocosm. *BioScience 34:*558–562.

Odum, E. P. 1985. Trends expected in stressed ecosystems. *BioScience 35:*419–422.

Odum, E. P., J. T. Finn, and E. H. Franz. 1979. Perturbation theory and the subsidy-stress gradient. *BioScience 29:*349–352.

Odum, H. T. 1982. Pulsing, power and hierarchy. In W. J. Mitsch, Rammohan K. Ragade, Robert W. Bosserman, and John A. Dillon, Jr., Eds. *Energetics and Systems.* Ann Arbor Science, Ann Arbor, MI, pp. 33–59.

Perombelon, M. C. M., and A. Kelman. 1980. Ecology of the soft rot *Erwinia. Ann. Rev. Phytopathol. 18:*361–387.

Rand, G. M., and S. R. Petrocelli, Eds. 1985. *Fundamentals of Aquatic Toxicology.* Hemisphere, New York. 666 pp.

Risser, P. G. 1985. Toward a more holistic management perspective. *BioScience 35:*414–418.

Saye, D. J., O. Ogunseitan, G. S. Sayler, and R. V. Miller. 1987. Potential for transduction of plasmids in a natural freshwater environment: effect of plasmid donor concentration and a natural microbial community on transduction in *Pseudomonas aeruginosa. Appl. Environ. Microbiol. 53:*989–995.

Sloof, W. 1983. *Biological Effects of Chemical Pollutants in the Aquatic Environment and Their Indicative Value.* University of Utrecht, The Netherlands, 191 pp.

Sydor, W. J., W. R. Miller, III, J. Cairns, Jr., Jr., and D. Gruber. 1982. Use of box and line plots to assess fish ventilatory behavior in biological monitoring systems. *Can. J. Fish. Aquat. Sci. 39:*1719–1722.

Taub, F. B. 1985. Toward interlaboratory (round-robin) testing of a standardized aquatic microcosms. In J. Cairns, Jr., Ed., *Multispecies Toxicity Testing.* Pergamon, New York, pp. 165–186.

Tebo, L. B., Jr. 1983. Effluent monitoring: historical perspective. In H. L. Bergman, R. A. Kimerle, and A. W. Maki, Eds., *Environmental Hazard Assessment of Effluents.* Pergamon, London, pp. 13–31.

Thompson, K. W., M. L. Deaton, R. V. Fourtz, J. Cairns, Jr., and A. C. Hendricks. 1982. Application of time-series intervention analysis to fish ventilatory response data. *Can. J. Fish. Aquat. Sci. 39:*518–521.

Vogl, R. J. 1980. The ecological factors that produce perturbation-dependent ecosystems. In J. Cairns, Jr., Ed., *The Recovery Process in Damaged Ecosystems.* Ann Arbor Science, Ann Arbor, MI, pp. 63–94.

Wall, T. M., and R. W. Hanmer. 1987. Biological testing to control toxic water pollutants. *J. Water Pollut. Control Fed. 59:*7–12.

Webster, J. R., J. B. Waide, and B. C. Patten. 1975. Nutrient cycling and the stability of ecosystems. In F. G. Howell, J. B. Gentry, and M. H. Smith, Eds., Mineral Cycling in Southeastern Ecosystems. ERDA CONF-740513. Reprinted in H. H. Shugart and R. V. G. Neill, Eds., *Systems Ecology.* Benchmark Papers in Ecology. Dowden, Hutchinson, and Ross, Stroudsburg, PA, pp. 1–27.

Westman, W. E. 1972. Some basic issues in water pollution control legislation. *Am. Sci. 60:*767–773.

Westman, W. E. 1981. Discussion of: a critique of assimilative capacity. *J. Water Pollut. Control Fed. 53:*1655–1666.

Woltering, D. M. 1985. Population responses to chemical exposure in aquatic multispecies systems. In J. Cairns, Jr., Ed., *Multispecies Toxicity Testing.* Pergamon, New York, pp. 61–75.

19

ECOLOGICAL ENGINEERING AND BIOLOGICAL POLISHING: METHODS TO ECONOMIZE WASTE MANAGEMENT IN HARD ROCK MINING

Margarete Kalin

Boojum Research Limited, Toronto, Ontario, Canada

19.1 INTRODUCTION

Mining has been one of the early interactions of humans with the environment, starting in the Bronze and Iron ages, circa 6000 and 1000 BC, respectively. Although environmental phenomena related to mining have been recognized for at least the past 100 years (Paine, 1987), it is only since the 1930s that awareness has increased to the stage where it is understood and accepted that mismanagement of our natural resources will result in dire consequences and that stringent safeguards are mandatory. Specifically, the effects of our interaction with the environment, both present and long-term, as we reap from the earth those natural resources it yields, are only now receiving the degree of attention they deserve.

As a result of this welcome awareness, waste management has become a common activity in base metal and precious metal mining. The reclamation of tailings, waste rock, and overburden are being practiced, and a wealth of information is available. However, waste management also includes the reclamation of water, the quality of which can be degraded as a result of mining.

Wastewater treatment technologies do exist. However, during operation and particularly following the cessation of mining activities, the treatment of effluents from the waste management area has become an ever-increasing financial burden for mining operations. New technologies are being sought that could result in a reduction in these expenditures during operation and could improve effluent quality. Most importantly, it is hoped that a system will be developed that will provide environmentally acceptable effluents at the time when mining operations cease.

It is recognized that the achievement of a cost-effective and competitive mining operation, from the commencement of exploration for ore bodies through to the day-to-day efficient operation of the mine, requires a certain expertise and sophistication. However, the environmental effects of these activities are frequently considered peripheral issues. Their importance as an integral part of the planning and development of the mine/mill is not always appreciated. In general, these environmental concerns are usually handled in the following manner:

1. An assessment of the potential impact on the environment of the mining operation is carried out as part of the mine development.
2. All concerned regulatory bodies eventually accept a condition under which the mine and mill will operate.
3. Monitoring programs are established to ensure compliance to the environmental regulations. In most cases, compliance requires an expensive wastewater treatment system.

The consideration of long-term and abandonment plans for mining operations is a more recent requirement of the overall environmental plan. These plans are based on long-term projections, sometimes covering several decades of operation. In reality, it is only when the ore body is approaching the end of its ability to produce an economic ore grade, forcing closedown of the mine and mill, that those measures necessary to protect the environment in the long term can correctly be ascertained and implemented. Part of this reality is the unpleasant specter of the cost of a program to achieve such an environmentally acceptable shutdown, especially when the waste material generates acid. In many cases, the effluents will require water treatment in perpetuity. With the rise in environmental concerns and the increasing awareness of the environmental problems associated with acid generation, the search for economic methods to deal with these concerns, both during operation and at closedown, has intensified.

19.1.1   Ecological Engineering and Biological Polishing: A Novel Approach

The technical basis of the processes that lead to acid generation has been extensively studied. Water, oxygen, and bacteria, all three of which are essential components of the natural environment, are the main factors driving the process (see also Chapter 12). Acid generation is therefore a natural chemical/biological oxidation process that is merely enhanced due to mining activities. In both tailings and waste rock, the surface area of pyritic material available for acid generation is increased, by comparison to the undisturbed ore and host rock, and owing to its change in location to above ground, its exposure to the atmosphere enhances the acid-generating process.

Natural processes, such as the weathering (oxidation) of pyritic rock, are part of the environment and are not the result of some technology developed by humans. In like manner, natural 'treatment processes' also exist, and it is generally accepted that biological activities are the most important components in self-purification of the natural recovery mechanisms of water (Hamilton, 1983). It follows, therefore, when one considers that acid generation is a natural phenomenon that is on occasion enhanced by humans, that its treatment should be possible by natural means. The understanding of these ameliorative processes should lead to their utilization in assisting us in our quest to protect the environment from degradation.

When the problem of acid generation is considered in this manner, a biological/ecological approach to its treatment is self-evident. Despite this, however, acid generation is currently regarded as a major environmental problem related to mining. Its management and control continue to require the expenditure of large sums of money, for which, at present, there are no guaranteed returns.

In an effort to increase our ability to assist the recovery process, research has been aimed at a thorough understanding of the natural recovery processes that occur in a small scale on waste management sites with acid-generating material. The only route to an environmentally and economically viable treatment system may be the expansion of these natural recovery processes to encompass the entire waste site, and the development of those ecologically engineered systems that display the characteristics of a long-term self-maintaining environment.

A brief summary is presented of the development of methods used, along with the basic concepts of ecological engineering and biological polishing as they apply to the establishment of self-maintaining systems for close-out of acid-generating wastes. Details are provided of the progress of various experiments that aim to establish reducing conditions, organic matter production, and promotion of the growth of biological filters that remove metals from the water and retain these substances in the waste management areas.

A description is provided of a project in which ecological engineering methods are being used for the first time on an acid-generating waste management area in northern Ontario, Canada. Using this approach, measures were implemented in 1986 to achieve, in 1990, a walk-away condition for the close-out of a massive sulfide tailings resulting from 10 years of operating a copper/zinc concentrator.

19.2   THE WORKINGS OF THE ECOLOGICAL APPROACH

The ecological processes that govern the natural recovery are extremely complex, and in-depth study and analysis of them have only recently attracted the attention of a few ecologists. Waste site ecology and the natural recovery of these sites is a young science in the ecological arena, and to date few ecologists have ventured into the type of extreme environments resulting from hard rock mining wastes.

The main components of a biological treatment for acid generation appear to be the following:

1. *Wetlands*. The promotion of wetland areas for waterlogged sections of the site would, owing to extensive evapotranspiration, intercept the infiltration of oxygen and water into the waste material. In order to fulfill such a function the wetland vegetation has to be rooted in the waste material. It is unlikely that the establishment of wetland vegetation with nutrient-rich or ameliorative amendments would provide the required beneficial effects on acid generation. Furthermore, wetlands of this nature are applicable only in those conditions where dam stability and water supply are not of concern in the long term. Wetlands, as a cover for raised tailings deposits where the water table would have to be maintained, will not likely result in self-sustaining systems, for the water supply would diminish with time.

To date, wetlands have received extensive attention mainly in the treatment of mining effluents from coal operations (see, e.g., Chapter 12). Only a few examples of actual working systems for seepage improvement have been documented (Barth, 1986).

2. *Sulfate Reduction*. Conditions that promote the chemical and microbial reduction of sulfate have to be developed. This might be possible by supplying organic materials that would serve as food sources for sulfate-reducing bacteria. To date, a microbiological setting that results in sulfate

reduction of acidic waste water has been identified. However, the rate-limiting criteria and the conditions required for self-maintaining systems are under investigation.

3. *Biological polishing agents.* The promotion of the growth of biological polishing agents, consisting of groups of attached algae, provide essential biofiltration of the waste water. These self-maintaining filters are tolerant to the extreme environments and have to be integrated to provide a complete biological treatment system.

4. *Terrestrial and aquatic moss covers.* Moss carpets can provide a complete cover for erosion control in areas where reclamation is not possible based on access. Such terrestrial carpets would also reduce infiltration of precipitation because mosses have no roots, but shallow rhizoid mats would store moisture during dry periods of the year. Aquatic moss, established in acidic ponds, will consume oxygen above the sediments to protect contaminated sediment from aerobic conditions. Research has recently begun to address the establishment of these covers, both terrestrial and aquatic.

The success of an engineering technique within the waste-management area, utilizing the biological processes described above, depends mainly upon a design of the system that ensures that the oxidation rates governing the release of the contaminants are exceeded by the growth rates and polishing capacity of the employed biota.

It follows then that a detailed knowledge of the growth behavior of the biota to be used is essential to such a program. Several years of experiments on precisely this aspect, together with the other components of the system, are required before the ultimate aim, namely, a self-sustaining system for acid-generating waste management areas, can be achieved. A summary of the biological and geochemical aspects that have been addressed to date is given in Kalin and van Everdingen (1987).

The application of the principles of ecological engineering and biological polishing on a waste-management site has been in progress since the spring of 1986. The concept was applied for the first time as an abandonment plan for the acid-generating wastes from a copper/zinc concentrator.

19.3 CASE STUDY

19.3.1 Site Description

Eighty-five kilometers northeast of Ear Falls, South Bay Mines, Ontario, Canada, a copper/zinc concentrator shut down in 1981 after 10 years of operation. The entire site, including the mine, townsite, and mill, covers about 75 ha. It is located between Confederation Lake, a recreational fishing lake, and Boomerang Lake, a smaller lake draining into Confederation Lake, all of which is part of the English River drainage basin.

The tailings cover a total area of 25 ha. Because 20 ha was dry, this area could be revegetated (1982), while 5 ha remains covered with water (Decant Pond). A dam was breached at the end of the operation, allowing seasonal runoff to drain into a muskeg-covered drainage area (Mud Lake), which finally reaches Confederation Lake (Figure 19.1).

About 760,000 metric tons are contained by dams built using waste rock and gravel on three sides of the tailings basin, whereas one dam (close to Boomerang Lake) was constructed through lifts with concrete poured onto bedrock. The estimated acid-generation potential of the tailings is 0.55 million metric tons of sulfuric acid.

The mine and mill site cover an area of approximately 15 ha, which includes a small pond collecting drainage from this area and from a revegetated waste-rock pile. During runoff, this area drains through a wetland lying in a 0.5-km-long ravine into Boomerang Lake. Along the tailings line, several small tailings spills occurred; these drain either toward Confederation Lake or into Boomerang Lake.

The location of the tailings deposit directly adjacent to Boomerang Lake (Figure 19.1) had resulted in acidification (from pH 7.2 to 4 by 1986) and increased zinc levels (from 0.02 to 8 mg/L by 1986) in the lake. Given the characteristics of the tailings (41% pyrite and 4% pyrrhotite), the problem of acid generation clearly must be addresssed.

19.3.2   The Development of a Close-out Scenario

The classical approach would have comprised a monitoring program agreement and a commitment to lime treatment in order to reduce the levels of zinc in Boomerang Lake. These measures would have had to continue for an undetermined period. Ultimately, more site locations may have required neutralization, for the direction and chemical nature of the contaminated plume from the tailings remained unknown. For the mining company, the most obvious drawback to this approach was that the site then required an undefined financial commitment (the degree of contamination and the length of time required effectively to treat the same being unknown quantities). Furthermore, the site continued to represent an environmental liability.

Between April and July 1986, Boojum Research Limited was given an opportunity to assess whether ecological engineering measures could be implemented on site to effect a walk-away abandonment. The site was found suitable and the project was started following the steps described in the next sections.

19.3.3   Phase I: Definition of the Waste-management Area

At the onset of the project, the investigation of the site took the form of an environmental inquisition, in that every attempt was made to pinpoint all

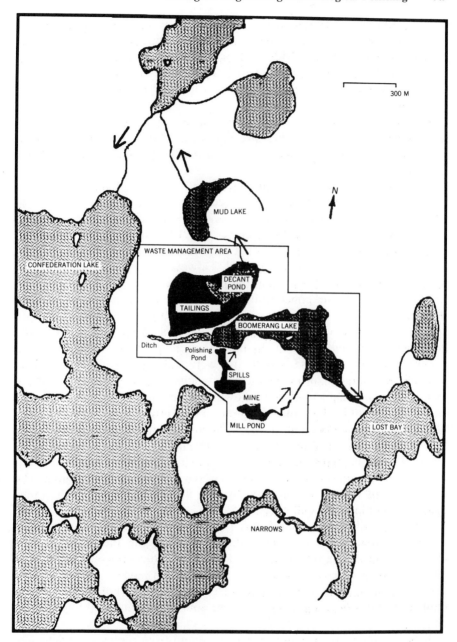

Figure 19.1 Overview of the South Bay Mine Site, including the waste-management area and the receiving water bodies (Mud Lake, Lost Bay, and Confederation Lake).

possible sources of contamination. This was not an easy task and was a particularly onerous undertaking when it was realized that most aspects of the past mining operation yielded evidence of unpleasant environmental consequences. In retrospect, from the standpoint of cost effectiveness, many of these consequences could have been avoided or at least lessened during operation.

It is possible to determine the way in which various regions of the waste-management area obtained their present chemical characteristics by an examination of the water quality trends during the period of mine and mill operations.

The increasing zinc concentration trend in Boomerang Lake (Figure 19.2) provides an example of the importance of background data analysis. In 1978, the zinc concentrations were increasing somewhat more rapidly than in the previous years (from 1 to approx. 3 mg/L), and continued to increase until operations shut down in 1981. In the year following shutdown (1982), a decrease in concentration occurred. However, this was followed by a gradual increase until 1986, when the project was started.

This pattern of zinc concentration changes in Boomerang Lake reveals that the present concentrations are possibly the result of runoff of tailings spills which may have occurred in 1978, as well as from a seep that slowly developed in the tailings dam. In 1986, the tailings dam was grouted, and water retention structures (small berms) were built on the spill areas in order to intercept contaminant loading events reaching the lake. These occurred mainly during spring runoff and following precipitation.

An analysis of background data alone is not sufficient, however, to identify all the characteristics of the site. Water sampling programs covering all four seasons are needed to determine the manner in which contaminants are released, assisting both the identification of all potential sources of contamination and the definition of the borders of the waste-management area. The importance of these aspects might best be illustrated by the observations made on the mine and mill site.

Although Mill Pond at South Bay is relatively small (1.2 ha) and subject to large fluctuations in water level throughout the year, the concentrations of copper and zinc in the pond are relatively high (0.7–109 mg/L, and 173–390 mg/L, respectively). These large concentration ranges, observed in the first year, indicate that the release of contaminants is not regular and is unlikely to be originating from the same source or influenced by seasonal changes.

It was evident, upon inquiry, that the sump from the mill building was still discharging water into the pond. Furthermore, an investigation of the mill area revealed that the concentrate, particularly in the storage area, was the main culprit in the production of high concentrations. Each precipitation event transports concentrate into gravel and waste rock; runoff from heavy rainfall further adds to this erratic behavior of the Mill Pond concentrations.

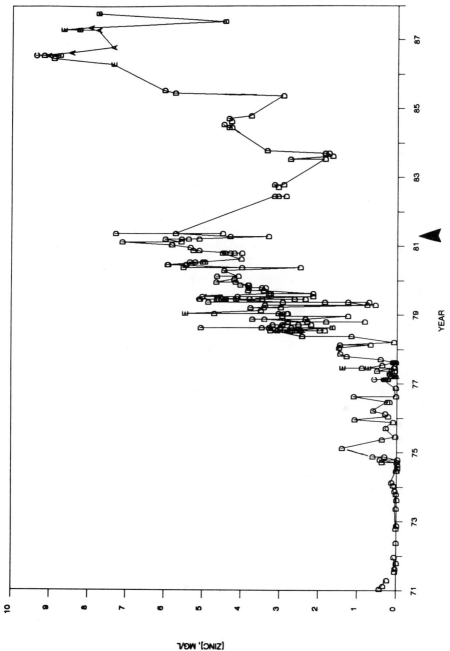

Figure 19.2 Concentrations of zinc in water sampled from the western portion of Boomerang Lake (sample stations B5–B10) between 1971 and October, 1987. Mining and milling ceased operation in May 1981 (arrow). B5 = A; B6 = B; B7 = C; B8 = D; B9 = E; B10 = F.

The concentrate pads were cleaned up; the area was covered with lime and a thin layer of borrow material was contoured so that the runoff would be retained in small trenches and puddles.

The discovery of completely unexpected sources/sinks of metals further highlights the necessity for a water sampling program. During spring runoff, Mill Pond water drains into a 500-m long ravine. A comparison of the copper and zinc concentrations in this ravine shows some interesting water chemistry characteristics. Water samples displayed concentration differences of copper and zinc between the east and west sides of the ravine; these differences defied explanation if both metal concentrations originated solely from the mill area. It was determined, from a discussion of the data with the operators, that aerial transport of copper concentrate downwind into this area could be different from that of zinc concentrate because of differences in moisture content, dryer operation, and storage area location.

Finally, in order to define background concentrations specific to the location for use as guidelines for the protection of water quality, the environmental state of the receiving environment was determined. Because Confederation Lake is a large recreational resort area, biological indicators are important, particularly because they are sensitive to subtle environmental changes. In many cases, these changes can be detected in metal concentrations only at a stage where environmental degradation has progressed to such a level that its control is cost prohibitive.

All water bodies on the site were analyzed for their phytoplankton communities, and these data indicated that, at all discharge points from the site, pristine water was supporting extremely pollution-sensitive desmids, for example, *Staurastrum* spp. These were absent from all wastewater bodies. These members of the phytoplankton community were identified as monitors of the ecosystem to be protected, and despite 10 years of operation and the increasing contamination of Boomerang Lake, their numbers had not been reduced.

The characteristics of the water sampled, together with an inventory of the indigenous species present on the site or to be tested for use on site, were used to define clearly the boundaries of the waste-management area, the contaminant(s) of concern, and the environment to be protected.

In the case of South Bay, the waste-management area consisted of the tailings, the mine and mill site, the spill areas, and Boomerang Lake (frame in Figure 19.1). The key environmental concerns are zinc and acidification, and the main indigenous biological polishing agents are periphytic green and blue-green algae, aquatic moss, and cattails. The biological polishing agents occur either as filaments or in mats. The characteristics (chemical and physical) of the locations in which they are growing within the waste-management area determine the measures that must be taken in order to utilize the biota best.

Table 19.1 Waste Inventory and Polishing Characteristics

Tailings

Tonnage: 760,000
Pyrite: 41.1%
Pyrrhotite: 4.1%
ZnS: 0.6%
SO$_4$:   555,000 t or 146 t/year
Fe (OH)$_3$:   315,000 t or 83 t/year
Zn: 3,000 t or 0.87 t/year

Hydrology Boomerang Lake

Zn loading: 5 t/year
Flow through: 300,000 m^3/year
Area: 250,000 m^2

Biological Polishing

Initial no. of shrubs/deadfall: 2,600
Area of lake occupied by trees: 10%
Zn removal green algae: 2 g/kg biomass-year
Moss Zn content old growth: 1.2 kg/m^2
Moss Zn content new growth: 0.04 kg/m^2
Zn removal by moss new: 0.1 g/m^2-year

19.3.4   Phase II: Waste-management Area Inventory and Implementation of Ecological Engineering Measures

In the same way that the operation of a mine and mill depends ultimately on the existence of proven ore reserves, the success of an ecologically engineered close-out and biological polishing measures depends on the total inventory of the waste site and the knowledge of its hydrologic and ecological behavior. Such a waste-site inventory is developed through hydrological, geochemical, and biological investigation. To illustrate the nature of the inventory, Table 19.1 gives a listing of some of the parameters obtained to date for the South Bay site.

The nature of the parameters listed clearly indicates that during this demonstration an understanding of the growth characteristics of biological polishing agents will be obtained on a scale needed for future applications of ecological engineering and biological polishing.

Equally important results will be obtained on the release mechanisms of the contaminants from the tailings. Based on the analysis of the piezometer water, the portion of the subsurface flow that contains the highest concentrations of metals will be intercepted by the construction of a polishing ditch and pond. This should divert the water into Boomerang Lake whereas otherwise it would reach Confederation Lake, the water body to be protected.

Through the determination of oxygen and sulfur isotopes, it became apparent that oxidation/acid generation proceeds under both saturated and unsaturated conditions. These findings are very important in that they lead to the assumption that the entire inventory of the tailings deposit may be transformed over time into its oxidation products. Estimates of this depletion time ranges from 180 to 35,700 years.

An essential component of the discharges will be iron hydroxide or jarosite precipitate. It is anticipated that the diversion ditch and its pond will serve as an initial precipitation area. Data that will be collected on the performance of the ditch during the pilot demonstration will aid in an understanding of the precipitation process that has previously been studied in seepage areas. The key factor controlling the mechanism appears to be a bromegrass area where annual biomass production exceeds decay, and a sufficient new precipitation area is added annually (Kalin, 1986).

With respect to the surface of the tailings area where the main problem is the establishment of reducing conditions and the accumulation of the metals released from the dry surface areas during runoff to Decant Pond, cattails and biological polishing with algal mats are utilized. Cattail growth can be established by various methods (Kalin, 1987). It is anticipated that the pond will ultimately be covered with cattails, which will provide reducing conditions below their root zone and supply nutrients to the algal polishing agents. The cattail stand, together with a biological polishing agent, should produce an acceptable water quality for that water leaving the pond.

Blue-green algal complexes are frequently indigenous to alkaline waste waters. In Decant Pond, thick algal mats grow on every dead stump and branch and completely cover shallow areas. The water characteristics of this pond since the tailings discharge ceased are given in Figure 19.3. In the first year, the zinc concentrations were low (Figure 19.3, Line K), owing to residual neutralization contained from the past and ongoing liming activities. However, as time progressed, the zinc concentrations were more erratic and increased somewhat, especially during the winter months, despite the continued liming (Lines N, O, L).

Liming was discontinued at the start of the project so that a determination of the natural behavior of Decant Pond could be made. In 1986, a very pronounced trend was noted (Line P), starting with high concentrations in the spring that rapidly decreased to concentrations of the same range (1–2 mg/L) achieved in the previous years with liming. This was followed, however, by an increase in concentrations again in October 1986. The same trend was observed in 1987, but by the end of the summer, the increase was considerably less pronounced (Line Q).

These water characteristics suggested that the growth of the algal mats could be related to the water quality. Analysis of these mats indicated that indeed the zinc concentrations are on average about 8%. An analysis of the

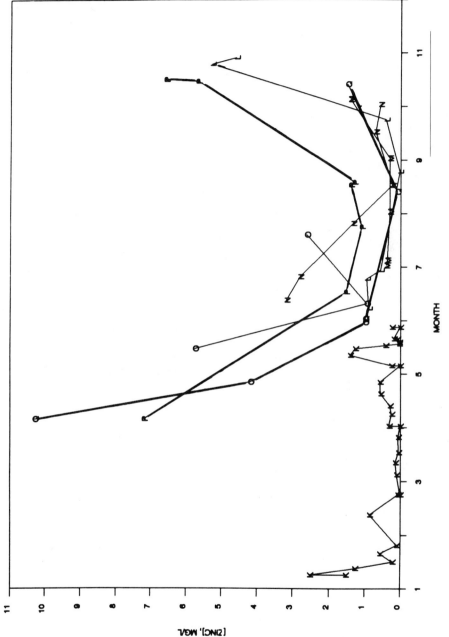

Figure 19.3 Seasonal concentrations of zinc in Decant Pond water sampled between 1981 and October 1987. K through Q represent concentrations in water determined in 1981 through 1987. (K = 1981; L = 1982; M = 1983; N = 1984; O = 1985; P = 1986; Q = 1987).

sediment, which yielded on average about 4% zinc, confirmed that the uptake was clearly from the water. Given that thick mats had also been observed on sticks suspended in Decant Pond water, it became clear that all waste wood material from the demolition of the buildings, and all other chemically inert materials (insulation, rubber tubing, etc.) should be placed into Decant Pond to increase the growth area available for the algal mats.

Growth characteristics and biomass volumes of the algal mats were collected during 1987. This facilitated some evaluation of the capacity of the algae. It can be assumed that 1 m^2 of dry algal mat contains about 460 g dry biomass weight. Using an average concentration of 80 mg/g of algae and 4 mg/L of Decant Pond water, an area of 4,700 m^2 would be required to take up the 172 kg of zinc contained in Decant Pond water, which has a volume of 43,000 m^3.

It follows, therefore, that if the observed algal growth pattern and the concentrations in the algal mats are maintained, there is ample evidence of the capacity of biological polishing. By increasing the surface area of growth of the algae by the addition of material from demolition during the summer of 1987, the seasonal trend of zinc concentrations should be less pronounced in autumn 1987.

Although only the ensuing years will provide confirmation of the mechanism, the concentrations measured in autumn 1987 are encouraging. The algal mats appear to have performed in the same manner as they did previously, when the lime truck method was used (Figure 19.3, Line Q). Assuming that the growth conditions for the algal mats can be sustained, Figure 19.3 may represent the first demonstration of a biological polishing agent at work.

19.3.5 Biological Polishing: Its Long-Term Performance

The preceding section has described the manner in which a biological polishing system is being developed and how this system is connected to or affected by the seasonal chemical behavior of the waste water body. In order to understand the long-term performance of such an ecological system, evidence must be provided to prove that environmentally acceptable conditions are indeed being achieved. Such proof is, of course, essential to the regulatory agencies involved, but is equally important during the project to evaluate the performance of the measures implemented. A computer model is a valuable tool in assessing long-term performance. CHINTEX is a waste-management model that was developed for the decommissioning of nuclear waste material. The modular nature of this model, described by Buchnea (1986), facilitated its adaptation to an assessment of the capacity of biological polishing.

Zinc concentrations in Boomerang Lake have been used (Figure 19.4) as an example of the manner in which the long-term performance may be evaluated. At the end of the project, the model will be used to evaluate the long-

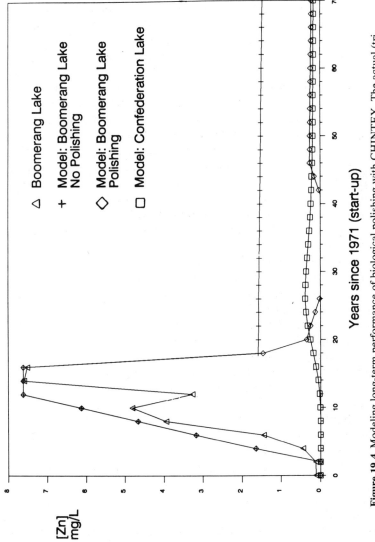

Figure 19.4 Modeling long-term performance of biological polishing with CHINTEX. The actual (triangles) and predicted (diamonds) concentrations of zinc in the western portion of Boomerang Lake between 1971 (0 years after shutdown) and October 1987 (16 years) are presented. In addition, predicted concentrations of zinc in Confederation Lake water (squares), and Boomerang Lake water with (diamonds) or without (crosses) biological polishing for the years 1988 through 2041 are extrapolated.

457

term performance of all ecological engineering measures and the biological polishing system implemented at the site. The predictions made by CHIN-TEX concerning the system's anticipated performance in the long term will be validated through on-site data collection, including a quantification of the growth rates of the biota and water characteristics.

Figure 19.4 illustrates how the results of the integration of the biological polishing module, which is presently, together with the hydrologic and geo-chemical parameters, at a preliminary stage of refinement, can assist in eval-uating the measures taken in Boomerang Lake. The hydrologic and biological investigations that were carried out as part of Phase II of the project, pro-duced a preliminary set of parameters, some of which are given in Table 19.1 and which were used as input parameters for CHINTEX.

The initial time at which discharge could have started from the tailings to Boomerang Lake is assumed, in the model, as being 1971, at the time of start-up of the mill. The increase in zinc concentrations in Boomerang Lake is represented in Figure 19.4 as the annual average concentration measured, whereas in Figure 19.3 the individual samples were plotted. The predicted zinc concentrations (shown by diamonds in Figure 19.4), based on an annual load of 5 tonnes of zinc contained in 30,000 m^3 of water originating from the tailings, with a pore water concentration of 170 mg/L, are very similar to those measured (triangles).

As part of Phase II, grouting of the tailings dam with the seep was carried out in an effort to divert the contaminated plume so that it flowed into the polishing ditch and pond, described earlier. This polishing ditch/pond was excavated specifically to provide chemical/biological polishing, and its lo-cation is given in Figure 19.1.

It is anticipated that the diversion of the annual load through the polishing ditch, according to model calculations, will produce a very significant de-crease in zinc concentrations in Boomerang Lake (diamonds in Figure 19.4) over the next 4 years. According to the predictions, however, the grouting and polishing ditch alone would decrease the concentrations in Boomerang Lake only to about 2 mg/L of zinc (crosses). With biological polishing, the concentrations in the lake may further decrease to below 1 mg/L, the same as those of Confederation Lake (squares).

When the trends of zinc concentrations in Boomerang Lake (presented earlier in Figure 19.2) are compared, a downward trend is indeed indicated, for the concentrations in 1987 are clearly lower. Whether the measures im-plemented will in fact perform in the manner indicated by the model's use of preliminary results depends largely on the growth/decay rates of the biota, as well as on the actual loadings to Boomerang Lake from the polishing pond, the latter of which is governed by the geochemical changes in the tailings.

The two means of biologically removing zinc from the water in Boomerang Lake consist of (1) the introduction of moss bags, and (2) promotion of green algal mats through enhancing the available surface area by the addition of cut shrubs and trees behind log booms in the lake. The removal capacity of these polishing agents is given in Table 19.1, but many aspects of their growth behavior have yet to be determined. Based on the present data, the aquatic moss is considerably less effective than the green algae. However, the anticipated moss carpet will also maintain reducing conditions in the sediments. Because the sediments are the sinks for the metals polished by the algae, maintenance of these reducing conditions is essential to the overall development of the polishing system.

19.4   DISCUSSION AND CONCLUSION

There are many questions yet to be answered in order to quantify the effectiveness of ecological engineering and biological polishing measures. As a first step, however, the results obtained to date are encouraging. It is indicated that this novel ecological approach taken to address the treatment of acid-generating wastes is a realistic avenue and ought to be pursued. As the project continues in the years to come, a more detailed understanding can be gained of the processes that occur.

At this stage of the development of ecological engineering, a conclusion is justified with respect to the concept in general. The fundamental difference between the conventional and the ecological approach to waste management is schematically presented in Figure 19.5. In the conventional approach, water reaches the mine, mill, and waste areas, where it is contaminated and, accordingly, requires treatment. The ecological approach would ideally, at the onset of activities, design the mining operation in light of a complete assessment of water flow both during and after operation. Such a design would, for instance, maximize diversion of water away from contaminant sources during and after the operation. This would minimize the contaminated water to be dealt with in the polishing systems.

Furthermore, biological polishing systems should be integrated into the treatment system at the outset of the operation, so that at the time of shutdown, chemical treatment can be gradually withdrawn. This is portrayed as the outermost box in Figure 19.5, the ecologically engineered component of the waste management area where biological polishing systems are used in conjunction with conventional water treatment.

The experience gained from this case study clearly suggests that had an ecological approach been implemented during the operation, the measures required at the time of shutdown could have been simplified. A polishing system, installed at the outset of mining, could have been utilized during operation. However, the South Bay site is providing an opportunity to assemble evidence to demonstrate that the concept is in fact a viable one in assisting the curtailment of environmental degradation from acid-generating waste sites using a self-maintaining, natural recovery system.

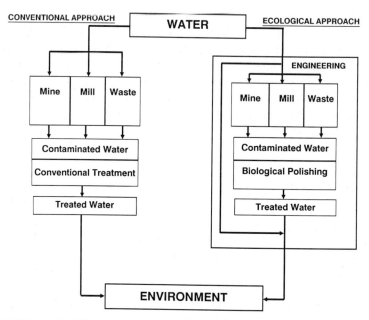

Figure 19.5 The main difference between the conventional and the ecological approach is represented by the water flow and by the integration of biological polishing into the waste-management area by engineering design.

ACKNOWLEDGMENT

This case study is supported by BP Resources Mining Division and was presented at the Canadian Mineral Processors Annual Operator's Conference in Ottawa, January 19–21, 1988.

REFERENCES

Barth, R. C. 1986. Reclamation technology for tailings impoundments. Part I: Containment. *Mineral Energy Res. Rev. Dev.* 29(1):1–25.

Buchnea, A. 1986. A systematic approach to the analysis of waste management systems. Paper presented at the CNS Annual Meeting, Toronto, Ontario, June 9, 1986.

Hamilton, C. E., Ed. 1983. Self-purification and other natural-quality recovery mechanisms. In *Manual on Water*, ASTM STP 442A, pp. 97–106.

Kalin, M. and R. O. van Everdingen, 1987. Ecological engineering: Biological and geochemical aspects, Phase I experiments. Acid Mine Drainage Seminar/Workshop. Proceedings of a conference held in Halifax, Nova Scotia, March 23–26, 1987, Environment Canada, Cat. No. EN 40–11–7/1987 E, pp. 565–590.

Kalin, M. 1987. Progress in the development of ecological engineering methods:

Curtailment of acid mine drainage. BIOMINET Annual Meeting. Proceedings of a conference held in Sudbury, Ontario, Canada, November 4, 1987.

Kalin, M. 1986. Biological polishing of acidic seepage creeks. BIOMINET Annual Meeting. Proceedings of a conference held in Toronto, Ontario, Canada, August 20–21, 1986.

Paine, P. J. 1987. Historic and geographic overview of acid mine drainage. In Acid Mine Drainage Seminar/Workshop. Proceedings of a conference held in Halifax, Nova Scotia, March 23–26, 1987, Environment Canada, Cat. No. EN 40–11–7/ 1987 E, pp. 1–46.

INDEX

463